BATTERIES

Volume 2

**Lead-Acid Batteries
and Electric Vehicles**

BATTERIES

Edited by
KARL V. KORDESCH
Union Carbide Corporation
Battery Products Division
Technology Laboratory
Parma, Ohio

Volume 2
Lead-Acid Batteries
and Electric Vehicles

MARCEL DEKKER, INC.   New York and Basel

Library of Congress Cataloging in Publication Data   (Revised)

Kordesch, Karl, 1922-
  Batteries.

  Includes bibliographical references.
  CONTENTS: v. 1. Manganese dioxide.--v. 2. Lead-acid
batteries and electric vehicles.
  I. Title.
TK2901.K67   621.35   73-82702
ISBN 0-8247-6084-0 (v. 1)
     0-8247-6489-7 (v. 2)

MARCEL DEKKER, INC.
270 Madison Avenue, New York, New York 10016

Current printing (last digit):
10 9 8 7 6 5 4 3 2 1

PRINTED IN THE UNITED STATES OF AMERICA

PREFACE

The lead-acid battery is commercially the most important rechargeable electrochemical system, with production values in the range of several billions of dollars per year, worldwide. In the United States alone, the needs of about 10 million new cars yearly and of the replacement market for 100 million older cars (at least once in 5 years) are obvious. It is truly a mass production business, considering that one starter battery may contain 50 lb of lead in, for example, 66 plates. Only the largest and most modern plants, capable of producing millions of plates per day, will be competitive and survive in such a business by keeping costs low. There are hundreds of thousands of lift trucks and golf carts in the United States, with at least half a ton of lead batteries in each. If the requirements of communication facilities and of the electric power companies are added, the result is that half of the lead industry production will find its way into batteries. The need for efficient utilization and improved construction is obvious; recycling will become even more important with the dwindling of raw materials.

We are grateful to Eugene Y. Weissman, the author of the first and principal chapter on lead-acid batteries, for his outstanding contribution. He leads the reader through the theory and practice of this galvanic system with an ease of which only a man deeply involved in the fundamentals and in the technology is capable; he avoids extreme detail.

The second chapter is written by the editor himself, professing to be an avid fan of electric automobiles. First-hand knowledge gathered from hand-built electric vehicle operation gives more credibility to performance studies and predictions than theory alone.

Nobody can exactly say when, but certainly sometimes before the year 2000, a considerable number of vehicles will be powered by batteries. The sentiments against pollution caused by the internal combustion engine were thought to provide the strongest push in that direction. However, the energy crunch has turned out to be the more powerful force. Saving oil products by using coal-derived or atomic fuels will be the theme of the coming decades.

The electric car as a supplement to long-distance transportation by electric trains is the probable answer for a hotly debated future scenario. The lead battery is at the present time the only power source we can use to develop electric vehicles for the road.

Some people may wonder why we go to the trouble of doing it, if we can see from a simple calculation that there will not be enough lead available to convert a sizable portion of our expected car population to electrics. The reply may be that even a 10% change would be a welcome relief in our predicament, and the production of 1 million cars using a 1/2 million tons of lead batteries per year would mean a mammoth venture in capital investment and technology.

It seems timely to have a new look at alternatives for the lead-acid batteries. They will be needed if electric vehicles do find acceptance and will also be the basis of improved efficiency in future large coal or atomic power plants, by serving as their load-leveling means. The crystal ball is cloudy, but some predictions may be realistic enough for us to believe in them.

Chapter 3 was added when it was realized that publicity dealing with the often claimed but not always achieved rapid progress of technology required authoritative confirmation from people close to manufacturers of lead acid batteries and designers of electric vehicles. Since developments in England and in the United States are covered in the accessible American-English literature and historic facts are found in available (classical) monographs, the contributions of German and Japanese authors were selected for this "addendum." Of course, some duplication of the material in the first two chapters could not be avoided; it was allowed to remain to make the contribution understandable within its own context.

Thanks are due to the publisher for his willingness to allow a considerable reworking of the manuscripts during the processing period for the sake of achieving a modern treatise on the subject.

KARL V. KORDESCH

CONTRIBUTORS

DIETRICH BERNDT, Varta Batterie A. G., Research and
Development Center, Kelkheim, Federal Republic of Germany.

KARL V. KORDESCH*, Union Carbide Corporation, Battery
Products Division, Technology Laboratory, Parma, Ohio.

AKIYA KOZAWA, Union Carbide Corporation, Battery Products
Division, Technology Laboratory, Parma, Ohio.

KLAUS SALAMON, Varta Batterie A. G., Research and
Development Center, Kelkheim, Federal Republic of Germany.

TOKUJIRO TAKAGAKI, Japan Storage Battery Co.,
Kyoto, Japan.

EUGENE Y. WEISSMAN[†], Globe-Union Inc., Corporate
Applied Research Group, Milwaukee, Wisconsin.

*Present Affiliation: Institut für Anorganisch Chemische
Technologie der Technischen Universitat Graz, Graz, Austria.

†Present Affiliation: BASF Wyandotte Corporation, Chemical Specialties
Business, Wyandotte, Michigan.

CONTENTS

CONTENTS OF VOLUME 1

ADDITIONAL VOLUMES OF THIS SERIES ARE IN PREPARATION

BATTERIES

Volume 2

**Lead-Acid Batteries
and Electric Vehicles**

Chapter 1

LEAD-ACID STORAGE BATTERIES

Eugene Y. Weissman[*]

Globe-Union, Inc.
Corporate Applied Research Group
Milwaukee, Wisconsin

*Present Affiliation: BASF Wyandotte Corp., Chemical Specialties Business, Wyandotte, Michigan.

1. DEVELOPMENT HISTORY OF THE LEAD-ACID BATTERY

It took approximately 50 years from the time of Volta's discovery of the galvanic cell principle in 1800 for its application to the lead/sulfuric acid/ lead dioxide system by Siemens and Sinsteden. This was followed 10 years later by Planté's reported results introducing the lead-acid battery as an electrochemical storage device [1].

Fittingly enough, Planté's name has been incorporated into the permanent lead-acid battery terminology with reference to the so-called Planté plates. These have their active material layers or surfaces formed electrochemically from the lead substrate, rather than applied to the substrate in a separate operation (see Sec. 4.1.4). Indeed, Planté's cells consisted of lead sheets wound spirally in a cylindrical configuration and separated by rubber strips, heavy linen, or felt. These sheets were formed electrochemically in situ to the Pb/PbO_2 couple.

Thereafter, progress continued systematically to the present day. A significant development, first described toward the end of the nineteenth century by Faure, was the utilization of pasted plate structures consisting of active material mixes applied to a variety of lead current collector grid configurations. Various types of separators, starting with such "exotic" materials as flannel, also found increasing usage.

The advent of a more technological era, with dynamos available for battery charging and electric motors for the utilization of the discharge energy, marked a gradual acceptance of the advantages of rechargeability in general, and of the lead-acid battery in particular. Most of the initial applications were stationary in nature, but by 1900, sufficient types of relatively compact and portable lead-acid batteries had been developed to enable their utilization for traction, starting, and lighting of automobiles, and military applications, in addition to the standby uses.

As with all other branches of technology, World War II provided a considerable impetus in terms of the extent and variety of applications to
which the lead-acid battery could be adapted. The industry grew, and with
it also the number of new developments and improvements in the field, including new materials of construction for separators and battery cases as
well as new methods of manufacturing and assembly of all internal battery
components, additional types of grid alloys for various applications, improved active material compositions, formulations, and processing techniques, and so on. Progress in all these areas has lately been made at an
accelerating rate, and the various aspects of interest are discussed in
some detail in the remainder of this chapter.

The resulting product grew lighter in weight and more reliable. Figure 1 exemplifies the reduction in lead and active material for a 6-V
100 Ah lead-acid battery from the 1900s to the 1960s [2]. An increase in
service life with time is also represented in the figure. Similar efficiency
and life improvement trends have been demonstrated for the contemporary

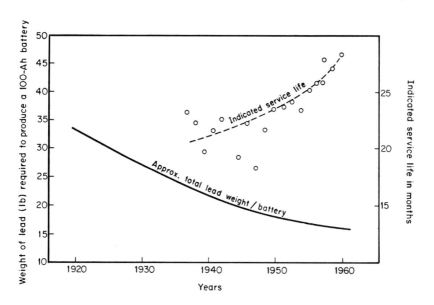

FIG. 1. Improvement in the reliability and active material utilization
efficiency of a typical 6-V 100-Ah lead-acid battery since the 1920s [2].

12-V systems. Specific examples are given in Figs. 20 through 23
(Sec. 3.1.) for the 12-V automotive battery, which replaced the 6-V battery nearly completely in automobiles built after the 1960s.

Table 1 is a summary of highlights in the development history of the
lead-acid battery, presented in chronological sequence [3].

TABLE 1

A Summary of Highlights in the

Development of the Lead-Acid Battery[*]

1860	Planté reduces the lead-acid system to practice using formed lead sheet plates
1881	Faure develops the pasted plate structure
1883	Tudor develops extended area (spun) lead plates
1880-1890	Grids are first designed into the lead-acid battery; antimonial alloys are first used
1882	Gladstone and Tribe's double sulfate theory
1890-1900	Wood separators in use
1900-1910	Lead dust and leady oxides as raw materials; "iron-clad" plate structure designed (hard rubber and asbestos)
1914-1920	Rubber separators; the role of expanders on capacity was first defined
1927-1937	Porous ebonite and microporous rubber put in use as battery separators
1935	Lead-calcium grid alloys first introduced for applications requiring low rates of self-discharge (float)
1948-1950	Cellulosic and man-made fiber bonded separators; lower density pastes (active material) find increased usage
1965	Maintenance-free batteries for portable devices
Mid 1960s	Lightweight construction, plastic battery cases; high automation of the production and battery assembly processes; high efficiency (low IR) designs
1970s	New applications: water-activated and maintenance-free batteries for automotive use; increased emphasis on traction batteries

[*]1860-1950 period adapted from Ref. 3.

2. THE ELECTROCHEMISTRY OF THE LEAD-ACID BATTERY

2.1. General

The theory of operation of the rechargeable lead-acid battery is based on the so-called double sulfate principle [4]. Basically, it accounts for the fact that the electrolyte in solution, sulfuric acid, is not only an ion-transport medium but a reactant as well. The discharge product of both the lead anode electrochemical oxidation reaction and the lead dioxide cathode electrochemical reduction reaction is lead sulfate, whence the terminology "double sulfate." The reactions can be represented as follows:

$$\text{Anode:} \quad Pb + HSO_4^- \underset{charge}{\overset{discharge}{\rightleftharpoons}} PbSO_4 + H^+ + 2e^- \tag{1}$$

$$\text{Cathode:} \quad PbO_2 + 3H^+ + HSO_4^- + 2e^- \rightleftharpoons PbSO_4 + 2H_2O \tag{2}$$

$$\text{Overall:} \quad Pb + PbO_2 + 2H^+ + 2HSO_4^- \rightleftharpoons 2PbSO_4 + 2H_2O \tag{3}$$

The half-cell reaction schemes are based on the electrolyte dissociation into H^+ and HSO_4^- ions, which is closer to the true picture. Complete electrolyte dissociation to H^+ and SO_4^{2-} ions, while represented as such in many instances, applies only in the case of extreme dilution.

During overcharge, water electrolysis into hydrogen and oxygen takes place.

The exchange current density for reaction (2), as determined by a Faradaic impedance method, was reported as 3.2×10^{-4} A/cm^2 in 10 N sulfuric acid [5].

Based on the free-energy change of $\Delta\underline{G}$ = -13.95 kcal/mole for the anodic reaction (E_0 = 0.303 V versus H_2/H^+) and $\Delta\underline{G}$ = -74.99 kcal/mole for the cathodic reaction (E_0 = -1.627 V versus H_2/H^+), one obtains the overall reaction values: $\Delta\underline{G}$ = -88.94 kcal/mole (E_0 = 1.93 V), specific energy = 161 Wh/kg, specific capacity = 83.5 Ah/kg [6].

In actual practice the lead-acid battery exhibits significantly lower specific energies, both relative to other systems and compared to its own theoretical values. Figure 2 shows the place of the lead-acid battery among other galvanic systems. This diagram may not be correct in all details,

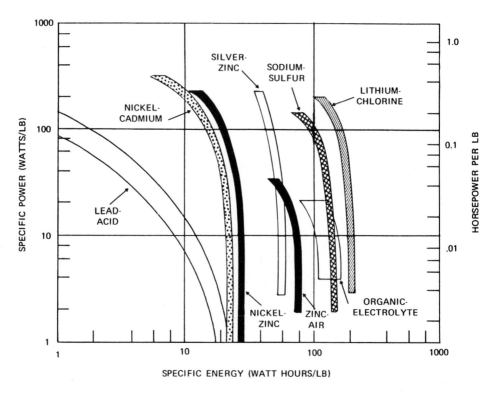

FIG. 2. A comparison between practical specific energy and power characteristics of the lead–acid storage battery and other systems. [Electric Vehicles, Chapter 2]

but it serves well for general comparisons and estimates of the capabilities of future power sources to be used in electric vehicles (see Chapter 2 of this book).

At typical electrolyte concentrations of the order of 35 wt % H_2SO_4 (sp gr = 1.265), the cell potential becomes approximately 2.1 V at room temperature. Table 2 lists calculated (average of two methods) and measured potential values as a function of electrolyte concentration, and Fig. 3 summarizes the potential versus electrolyte concentration loci of all the stability domains of lead sulfate (between curves 1 and 10), lead (below curve 1), and lead dioxide (above curve 10).

With lead sulfate having a specific volume roughly 1.5 times larger than lead dioxide and 1.8 times larger than lead, the electrodes—or plates

TABLE 2

Experimental and Calculated (Average of Two
Methods) Room Temperature EMF of the Lead-
Acid Cell as a Function of Electrolyte Concentration[*]

H_2SO_4 concentration		EMF V	
M	wt %	Experimental	Calculated
0.05	0.488	1.762	1.764
0.10	0.970	1.796	1.799
0.20	1.925	1.831	1.831
0.50	4.675	1.881	1.880
1.0	8.933	1.919	1.920
2.0	16.400	1.971	1.974
3.0	22.737	2.014	2.016
4.0	28.179	2.053	2.056
5.0	32.901	2.090	2.091
6.0	37.047	2.124	2.134[†]
7.0	40.707	2.155	2.165[†]

[*]Adapted from Ref. 7.
[†]Single value.

as they are usually known—in a lead-acid battery are subjected to significant stresses and morphology changes during charge-discharge cycling.
As a result of a combination of the common ion effect and sulfate/bisulfate
equilibrium variations, the solubility of the reaction product—lead sulfate—
is highest in water, passes through a minimum followed by a maximum at
about 10 wt % sulfuric acid, and then decreases again (Table 3) [8]. These
solubility characteristics are important from a performance standpoint, as
will become clearer in what follows.

Lead sulfate is highly resistive (approximately 10^8 Ω-cm) and has an
orthorhombic crystal structure readily identifiable by X-ray diffraction.
It is present in a fine-grained structure or as large crystals, constituting

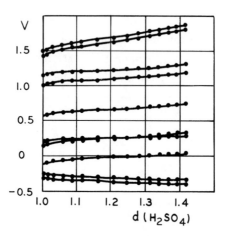

FIG. 3. Dependence of the equilibrium potentials of anodic oxidation reactions of lead on the concentration of sulfuric acid [7]:

(1) $Pb + SO_4^{2-} \rightarrow PbSO_4 + 2e^-$

(2) $Pb + HSO_4^- \rightarrow PbSO_4 + H^+ + 2e^-$

(3) $H^+ + e^- \rightarrow (1/2)H_2$

(4) $Pb + 2H_2O \rightarrow Pb(OH)_2 + 2H^+ + 2e^-$

(5) $Pb + H_2O \rightarrow PbO + 2H^+ + 2e^-$

(6) $Pb + 2H_2O \rightarrow PbO_2 + 4H^+ + 4e^-$

(7) $PbO + H_2O \rightarrow PbO_2 + 2H^+ + 2e^-$

(8) $2H_2O \rightarrow O_2 + 4H^+ + 4e^-$

(9) $PbSO_4 + 2H_2O \rightarrow PbO_2 + HSO_4^- + 3H^+ + 2e^-$

(10) $PbSO_4 + 2H_2O \rightarrow PbO_2 + SO_4^{2-} + 4H^+ + 2e^-$

TABLE 3

Solubility of Lead Sulfate

in Solutions of Sulfuric Acid [8]

Concentration of H_2SO_4			Weight of $PbSO_4$ per liter	
Percent	Molarity 25°C	Specific gravity $\frac{25°C}{25°C}$	At 25°C (mg)	At 0°C (mg)
0.0	0.00	1.000	45.0	28.0
0.5	0.05	1.003	4.60	2.06
1.0	0.10	1.006	4.91	2.10
5.0	0.52	1.033	6.15	2.46
10.0	1.08	1.067	6.68	2.86
15.0	1.68	1.102	6.28	2.63
20.0	2.32	1.140	5.18	2.21
25.0	3.00	1.179	3.76	1.76
30.0	3.72	1.219	2.75	1.27
35.0	4.48	1.260	2.02	0.84
40.0	5.30	1.303	1.52	0.53
45.0	6.16	1.348	1.23	
50.0	7.10	1.395	1.08	

the so-called hard sulfate, which is not regenerable to the electrochemically active species of lead and lead dioxide. Agglomerations of significant amounts of hard sulfate usually typify end of life (Fig. 4).

Recent electron microscopy and X-ray diffraction studies of Chiku and co-workers [9] have revealed a morphological difference between the rechargeable and the refractory ("hard") lead sulfate species present on pure lead. The rechargeable lead sulfate was defined by them as "active material" and was shown to nucleate preferentially in the early stages of the anodic oxidation of lead on the low index lead crystal surfaces [in partic-

FIG. 4. Scanning electron microscopy view of failed positive plate showing large agglomerates of hard sulfate (grid: Pb-0.08 wt % Ca; magnification: 700×).

ular the (001) and (111) planes] and grow dendritically as a function of the discharge conditions.

Not all the active material, whether lead or lead dioxide, can be utilized electrochemically and the faradaic efficiency, depending on battery age, temperature, and rate, will usually vary between 25 and 60%. This efficiency is lower than in most other types of storage batteries, and constitutes one of the major lead-acid battery R&D problems, the ultimate goal being a higher energy content per unit weight and volume. The unused active material provides the necessary electrically conducting network in the plates as well as the recharge sites. As expected, thin (1-mm) and highly porous plates (65 to 70%), discharged at low current densities such as those corresponding to the 10- to 20-h rates, will yield the highest faradaic efficiencies. This advantage is offset by a reduced cycle life as one goes beyond the practical extremes of plate thickness or porosity.

The current efficiency in a standard battery using lead-antimony grids (see Sec. 4.1.2.) is of the order of 85 to 95%, and it approaches 100% in lead-calcium systems owing to the significantly lower rate of gas evolution. Burbank et al. [10] used the term "virtual resistance" to include all polarization effects in a pulse-discharged battery. Typical values are of the order of 0.1 to 0.4 Ω, with short circuit currents of 5 to 20 A per 1-Ah plate.

2.1.1. Self-discharge Effects

Self-discharge phenomena specific to the positive and negative plates are discussed in the following sections. A general representation of a typical lead-acid battery capacity loss, in terms of sulfuric acid specific gravity units as a function of temperature, is given in Fig. 5.

The relationship between specific gravity and available battery capacity is often used as remaining state of charge indication. However, consulting later sections (4.16, Table 8, and 4.3.3., Table 13) will show that such indications are only useful if the design specifications of the manufacturer are known.

Potential-pH diagrams are useful for the representation and interpretation of self-discharge or passivation effects, although they are theoretically significant only for systems at thermodynamic equilibrium. Figures 6 and 7 represent the lead-lead oxides-lead sulfate potential-pH equilibrium domains for a lead-0.075 wt % Ca grid system [12] (which is becoming increasingly pertinent with the increased applicability of the

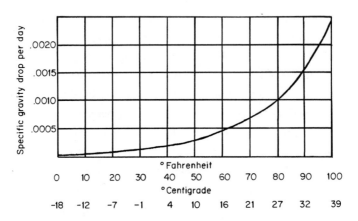

FIG. 5. Self-discharge of new, fully charged lead-acid batteries [11].

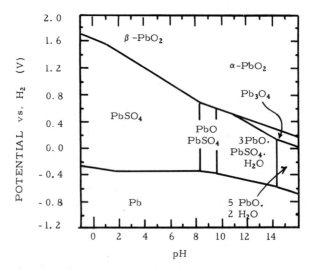

FIG. 6. Potential (versus H_2/H^+)-pH diagram of the system
$Pb-H_2O-H_2SO_4$ for $(a_{SO_4^{2-}} + a_{HSO_4^-}) = 1$. From Ref. 12.

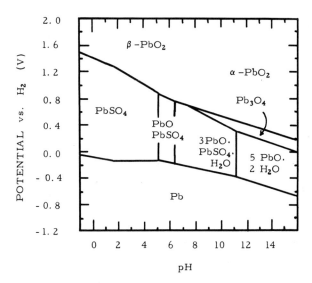

FIG. 7.   Potential (versus H_2/H^+)-pH diagram of the system $Pb-H_2O-H_2SO_4$ for $(a_{SO_4^{2-}} + a_{HSO_4^-}) = 10^{-7}$.   From Ref. 12.

maintenance-free battery—see Sec. 3.4).   The sulfate ion activities are unity and 10^{-7}, respectively, with the latter condition prevalent in lead-acid battery pastes [13] (Sec. 4.2.2) or at locations of complete acid exhaustion such as the interior of grid corrosion films and the vicinity of alkaline corrosion products (PbO, Pb_3O_4, α-PbO_2).   More elaborate potential-pH diagrams, also listing soluble lead-ion domains and limits of thermodynamic stability of water, have also been published [14] as well as a detailed schematic representation of passivation films at various potentials including their relative thicknesses and porosities (Fig. 8) [15].

2.1.2.   Performance and Structural Aspects

Section 3 of this chapter deals with battery applications and includes pertinent performance information.   Therefore, only aspects of special technical interest are discussed here.   One of them involves the voltage transients encountered in a lead-acid battery at the onset of charge or discharge [16].

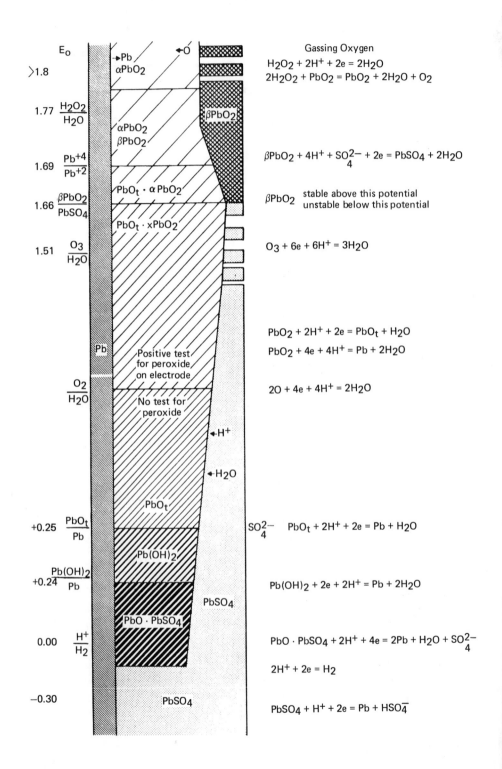

E_0

> 1.8

→Pb   ←O
αPbO₂

βPbO₂

$1.77 \quad \dfrac{H_2O_2}{H_2O}$

αPbO₂
βPbO₂

$1.69 \quad \dfrac{Pb^{+4}}{Pb^{+2}}$

PbO$_t$ · αPbO₂

$1.66 \quad \dfrac{\beta PbO_2}{PbSO_4}$

PbO$_t$ · xPbO₂

$1.51 \quad \dfrac{O_3}{H_2O}$

Pb

Positive test
for peroxide
on electrode

$\dfrac{O_2}{H_2O}$

No test for
peroxide

←H⁺

←H₂O

PbO$_t$

$+0.25 \quad \dfrac{PbO_t}{Pb}$

$+0.24 \quad \dfrac{Pb(OH)_2}{Pb}$

Pb(OH)₂

PbSO₄

PbO · PbSO₄

$0.00 \quad \dfrac{H^+}{H_2}$

-0.30

PbSO₄

Gassing Oxygen

$H_2O_2 + 2H^+ + 2e = 2H_2O$

$2H_2O_2 + PbO_2 = PbO_2 + 2H_2O + O_2$

$\beta PbO_2 + 4H^+ + SO_4^{2-} + 2e = PbSO_4 + 2H_2O$

βPbO_2 stable above this potential
unstable below this potential

$O_3 + 6e + 6H^+ = 3H_2O$

$PbO_2 + 2H^+ + 2e = PbO_t + H_2O$

$PbO_2 + 4e + 4H^+ = Pb + 2H_2O$

$2O + 4e + 4H^+ = 2H_2O$

SO_4^{2-} $PbO_t + 2H^+ + 2e = Pb + H_2O$

$Pb(OH)_2 + 2e + 2H^+ = Pb + 2H_2O$

$PbO \cdot PbSO_4 + 2H^+ + 4e = 2Pb + H_2O + SO_4^{2-}$

$2H^+ + 2e = H_2$

$PbSO_4 + H^+ + 2e = Pb + HSO_4^-$

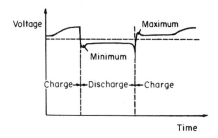

FIG. 9. Transient voltage effects in the lead-acid battery [16].

The beginning of discharge is characterized by a minimum-shaped voltage drop known in the trade as "Spannungssack," "coup de Fouet," or "whip stroke." The reverse occurs at the beginning of charge ("Spannungsberg"). These phenomena, which are represented schematically in Fig. 9, occur mainly at the positive plate.

The voltage drop at the onset of discharge has been reported to be current density-dependent, with typical values being about 20 mV from 25 to at least 75 mA/cm^2, and lower polarization values at higher and lower current densities [16]. Although some of the voltage drop is ohmic in character, most of it appears to be the result of a charge transfer in the double layer. The main rate-limiting step is the formation of microdendritic lead sulfate crystals under conditions of local supersaturation with respect to bivalent soluble lead. This phenomenon has been dubbed "crystallization overvoltage" and verified by optical microscopy [17].

The Spannungsberg observed at the onset of charge is ohmic in character owing to the insulating properties of the discharged species, i.e., lead sulfate. As such, the maximum overvoltage values will be current density-dependent, for example, 45 mV at 5 mA/cm^2 and 240 mV at 40 mA/cm^2 [16]. Nevertheless, a crystallization overvoltage component also appears to play a role during the electrochemical formation of PbO_2, that is, during charging of the positive plate, and it is reported as the main reason for discrepancies between predicted and experimental overvoltage values [18].

FIG. 8. Schematic representation of the anodic passivation films on lead [15].

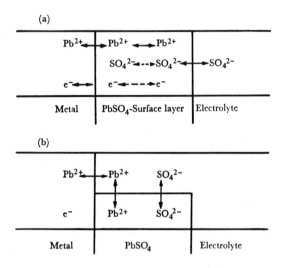

FIG. 10.   Schematic reaction pattern of the Pb/PbSO$_4$ electrode [20].
(a) Solid state reaction. (b) Reaction including solved Pb^{2+}-ions.

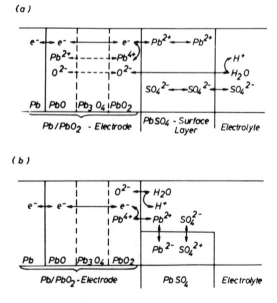

FIG. 11.   Schematic reaction pattern of the PbSO$_4$/PbO$_2$ electrode [20].
(a) Solid state reaction. (b) Reaction including solved Pb^{2+}-ions.

Thus, at a dilute acid concentration of $2\underline{N}\ H_2SO_4$, corresponding to electrochemical formation or discharged state conditions, the crystallization overvoltage component was measured at a limiting value of 0.02 V—close to the theoretical—for a current density of 10 mA/cm^2, beyond which it remains constant. On the other hand, the ohmic component will rise continually with an increase in the formation current density.

Berndt [19,20] investigated the effect of temperature and current density on the utilization of lead and lead dioxide electrodes and pointed to the soluble bivalent lead ions as defining an intermediate charge or discharge step (Figs. 10 and 11; note that the sulfate/bisulfate ion equilibrium is ignored in this representation, for the sake of simplicity). This is evidenced by diffusional rate limitations at the positive plate, at low temperatures and high discharge currents, because of the inability of the reaction to penetrate into the interior of the plate. These performance limitations were recently confirmed by means of flowing electrolyte studies and X-ray microprobe analyses with labeled sulfur ^{35}S [21].

The solid state reaction representation [parts (a) of Figs. 10 and 11] were questioned by Berndt and considered possible only across films not thicker than a few angstroms. Carr and co-workers [22] used linear-sweep voltammetry on pure lead and electrodeposited lead dioxide test electrodes to show that both types of transport are controlling, with a difference: in the electrolyte only the SO_4^{2-} ions are of consequence; the Pb^{2+} ions play a charge-carrying role, together with SO_4^{2-}, in the solid film which develops on the electrode. At low sweep speeds and low concentrations of sulfuric acid (i.e., close to fully discharged conditions), the rate-controlling step is in the solution phase. It transfers to the solid film phase at higher sweep speeds and/or high electrolyte concentrations.

The differences in the electrochemical performance of α- and β-PbO_2 (see Sec. 4.1.1 for a definition of the two dioxide species) during cathodic reduction, and the fact that the α species is progressively converted to the β form during cycling, have also been confirmed in the study of Carr et al. [22]. One should keep in mind, however, that results based on test electrodes of the type used by Carr can only be used qualitatively. The reader is directed, in this respect, to a pertinent quotation in Sec. 2.2.

Skalozubov used radioactive ^{35}S to demonstrate how the effective surface area of the active material in the lead-acid battery plates first increases to a certain extent during cycling, and thereafter decreases owing
to lead sulfate crystal growth [23]. This confirms the known fact that the
ability of a typical lead-acid battery to "hold its charge" will increase to
roughly 120% of nominal capacity and then gradually decrease to the end of
life point, which is taken as approximately 80% of nominal (Fig. 12).

At the negative plate an added resistance to effective utilization of the
active material is contributed by the insulating properties of the lead sulfate layers formed during discharge. A detailed discussion on the subject
has recently been published concerning plates made with lead-antimony
grids of various compositions [24].

Gillibrand and Lomax also defined a diffusional limitation of the discharge step, in terms of the transport of sulfate ions from the bulk of the
electrolyte to the lead electrode surface [25,26]. Based on their results,

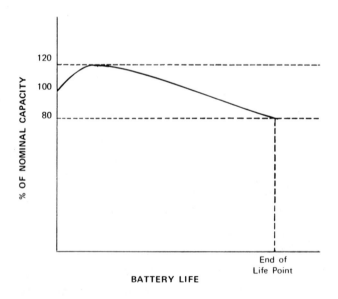

FIG. 12. "Charge-holding" capability during lifetime of lead-acid
battery.

they suggested a Peukert [27] type of exponential relationship of capacity versus discharge rate having the form

$$t = \frac{K}{I^n} \tag{4}$$

where t is the duration of discharge, I is the current density, and K and n are constants.

In a more recent investigation [28] involving a temperature range of -18 to +16°C and current densities between 21 and 165 mA/cm^2, the following relationship was established between the effective plate capacity C at θ°C and the current density:

$$C = \frac{K_0(1 + \alpha\theta)}{I^{(n-1)}} \tag{5}$$

where n = 1.4 for both the positive and the negative plate, K_0 = 0.32 and 0.24 min, and α = 0.021 and 0.015 min/°C, for the negative and positive plates, respectively.

A variety of studies have been conducted, by means of radiotracers, in order to define current and active material distributions in the plates as a result of discharge [29-34]. It was found that at low current densities (say ≤10 mA/cm^2), lead sulfate is uniformly distributed throughout the plate thickness, whereas at high current discharges, concentrations at the surface and near the grid (for antimony-free grids) are higher than in the bulk of the plate. The buildup of sulfate near the antimony-free grid is responsible for subsequent active material shedding and premature capacity losses, especially at the positive plate. Furthermore, current maldistributions are particularly evident at the beginning and end of discharge, with more pronounced irregularities and more rapid changes at the positive plate. Bode and co-workers interpreted this phenomenon—at the usually electrolyte-limited positive plate—in terms of concentration gradient effects involving the transferring ionic species. One notes that at very low temperatures, such as -18°C, the diffusional limitation loses its significance and a constant lead sulfate content is observed throughout the positive plate thickness [35].

This type of uniform lead sulfate concentration was also confirmed by electron microprobe [36] and X-ray diffraction [37] studies. Furthermore, it was pointed out that the intensity ratio $Pb/PbSO_4$ is a useful index of battery life; it gets smaller, even for a fully charged battery, as the battery nears end of life [37].

2.1.3. Theoretical Electrode Models

Lead-acid battery plates can be treated theoretically as porous electrodes with sparingly soluble reactants. Stein looked at the problem from the standpoint of high current density discharges [38,39], which are invariably accompanied by a reduction in battery capacity. His analysis, though debatable in certain respects (e.g., his assumptions regarding current versus electrolyte concentration dependence inside a pore), is interesting in that it shows how electrolyte starvation in the pores is not the primary capacity decrease mechanism. Rather, electrolyte concentration changes at the pore entrance are responsible. Similar conclusions have also followed from a theoretical analysis by Lehning [40], suggesting thin plates for optimum electrochemical polarization.

Stein's calculated current distribution in the pores comes out fairly uniform except at the pore entrance (Fig. 13) and, beyond a limiting value, the plate thickness no longer plays a capacity-increasing role. This fact

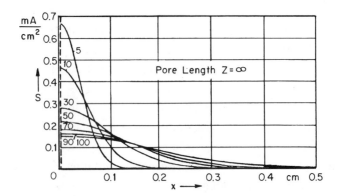

FIG. 13. Current density at the pore wall, S, as a function of the depth of penetration, x. Parameter: discharge time in seconds [38].

has been corroborated in practice. In Stein's model the electrolyte conductivity plays a significantly more important role than the diffusivity of
the transferring components.

Dunning et al. [41] recently proposed an idealized "sparingly soluble
reactants" model which considers the transport of active species within the
porous electrode itself and does not get involved with the transport between
the bulk of the electrolyte solution and the interface. The active material
is assumed to be available within the electrode as low-solubility dispersed
crystallites, and diffusion of the electrochemical reactants and products
proceeds at very low rates. This approach belongs to the general category
of "macrohomogeneous" or "continuous" models [42-44], which disregard
the actual pore geometry of the plate matrix and, instead, assume a ran-
dom structure which can be characterized in terms of averages of surface
area per unit volume, matrix and electrolyte resistivity, effective plate
porosity, and so on. There is reason to believe that the continuous pore
model may yield practical results, in particular for the analysis of the
more complex positive plate performance which usually is limited primarily by the availability of sulfuric acid reactant to its active porous
structure.

Experimental substantiation of performance versus structure theories
and generalized models applied to the lead-acid battery system can be expected in the near future. Some initial inputs have been provided by Voss
and Freundlich [45] in their attempts to compare the discharge capacities
of α- and β-lead dioxide electrodes; as opposed to Stein's work [38,39],
they were concerned with low-rate phenomena where changes attributable
to surface area or lead dioxide morphology effects would be more readily
detectable (the following section treats the general aspects of lead dioxide
electrode performance in more detail).

It can safely be stated that considerably more insight and possible
breakthroughs in the general area of theoretical performance models will
be forthcoming in the next 5 to 10 years.

2.2. The Positive Electrode

The life-limiting component of any lead-acid battery system is the lead
dioxide electrode. Owing to the complexity of the charge, discharge, and
self-discharge phenomena taking place at the positive lead dioxide battery
plate, they have formed the subject of numerous studies and some reviews
[10, 46-49]. Ruetschi et al. [50] described the remarkable behavior of the
lead dioxide electrode in the lead-acid system as indicative of a high degree
of metastability. Otherwise, its significantly higher potential relative to
the decomposition potential of water would lead to spontaneous oxygen evolution.

   The following quotation from a recently published review on the lead-
acid cell describes some of the difficulties associated with the study of the
lead dioxide electrochemical system:

> Most attempts to analyze the reactions of charge and discharge from a
> kinetic standpoint have been made on electrodes formed by anodic
> treatment of lead metal or of electrodeposited PbO_2. Such an elec-
> trode is not, of course, an equilibrium mixture of PbO_2 and $PbSO_4$
> and cannot be expected to behave in the reversible manner. To polar-
> ize a monophasic electrode, so that its potential crosses into a range
> of thermodynamic metastability, is much the same as cooling a liquid
> below its freezing point in the absence of nuclei. The resulting super-
> cooling has no relation to the bypassed equilibrium condition. Equi-
> librium is not established until all phases are present in sufficient
> quantity and in suitable physical juxtaposition. It is thus more likely
> that the kinetics of the reversible plate reaction will be solved when a
> suitable electrode is designed. A kinetic study of the massive porous
> battery plate is complicated because the reactions are diffusion-
> controlled, and the constantly changing porosity and composition ren-
> der a stabilized surface difficult to maintain for kinetic evaluation of
> the reversible plate itself [10].

Indeed, available material on the electrochemical mechanisms of
charge and discharge at the lead dioxide electrode goes little farther than
pointing out its reversibility. Most of the available data discuss mor-

phology changes rather than charge transfer processes, and a great deal of uncertainty and disagreement still prevail in this general area.

2.2.1. Self-discharge Effects

To the extent that a positive plate self-discharges, it will do so as a result of local galvanic action between the grid metal and lead dioxide, or reactions between lead dioxide and the adjoining electrolyte and other species. The lead sulfate "films" ultimately formed afford a certain amount of passivation, albeit at a penalty in terms of increased internal plate resistance.

Ruetschi and Angstadt [51] proposed the following five self-discharge schemes, with reaction (8) characterizing the most common type of battery, using lead-antimony grids:

$$PbO_2 + 2H^+ + SO_4^{2-} \rightarrow PbSO_4 + H_2O + \frac{1}{2} O_2 \tag{6}$$

$$PbO_2 + Pb + 2H_2SO_4 \rightarrow 2PbSO_4 + 2H_2O \tag{7}$$

$$5PbO_2 + 2Sb + 6H_2SO_4 \rightarrow (SbO_2)_2\, SO_4 + 5PbSO_4 + 6H_2O \tag{8}$$

$$PbO_2 + (\text{oxidizable separator material}) + H_2SO_4 \rightarrow$$
$$PbSO_4 + (\text{oxidized material}) \tag{9}$$

$$PbO_2 + H_2 + H_2SO_4 \rightarrow PbSO_4 + 2H_2O \tag{10}$$

The presence of antimony in the grid reduces the passivation action of the lead sulfate layer by opening pores in it as a result of the formation of soluble antimony sulfates (see also Sec. 4.1.2). It is interesting to note that at high acid concentrations, protection against self-discharge is due to lead sulfate formation, whereas at low acid concentrations, antimony corrosion products take over the same function. The result is a self-discharge versus acid concentration curve exhibiting a maximum over the sulfuric acid specific gravity range prevailing in the lead-acid battery at various states of charge.

Because of the shape of recently obtained sweep voltammograms [52], there remains, however, some doubt regarding the antimony-induced pore formation mechanism. Furthermore, there is increasing evidence to the effect that the lead oxides also play a significant corrosion protection role, as will be seen in what follows.

In studies involving pure lead, Tarter and Ekler [53, 54] demonstrated the sensitivity of $PbSO_4$ to PbO_2 conversion data to the time of immersion of the electrode in the sulfuric acid solution. For very short immersion times ("fresh" electrodes), a typical potential versus time curve is shown in Fig. 14. Noteworthy is the self-discharge plateau, H-K, whose maximum length corresponds to a 10.5$\underline{N}$ H_2SO_4 concentration. This behavior has been interpreted as a trade-off between an increase in PbO_2 stability at higher electrolyte concentrations and a decrease in the corresponding amounts of PbO_2 formed on charge.

Kabanov and co-workers [55] suggested a self-discharge mechanism occurring by oxygen transport from the oxide layer to the lead grid. A solid-phase reaction of the type PbO_2 + Pb $\rightarrow$ 2PbO was also discussed by Lander [56-60]. Furthermore, his findings suggest that in ordinary service PbO_2 will be present adjacent to the grid, more than PbO, and the rate of corrosion will be determined accordingly. Deep discharge or overcharge states would correspond to maximum corrosion conditions, and a fully charged float mode approximating automobile voltage regulation would represent an optimal situation. For such conditions, a lead sulfate barrier

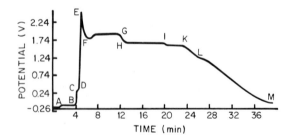

FIG. 14. Typical potential-time recording for short immersion electrode. A to G, charging; G to M, discharging; A, Pb/$PbSO_4$ potential; F to G, "charging plateau" (PbO_2/O_2 potential); H to K, "first discharge plateau" ($PbSO_4$/PbO_2 potential); L, mixed potentials. For other steps: see [54].

prevents grid attack by sulfate ions but not by the water molecules present in the electrolyte solution.

This selective permeability of $PbSO_4$ corrosion layers on Pb in sulfuric acid, first suggested by Pavlov and co-workers (see Sec. 2.3), has been shown by Ruetschi to resemble that of precipitation membranes prepared by the interdiffusion of Pb^{2+} and SO_4^{2-} ions into cellophane foil [61]. The lower limit of the potential range where PbO and/or α-PbO_2 form in the interior of a corrosion film, beneath the permselective layer of lead sulfate, that is, +0.25 or +1.15 V versus H_2/H^+, respectively, can be predicted theoretically.

The lead sulfate crystals in the outer layer are of the order of 0.1 to 10 μm in size and the layers become impermeable to SO_4^{2-} and HSO_4^- ions when they are at least 1 μm thick. Further corrosion of lead then occurs by water dissociation, Pb^{2+} precipitation as PbO, and migration away of the H^+ ions. It should be noted, however, that in the case of lead-antimony alloy grids the sulfate films are liable to be less dense owing to the higher solubility of the basic antimony sulfates. Ruetschi suggests that the lead sulfate films may then be less permselective, with a corresponding effect on the diffusion potential, and α-PbO_2 would be able to form even at lower potentials than +1.15 V versus H_2/H^+.

The grid protection mechanism by apparently dense lead dioxide films has also been mentioned recently in connection with a novel way of manufacturing "pseudo-dry charged batteries" [62] (see Sec. 4.2.5.3 regarding the "power spin" processing technique). In this case the lead-acid batteries are kept in storage with a certain amount of electrolyte solution retained in the plate-separator assemblies.

2.2.2. Structural Aspects and Theoretical Models

Simon and co-workers have carried out detailed microscopic studies of morphology changes in cycling lead dioxide (positive) plates [63, 64]. These were various commercial samples using antimonial lead grids. Basically, the discharge process was shown to be associated with lead dioxide dissolution and nucleation of lead sulfate crystals at preferred sites on the lead dioxide-active material. At the end of discharge, even the lead dioxide portion which has remained uncovered can be considered essentially encapsulated within the lead sulfate crystals. This is representative of the typically limited utilization of active material in the lead-acid battery, of the order of 50 to 60% at the most. While not providing any more discharge capacity, the remaining lead dioxide structure can be visualized as providing a conductive path (lead sulfate is, as we have seen, highly resistive) during the recharge process.

Recharge formation of lead dioxide proceeds, as expected, on the existing nonsulfated lead dioxide particles. Eventually, some of the residual lead sulfate crystals become dioxide encapsulated and, upon final dissolution, leave hollow lead dioxide shells, i.e., a weakened structure.

While the lead dioxide plate structure continues changing with time, Simon and co-workers reported that plates produced by various manufacturers appear to become morphologically indistinguishable after only a few initial charge-discharge cycles [64]. This is so because during the initial electrochemical formation process, lead dioxide originates from a matrix whose composition (in terms of the basic sulfates and lead oxides) varies from manufacturer to manufacturer. After cycling, however, lead dioxide is always derived from a matrix of lead sulfate crystals whose microstructure and composition will be identical, irrespective of the nature of the starting material in the fresh, unformed plate, provided that the discharge conditions are comparable.

In similar microscopy studies by Simon et al. it was also shown that lead dioxide forms by two separate mechanisms during charging: nucleation and growth upon any lead dioxide left over from the preceding discharge reaction, producing a layer of very small and compacted crystals, and growth from the lead sulfate-encapsulated material forming a more reactive and fragile network of porous lead dioxide. After a high rate discharge, a multilayered dense coating of small lead sulfate crystals is formed and the lead dioxide utilization efficiency predictably drops to lower values. A correspondingly dense layer of lead dioxide is formed upon recharge, owing to the lack of the less dense lead sulfate-encapsulated species described above. Some reactivity loss ensues with time, and several lower-rate charge-discharge cycles are necessary to restore a more active structure (see also Sec. 4.1.2 regarding the effect of antimony on the lead dioxide structure).

Theoretical electrode modeling has been applied specifically to the positive lead dioxide electrode. Euler [65] used electrical analog analyses to suggest that, because of the heterogeneity of the lead dioxide system, the initial uneven current distribution across the electrode thickness is maintained for a relatively long discharge time, with a reaction layer proceeding inwards from the electrode surface. His analysis is limited, however, to small current densities owing to his assumption of a linear rather than exponential polarization equation. The more general macrohomogeneous model, which regards the electrode as a combination of a solid and a liquid phase and disregards the actual geometry of each pore, has recently been applied to the positive plate current distribution problem by Simonsson [66] and checked experimentally. The model was used to show how, in the case of discharges at high current densities (100 mA/cm^2), the concept of a reaction layer moving from the electrode surface to its interior can be used. At low current densities (2.5 mA/cm^2) current distribution, and the corresponding lead sulfate concentration, are uniform at the beginning of discharge with the outer parts adjacent to the surface being preferentially utilized under deeper discharge conditions.

2.2.3. Cycle Life

The positive lead dioxide plate is the life-limiting component in a lead-acid battery. As such it has formed the subject of continuing studies aiming at a correct definition of the process whereby the plate performance and structure degrade with time to the point of eventual failure. Active material shedding and grid corrosion are the main phenomena targeted in those studies.

Voss and Huster [67] suggested, based on empirical studies, the existence of a linear relationship between the depth of discharge to which a positive plate is subjected and the logarithm of the number of charge-discharge cycles measured to the failure point. The latter is defined as the point at which the capacity corresponding to the depth of discharge under evaluation can no longer be reached. This finding had previously been reported also for alkaline batteries. Furthermore, the authors obtained a linear relationship between the cycle life to the 40% capacity point and the reciprocal of the depth of discharge. Their experiments were conducted "all other things being equal," and they represent merely one exploratory step in the direction of cycle life prediction for defined battery systems.

Active material shedding is intimately associated with the performance decay and end of life phenomenon. Radioactive tracer studies using ^{210}Pb have been used to postulate that shedding is a result of slow lead transfer through the pores of the active material, in the form of soluble lead sulfate [68].

2.2.4. Float Conditions

There are numerous applications where a lead-acid battery is maintained fully charged, under voltage-controlled float conditions, for a significant portion of its service life. This is tantamount to a low rate of intermittent charging at potentials slightly above the open circuit value. Normally the process is dominated by the voltage of the negative plates, and the voltage fluctuation of the battery is reflected by that of the negative plates.

In a battery using lead-calcium grids, it has been shown that the charge acceptance efficiency of the positive plates is only about 78 to 85% compared to almost 100% for the negative plate (see also Sec. 4.3.1). There are marked differences in the morphology of good and failing ("soft" active material) positive plates, with the former possessing prismatic crystals with many side arms and the latter exhibiting nondescript agglomerates with smooth contours and rounded edges [69]. Little crystalline $PbSO_4$ is present in the floating plate and thus cannot contribute by itself to structural stability. The texture and capacity of the active material can be improved by prolonged discharge cycles with prismatic $PbSO_4$ crystals formed as radial spines on the active material particles.

To these observations Simon [70] has added the results of studies showing that, under certain conditions, a reticulate structure of α-PbO_2 adds a mechanical reinforcing effect to that of the branched prismatic

β-PbO$_2$ crystals. The reticular structure actually appears to enclose the β-PbO$_2$ crystals within its mass.

2.2.5. Low Current Processes

A. Ragheb and co-workers [71] reported that at very low current densities, of the order of less than 1 mA/cm^2, lead dioxide-covered pure lead will dissolve anodically at approximately +1.60 V versus H$_2$/H$^+$. During the recharge process the lead ions are then converted directly to lead dioxide within the pores of the already formed active mass of lead dioxide. At approximately -0.12 V, the reduction of lead dioxide to lead sulfate also takes place via a soluble lead ion mechanism.

These very low current processes can be responsible for a higher electrode activity through an increase in the number of the active sites by the soluble lead ion processes described above.

The lead sulfate crystals formed during discharge are such that they grow individually to a definite size without passivating the lead dioxide. This is quite different from the situation prevailing at high cathodic current densities when partially passivating lead sulfate films, rather than distinguishable larger crystals, are deposited on aggregates of lead dioxide particles [72].

2.3. The Negative Electrode

The negative plate can self-discharge according to the reaction

$$Pb + H_2SO_4 \rightarrow PbSO_4 + H_2 \tag{11}$$

The relatively high hydrogen overvoltage of pure lead (or lead-calcium alloys) can be significantly decreased by the presence of other contaminants. In the case of antimony as a grid-alloying constituent (see Sec. 4.1.2), the decrease amounts to approximately 0.2 V at room temperature and results in a greatly accelerated self-discharge reaction (an average of 15 to 30% of the capacity per month, versus roughly 3% for lead or lead-calcium systems).

The self-discharge is a result of antimony migration from the positive to the negative plate; this migration can be retarded by the use of effective microporous separators. The same holds true for another type of negative self-discharge reaction due to oxygen transport and diffusion-limited [51]:

$$Pb + \frac{1}{2} O_2 + H_2SO_4 \rightarrow PbSO_4 + H_2O \qquad (12)$$

In the absence of lead sulfate nucleation sites, such as on electropolished lead surfaces, a better insight can be obtained regarding the mechanism of lead anodization to the Pb^{2+} sulfate form. Archdale and Harrison [73-75] have shown that at low anodic potentials, between -0.307 and --0.278 V versus H_2/H^+, Pb dissolves as Pb^{2+} and $PbSO_4$ ions. At potentials higher than -0.278 V, $PbSO_4$ is formed either by a solid state reaction or by a solution-precipitation mechanism.

As the discharge of a negative plate proceeds at progressively higher current densities, the lead sulfate distribution across the plate thickness becomes nonuniform and the active material converts to sulfate to a greater extent at or near the electrode surfaces. This transition from even to uneven utilization occurs at about 30 mA/cm^2 (approximately the 1-h rate for "standard" commercial plates) [76], whereas in the case of a positive plate it can occur significantly sooner, i.e., between 6 and 30 mA/cm^2 [33]. Generally speaking, the negative plate is performance-limited by its available surface geometry and its capacity for lead sulfate passivation and, in those terms, can have a calculated optimum thickness of the order of 1.8 mm [77] or less, depending on its required discharge regime. The passivation itself occurs by lead sulfate plugging of the active material pores rather than by coating of the lead particles [78].

In an optical microscopy study of various commercial plate samples made with lead-antimony grids, Simon [79] pointed out that lead sulfation or lead sulfate reduction phenomena at the negative plate are different during the initial electrolytic formation step (see also Sec. 4.2.5) and during subsequent charge-discharge cycling. During formation one deals with a sulfate solution and lead reprecipitation mechanism ending with well-defined morphologies, such as dendrite or platelet-type crystals. During cycling lead will form, upon charge, on top of existing lead sulfate structures conforming to their shape.

When the lead anode discharges, as the layer of freshly formed sulfate increases in thickness, new crystals are generated; these are usually equiaxial polyhedra. After a finite number of charge-discharge cycles, the electrode structure becomes granular and loses its original needle-like morphology. With time, and even more so as a result of deep discharging, significantly larger portions of the originally available lead become structurally encapsulated by the sulfate and unavailable for further discharges.

Pavlov and Popova [80] investigated the lead/lead sulfate system for pure lead in 1$\underline{N}$ sulfuric acid, i.e., at formation solution concentrations. They defined two kinds of passivation: stable and unstable. Stability is the result of distances between the lead sulfate crystals equivalent in magnitude to the ionic diameters. This makes the anodic lead sulfate layer act as a "permselective" membrane allowing free access of the H_3O^+, OH^-, and Pb^{2+} ions to the lead interface but hindering the transport of SO_4^{2-} ions. Passivation then follows as a result of intercrystalline precipitation of lead monoxide and basic lead sulfates. Instability, or self-depassivation, is the result of larger intercrystalline distances within the lead sulfate layer with passivation possible only during anodic polarization. Upon interruption of the polarization, the protective precipitate of lead monoxide and basic lead sulfates dissolves. As described in Sec. 2.2.1, the lead sulfate selective permeability concept was more recently applied by Ruetschi [61] to the positive electrode as well.

2.4. Phosphoric Acid Effects

2.4.1. General

The utilization of phosphoric acid as an electrolyte additive has been patented as far back as 1929 [81, 82] for the claimed purposes of strengthening of the positive active material and preventing "harmful sulfation," i.e., "hard" or refractory sulfate formation, during long discharged stand conditions. In the case of tubular positive plates (see Sec. 4.1.4), a reduction in active material shedding and in the rate of positive grid corrosion have also been claimed [83].

More recently, Tudor and co-workers [84-86] used autoradiography with labeled sulfur, ^{35}S, and phosphorus, ^{32}P, to demonstrate that the cycle life of lead-acid batteries using plates with lead-calcium grids is increased when the electrolyte contains small amounts of phosphoric acid, at the cost of some reduction in the positive plate capacity, of the order of 5 to 10% [87,88]. In essence, the mechanism by which phosphoric acid increases cycle life was interpreted as one of modification of the pattern of lead sulfate distribution at the active material-grid interface [86]. The result is a more conductive interface, with less of a $PbSO_4$ barrier and less interference with the discharge process. It would also appear that a chemical and/or electrochemical reaction occurs between phosphoric acid and the positive active material, with phosphoric acid incorporation during charge and release during discharge [85]. Some substantiation of this phenomenon is expected in the near future, as a result of electron microprobe analysis techniques coupled with scanning electron microscopy.

The possible formation of lead phosphate compounds in the positive plate during charge is also likely to be responsible for the observed reduction in capacity in such cases. This phenomenon is also evident in batteries using lead-antimony grids such as, for instance, water-activatable types based on the utilization of a boron phosphate-gelled sulfuric acid electrolyte [87,88] (Sec. 4.1.6). Other effects due to the presence of phosphoric acid in the electrolyte solution are lead "mossing" and dendrite formation on the negative electrode.

Lead mossing at the positive plate has also been reported in the case of electrolytes containing more than 10 g of phosphoric acid per liter of solution [83], although this finding has not been corroborated by other investigators [87,88].

2.4.2. Soluble Lead(IV) Species

Moss and dendrite formation depend on the presence of soluble lead in the electrolyte solution. The Pb(IV) species have lately received more attention, particularly with regard to their possible effect on the performance of the positive lead dioxide electrode. Thus, dissolved β-PbO$_2$ was reported in sulfuric acid solutions [89] (for example, 0.15 mM/1,000 g H$_2$O in approximately 33% H$_2$SO$_4$).

In phosphoric acid-containing sulfuric acid solutions, the existence of two relatively stable tetravalent lead phosphate species has been reported, based on chemical and electrochemical studies [46, 90-92]: yellow 2PbO$_2$·P$_2$O$_5$·XH$_2$O and white PbO$_2$·P$_2$O$_5$·XH$_2$O. Recent polarographic measurements have confirmed the existence of soluble Pb(IV) species [93] and their generation, at the onset of overcharge, at the lead dioxide electrode. These are strong oxidizing agents, hence some separator degradation effects are observed.

3. APPLICATIONS AND PERFORMANCE CHARACTERISTICS

3.1. Automotive and Related Uses—The SLI Battery

The starting-lighting-ignition or SLI battery (Fig. 15) [94] has several functions clearly defined by its name (see also the performance criteria defined in Sec. 4.3.2):

1. Engine starting, which requires a high rate—several hundred amperes—discharge capability over a wide temperature range. Particularly critical is the performance, i.e., available capacity above a minimum voltage, at winter temperatures.
2. Operating the lights without the benefit of battery recharge for sufficient periods of time. This requires an adequate low rate capacity. Also, powering all required electrical accessories even under engine idling conditions.
3. Maintaining the electrical engine system (ignition) operational.

Since these various types of requirements would each demand a different optimized battery design, the available product inevitably represents a compromise.

With the advent of higher compression ratio and higher displacement engines in the early 1950s, a transition was made from 6- to 12-V battery

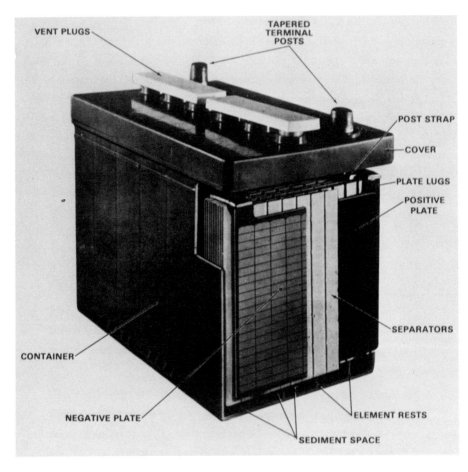

FIG. 15. SLI battery, courtesy of Globe-Union, Inc. [94].

designs with a corresponding, though not proportional, reduction in rated
capacities from 70 to 135 Ah down to 50 to 110 Ah.

Figure 16 is a family of capacity-voltage discharge curves for a typical 100-Ah (20-h rate) SLI battery, with the discharge current as parameter [95]. The data are for room temperature operation, and the accepted
voltage cut-off line is also indicated in the figure. Room temperature instantaneous current-voltage relationships for several sizes of SLI batteries

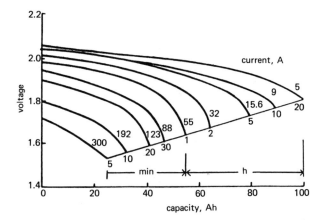

FIG. 16. Capacity-voltage curves for typical 100-Ah (20-h rate) SLI battery [95].

are given in Fig. 17, and current versus available capacity of a 100-Ah battery as a function of temperature is given in Fig. 18.

Lately, increased attention is being focused on high-rate, low-temperature discharge capabilities for which an SAE specification (see Sec. 4.3.2) would require 3 min discharge at three times the C_{20} rate at $-18°C$ (C_{20} is the capacity at the 20-h rate).

With the appearance of new engine designs such as the rotary type, a new set of SLI battery performance requirements has surfaced. Even higher starting capacities would be needed for aircraft engine starting at subzero temperatures, whereas low rate capacities (e.g., at the 20-h rate) are of less consequence. Thus, one can anticipate increased activity involving designs with more and thinner plates per element and minimum-resistance separators such as those in the microporous polymer category.

Significant improvements have been made over the years in the materials of construction and assembly methods for lead-acid batteries in general, and SLI batteries in particular. These are discussed in some of the other sections in this chapter. A typical weight breakdown for a modern

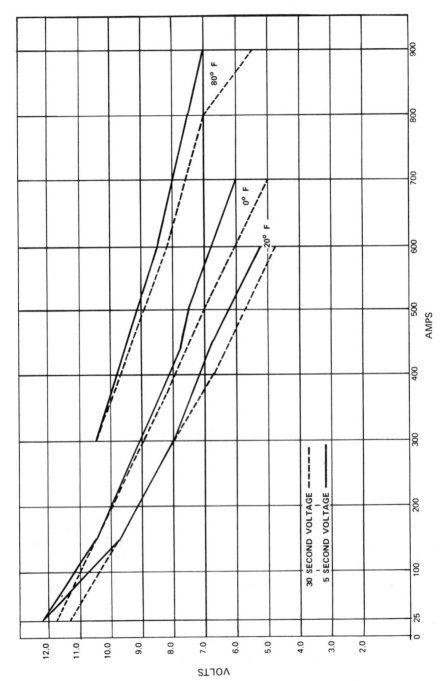

FIG. 17.   Typical engine-starting discharge characteristics of original-equipment, 68-Ah, 12-V SLI battery.

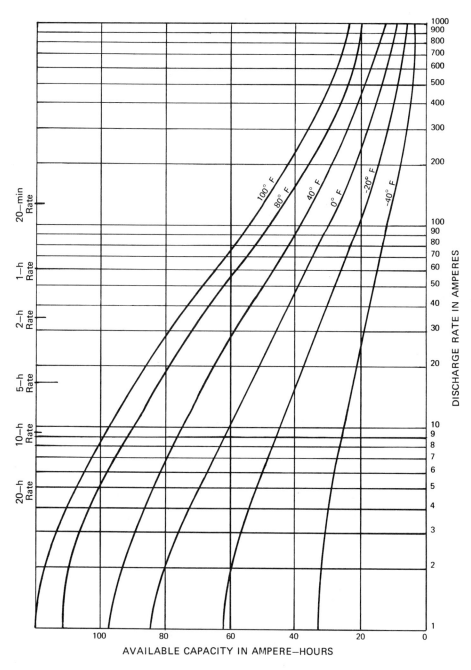

FIG. 18. Current versus available capacity at various temperatures for a typical 100-Ah (20-h rate) SLI battery.

SLI battery is detailed in Fig. 19. It is also interesting to note the effect
of an improvement such as the thin-walled polypropylene case battery (e.g.,
the Sears Die Hard variety) on specific energy, high rate capability, and
utilization efficiency of the positive active material which is normally
electrolyte-limited in high-capacity lead-acid batteries (Figs. 20-22 [96],
23 [97]).

3.2.   Traction and Industrial Batteries

There exists a variety of specialized applications involving on-the-road-
type vehicles—delivery vans, small limited-range passenger cars, golf
carts, motorbikes, rail cars—and industrial vehicles—warehouse train vehicles, forklift trucks, mine vehicles, etc.—and this specialization is reflected in the types and characteristics of the corresponding batteries.

Lead-acid batteries are used almost exclusively for electric on-the-
road vehicles and have formed the subject of numerous studies and tests
involving all-electric as well as hybrid electric power plants.   The latter

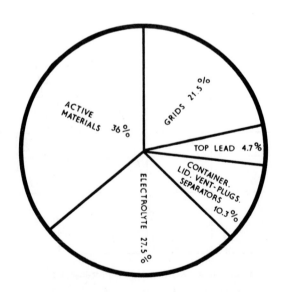

FIG. 19.   Weight analysis of a typical SLI battery [95].

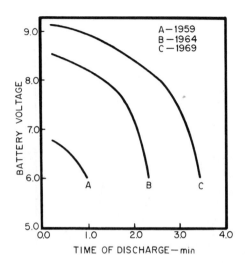

FIG. 20. Improvement of the high rate capability of a group 24 SLI battery [96] (container material: A, hard rubber; B, resin rubber; C, polypropylene).

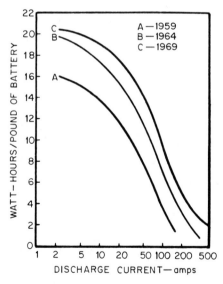

FIG. 21. Improvement of the weight-specific energy of a group 24 SLI battery [96] (container material: A, hard rubber; B, resin rubber; C, polypropylene).

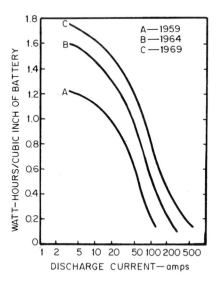

FIG. 22. Improvement of the volume-specific energy of a group 24 SLI battery [96] (container material: A, hard rubber; B, resin rubber; C, polypropylene).

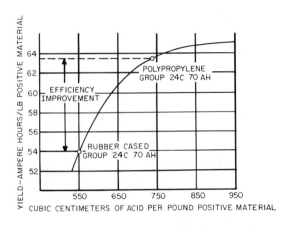

FIG. 23. Improvement in the positive active material utilization efficiency of group 24 lightweight case batteries [97].

normally use combinations with constant-rpm internal combustion engines, although other schemes, such as those based on hybridization with fuel cells or other types of batteries, have also been considered. (See Chapter 2, Electric Vehicles.)

The development of lightweight polymeric battery cases and improved separators and intercell connectors has been highly significant for the technology of on-the-road electric vehicles, especially in areas of widespread use, such as Great Britain, where the use of electric delivery vans is well established and documented [98], or Germany, where extensive use has been reported for passenger railcars [99]. Both tubular and flat-plate batteries are used, and significant progress has been made possible by increases in the volumetric capacity of tubular-plate batteries by using woven tubes that make it possible to reduce the design pitch (see Sec. 4.1.4). Figure 24 illustrates both types of traction batteries mentioned above [100].

Industrial vehicle batteries normally reflect more conservatism in their design. Hard rubber cases are frequently used, with a resulting waste of at least 10% of the potentially available active material volume. There is a large range of discharges that truck batteries must be able to deliver in practice. A pedestrian pallet truck, for instance, will usually require a 3-h discharge rate or less whereas, at the other extreme, a high-lift narrow-aisle truck will often use up to 75% of the battery capacity at the 1-h discharge rate [101].

Many types of industrial trucks drain their batteries at a relatively low rate, because they are used only at low speeds. This, in turn, causes the battery to deliver its available electrochemical capacity at a higher level of utilization efficiency. Appropriate designs, such as those based on tubular positive plates, will provide 1,500 cycles and more at 80% depth of discharge and 1 cycle/day. Other types of industrial batteries, using a rugged, thick-plate design, will typically run 5 to 7 years at low rates of discharge. Specific energy per weight is of no great consequence; as a matter of fact, a finite dead weight (counterweight) is required in many types of industrial trucks. Space and cost are of prime importance followed by long service life and low maintenance requirements.

FIG. 24. Examples of flat and tubular-plate lead-acid cells used in traction batteries [100]. Left, flat positive plate cell showing: 1, positive plate (antimonial lead); 2, glass wool separator; 3, porvic separator; 4, wood veneer separator; 5, negative plate (antimonial lead); 6, bottom block of mud rib; 7, spray arrestor; 8, cell cover; 9, intercell connector; 10, PVC rod; 11, cell jar; 12, vent plug and gasket; 13, terminal post and bus-bar; 14, post gasket. Right, tubular positive plate cell showing: 1 and 3, perforated PVC tubes; 2, glass fiber sheath; 4, polyethylene foot; 5, active material; 6, lead alloy spine; 7, separation comprising microporous dia-phragms and perforated PVC spacers; 8, supports; 9, negative plate; 10, acid baffle and separator shield; 11, ebonite lid; 12, vent plug; 13, hard rubber or plastic container; 14, acid-tight post seal.

In the case of traction batteries for on-the-road vehicles or leisure-
type applications (e.g., golf carts), adequate specific energy per weight is
crucial if the capability for longer distances between recharges is to be attained. High rate pulse capability on shallow depths of discharges, of the
order of 10 to 25% of the capacity combined with a good efficiency of high-
rate charging until an almost fully charged state is reached (very difficult
to achieve) is a requisite for some types of battery-internal combustion
engine hybrid systems. For all-electric traction, the battery requirements change to deep discharge capability at high efficiencies of active
material utilization and long cycle lives, at a great variety of discharge
and charge rates.

Standardized traction battery capacity ratings are provided by the
manufacturers on the basis of a 5-h constant current discharge at $27°C$. A
typical set of discharge curves is given in Fig. 25.

No single type of lead-acid battery can satisfy such a diversity of applications, and different designs are used as needed. For on-the-road vehicles a balance has to be struck between achieving high rate capability, particularly for the hybrid designs, and adequate service life. High rates can
be obtained by the use of high-area multiplate cells, specialized designs
such as bipolar structures [102-107], which have yet to be reduced to mass
production, or the less significant so-called tripolar design—current col-

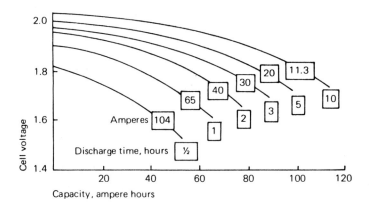

FIG. 25. Typical traction battery discharge characteristics [100]
(base: 100 Ah at the 5-h rate).

lection from two bottom and one top tab at each plate—which has entered
production [108, 109].

Special grid designs to decrease the internal resistance as much as
possible have also been developed; the laminar type is an example in point
[110, 111]. In this case, a plate uses several grids superimposed in such
a way that the conducting branches of one grid are offset with respect of
those of the other grid or grids. Some of the grids use radial rather than
horizontal and vertical conducting members. The ratio of active material
to grid weights is extremely low. An example of such a design [111] uses
6 wt % Sb grids 1 mm thick, with a triple grid laminate for the positive
plate and a double grid laminate for the negative plate. The active mate-
rial/grid weight ratios are approximately 1. Twenty-two watt-hours per
kilogram and 60 Wh/dm^3 are claimed at the 1-h rate. These are high specific energy values by lead-acid battery standards.

A general discussion on the data and performance required of the lead-
acid battery by the electric vehicle designer has recently been presented
[112]. In the experimental area, Agruss evaluated various storage battery types for traction applications [113, 114]. For lead-acid batteries he
showed that, for varying power discharges, the principle of additivity was
obeyed. Accordingly, the capacity removed at any power level may be
added to that removed at a different power level to determine total amount
consumed and amount remaining before the cut-off voltage is reached.

Kordesch reported on different galvanic hybrid systems [115] and
passenger car field test results involving fuel cell/lead-acid battery hybrid
power trains [116, 117]. The batteries were commercial 84-Ah SLI types
of the Sears Die Hard variety, in lightweight polypropylene cases, and they
yielded 24 Wh/kg at the 50-A discharge rate. This type of result shows
that even SLI batteries can be subjected to a hybrid-traction duty cycle
within acceptable life and performance limits, if one uses a taper charge
mode.

In the United States there is available an Interim Federal Specification,
No. W-B-00131J, dated March 13, 1973, for industrial and traction batteries and a Provisional Electric Vehicle Specification (SAE, 1976). Much
work remains to be done in this area, though, as attested by the rather

active literature dealing with electric traction in its varied aspects. A recent communication of interest [118] deals with the dynamic characteriza-
tion of lead–acid batteries for vehicular applications. The "capacity recuperation phenomenon" is referred to. This terminology describes the fact
that, within limits, the longer one maintains the period of open circuit
stand between discharge pulses, the higher the available discharge capacity
which is obtained. In most instances, and for a narrow range of open circuit stand times, e.g., 1 to 2 min, the normally ascending capacity versus
OC time actually passes through a minimum (Fig. 26).

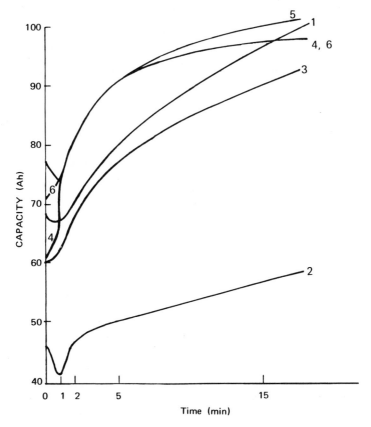

FIG. 26. Illustration of the capacity recuperation phenomenon for
200-Ah golf cart batteries (1.280 sp gr electrolyte). Batteries 1 and 3,
and 4 and 6 are identical; 2 and 5 are different models from a third manufacturer; results are condensed from Ref. 118.

For hybrid power trains, where the battery is normally called upon to deliver shallow discharge pulses of high current, Silverman and co-workers [119] have carried out an optimization study for an improved specific power battery capability. High-porosity thin plates, with a grid having a wider spacing between the lead members, yielded in excess of 440 W/lb and sustained power of 415 W/lb for 20 sec. This compares favorably with some of the better results realized in practice for plastic case SLI batteries (Sec. 3. 1).

3.3. Stationary Batteries for Standby Power Supply

Lead-acid batteries are in widespread use as stationary sources of standby power for a variety of requirements, including: telephone exchanges, computer installations (where the feature of UPS, i. e., uninterrupted power supply, is particularly critical for the latest or third-generation models), remote location communication stations, etc.

While the negative plates in such batteries are of the pasted Faure type or of the box type where the active material is sandwiched between pairs of grids riveted together [120], the positive lead dioxide plates are of various designs: Planté, Faure, or tubular (see Sec. 4. 1. 4 for a complete description of the various types of lead-acid battery plates).

Telephone exchanges, which are the principal users of stationary batteries both in terms of installed ampere-hours and annual expenditures, normally require 46 to 52 V dc. The standby batteries provided for that purpose in Europe often use Planté [120-124] or tubular [125] positives. In the United States the Faure positives are extensively employed [126, 127]; their greater voltage range between the fully charged and discharged states are compatible with the somewhat wider exchange voltage limits (44 to 54 Vdc) and furthermore they offer, together with tubular plate designs, the advantage of lower cost, weight, and volume.

In the Bell System, which uses cells in the capacity range 50 to 7,000 Ah, with an estimated total of 1,000,000 to 1,250,000 cells in the 180- to 1,680-Ah range in present service [126], three functions are served by the batteries: rectified dc filtering (the batteries are continuously

trickle-charged, or floated, across the rectifiers to provide an uninterrupted source of dc power), instantaneous power reserve for short outages
in the millisecond to 20-min range, and extended reserve for periods of
3 to 48 h in the case of engine generator failure [127].

Even larger capacities, of up to 15,000 Ah, are standard for Planté
cells [121].

Lead-antimony and, increasingly, lead-calcium alloys or pure lead
are used for the grids with the lead-antimony alloys presenting the known
disadvantage of a higher tendency to self-discharge, as discussed in other
sections of this chapter.

Normally a dilute electrolyte solution is used (specific gravity 1.210
to 1.215, and occasionally as high as 1.240) microporous polymer (e.g.,
PVC) separators are provided between the plates, and the cells are contained within jars normally made of glass or a suitable polymer such as
polypropylene or polystyrene. Figures 27 and 28 depict typical pasted and

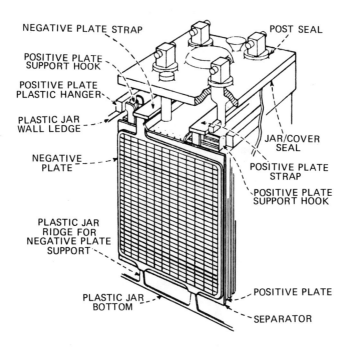

FIG. 27. Pasted-plate stationary battery [126].

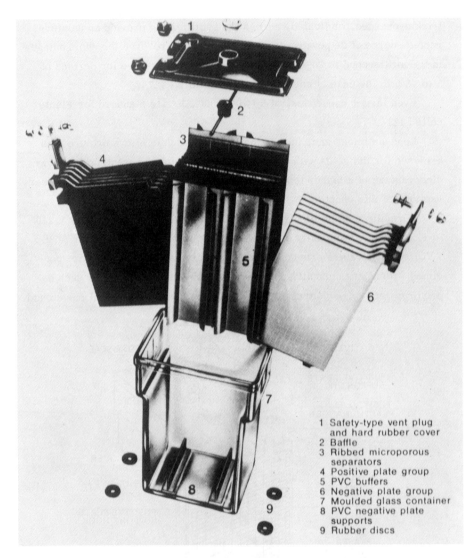

1 Safety-type vent plug
  and hard rubber cover
2 Baffle
3 Ribbed microporous
  separators
4 Positive plate group
5 PVC buffers
6 Negative plate group
7 Moulded glass container
8 PVC negative plate
  supports
9 Rubber discs

FIG. 28. Stationary battery (cell) using positive Planté plates—exploded view [128].

Planté-type cells, respectively, while Fig. 29 is a photograph of a stationary battery installation [128].

The life expectancy of lead-antimony batteries is of the order of 15
years, while that of lead-calcium batteries may be 30 years or more.
Standard life specifications (e.g., U.S. Federal Specification, per Table 4)
are available [129]. Also, various float voltages have been reported in the
literature for stationary batteries maintained in a state of readiness, and a
typical figure is 2.17 V per cell [130]. Optimum floating charge voltages
of 2.16 V for cells using 1.215 specific gravity electrolyte and 2.18 V for
1.240 specific gravity electrolyte, with an allowable voltage range of ±2%,

FIG. 29. Battery of pasted positive plate cells showing transparent
cell containers, safety-type vent plugs [128].

TABLE 4

Minimum Useful Life Specifications

for Stationary Lead-Acid Batteries [129]

Battery type	Life (years)
Lead-calcium, flat pasted	20
Lead-antimony, flat pasted	14
Planté design*	25
Multitubular design*	20

*In terms of positive plate characteristics; lead alloy composition unspecified.

have recently been discussed [131]. New lead-antimony units remain fully charged when properly floated but, as they age, so-called equalizing charges are required every 3 to 6 months, and more frequently in warmer ambients, at higher than float voltages. Lead-calcium batteries can be floated at the higher voltages of 2.20 to 2.25 in order to avoid the need for equalizing charges [130]. A detailed discussion of the float behavior of the lead-acid battery system has recently been published [132].

Figure 30 gives a set of typical discharge curves for a 100-Ah pasted-plate stationary battery with the cut-off voltages as the parameter [133].

An innovation in the field of long-life stationary batteries is the Bell conical grid design [126, 127, 134-137] (Fig. 31). This structure has been incorporated in a radically new cylindrical cell design also incorporating the use of pure lead and positive pastes prepared from tetrabasic lead sulfate and water. The novel grid design is claimed to accommodate the normally destructive long-term grid growth effects, particularly in the positive plate. The presulfated lead oxide used for the positive pastes results in an interlocking structure of elongated PbO_2 crystals, conserving the original morphology of the raw material, $4PbO \cdot PbSO_4$ (see also Sec. 4.2.2). This is claimed to minimize shedding in service [127, 138].

Significant increases in service life are projected for the Bell cylindrical battery, and time will tell if the promising results from accelerated life tests are indeed corroborated in actual use. A complete representa-

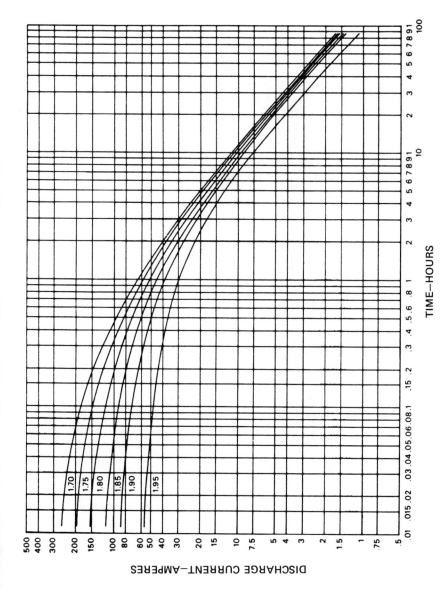

FIG. 30. Discharge characteristics for 100-Ah (at 8-h rate) pasted-plate stationary battery with sp. gr. 1.210 electrolyte [133].

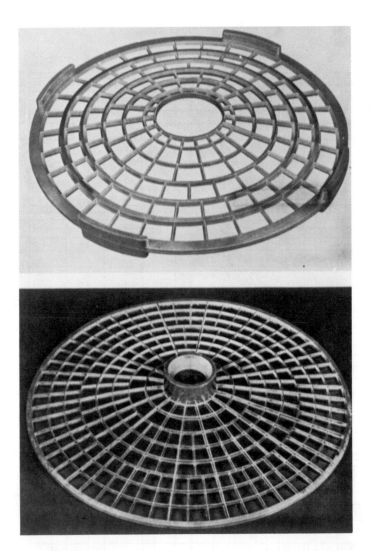

FIG. 31. Conical grids—Bell System [126]. Above—positive grid.
Below—negative grid.

tion of the battery design presently being readied for production and use is given in Fig. 32.

3.4. Maintenance-free Batteries

As lead-acid storage battery use continues to grow, in terms of the number of applications as well as the quantities and types of equipment for each application, the need for maintenance-free designs is becoming more and more pronounced (see also discussion on lightweight plates in Sec. 4.1.4). This requirement has become obvious not only for traction, stationary, and miscellaneous appliance use, but even for SLI applications. In the latter case this is due at least in part to the need for crowding additional components under the car hood in order to meet increasingly more stringent anti-pollution standards. Space and the higher engine operating temperature (owing to pollution abatement-induced inefficiencies) are thus becoming a problem. This may require the battery to be designed into another location on the car body where periodic maintenance, i.e., "topping off" with water, will become impractical.

As will be shown in what follows, an almost generally used approach to the maintenance-free design requires replacing the conventional lead-antimony grid alloy with one less prone to self-discharge and gassing (e.g., lead-calcium; see also Sec. 4.1.2), though at the cost of a higher tendency for positive active material shedding in service. In some instances attempts have been made to compromise by selecting what would appear to be the best of both worlds, i.e., lead-antimony positives (shedding resistant) and lead-calcium negatives (higher hydrogen overvoltage). This approach has been reported by C&D Batteries (U.S.) at the 4th International Conference on Lead in Hamburg, in 1971, for industrial batteries using pasted plates of the Faure variety [139]. Nevertheless, it would seem just a matter of a relatively short time before enough antimony will have migrated from the positive to the negative plates during cycling, even through the best of separators, to eliminate the advantage of the originally antimony-free negative plate. Furthermore, a voltage-sensing type of charge control would require periodic adjustments in such a system.

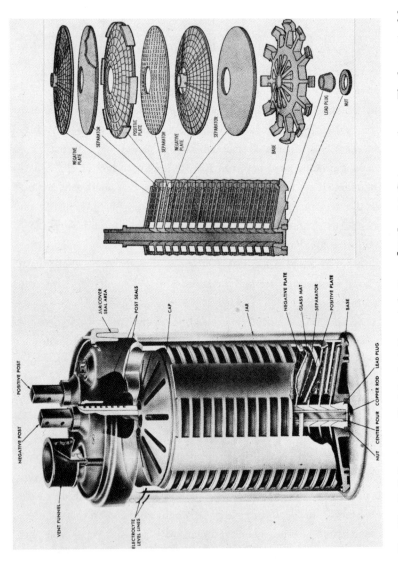

FIG. 32. The Bell System cylindrical stationary battery [137]. (a) Cutaway view. The largest of four proposed sizes of the round-cell battery is shown here (2.06 V, 1450 Ah at the 5-h rate, 14 in. diameter, 30 in. tall). Battery jar and cover consist of clear plastic. (b) Plate, separator, and base assembly.

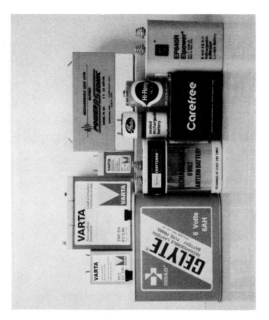

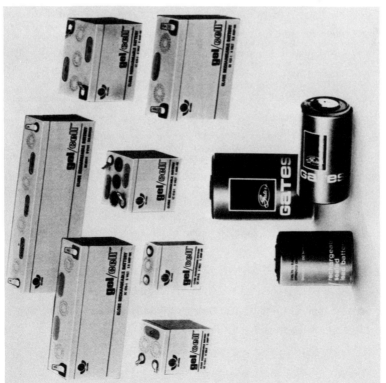

FIG. 33. Various types of nonautomotive maintenance–free lead–acid batteries and cells.

3.4.1. Nonautomotive Batteries

The most commonly used types are relatively small-size attitude-insensitive units for emergency standby applications, cordless power equipment, and various portable and moving appliances [140-147]. Representative examples are shown in Fig. 33.

The electrolyte must be in an immobilized state; this is achieved by conversion to a thixotropic gel with additions of up to 5 wt % silica having submicrometer particles [140,142,144,146,148,149], with or without additional support provided by glass fiber mats, by immobilization with sodium silicate [8,150], or with absorbent separator-matrix components usually made of cellulosic or glass fibers [141,143;151-153]. Usually, though not always, the plate grids are made of lead-calcium alloys in order to increase the hydrogen overvoltage and reduce self-discharge on stand. Often the electrolyte solution also contains phosphoric acid in order to minimize positive active material shedding and thus increase cycle life (see also Sec. 4.1.6.2).

The vents of such batteries are designed to operate as check-valves allowing excess pressure to escape and, at least in theory, preventing the ingress of atmospheric constituents. (We recall here that moist lead negatives will react spontaneously with oxygen and thus self-discharge according to the scheme: $Pb \xrightarrow{O_2} PbO \xrightarrow{H_2SO_4} PbSO_4$.) Usually, the internal gas pressure is equilibrated close to the atmospheric. "Sealed" cells often use emergency vents (weakened spots) for safety reasons.

No provisions exist for additions of make-up water. In case of slow venting the end-of-life can be the result of water losses rather than the usual failure modes, namely: positive plate grid corrosion and active material shedding, or internal shorting. Usually the amount of electrolyte is capacity limiting; this is especially true if the system operates only semi-wet.

Most small maintenance-free batteries come in plastic cases. There are some exceptions, such as the Gates Rubber Corp. cells (Fig. 33), which are encased in metal. The latter type is capable of venting at significantly higher overpressures, of the order of several atmospheres. This

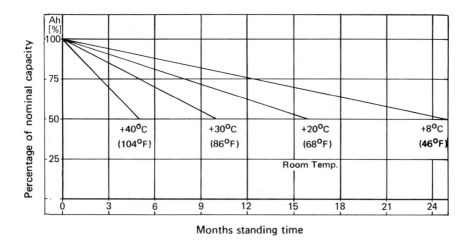

FIG. 34. Self-discharge characteristics of gelled electrolyte
maintenance-free battery using lead-calcium grids [154].

feature brings this type into the group of sealed systems (described in Sec.
3.4.3) with low rate overcharge capability.

The operation of some types of maintenance-free systems has achieved
a high degree of reliability, with 300 to 500 deep discharge cycles possible.
Those with nonantimonial plate grids self-discharge at rates significantly
below the ones corresponding to a standard automotive battery (Fig. 34).

Figures 35 and 36 represent typical discharge capacity curves and the
effect of temperature on battery capacity, respectively. Recharge is
usually carried out in two ways. For "float" or standby service the battery
is held continuously across an appropriate voltage source, for example,
2.25 to 2.30 V [154]. For deep discharge applications a tapered current
controlled voltage charging mode is typical; furthermore, the initial charging current is limited to, say, three to four times the 20-h rate, and at the
end of charge voltage, for example, 2.4 V [154], charging is maintained
until the current drops to a prescribed value determined by the capacity
rating of the battery. After that, the charger is disconnected or switched
to a float mode.

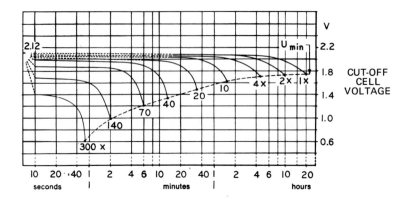

FIG. 35. Operating discharge times at various multiples of the 20-h rate for gelled electrolyte maintenance-free batteries [154].

3.4.2. Automobile Batteries

1973 marks the first year when increased activity has been observed (in the United States) in the manufacturing of a complete line of maintenance-free SLI batteries. They incorporate some of the features discussed above, namely: plates using nonantimonial grids, e.g., made of Pb-0.05 to 0.10 wt % Ca, of a conventional design, or specialized configurations such as the Delco "wound wire" electrode consisting of a fan-shaped array of lead-

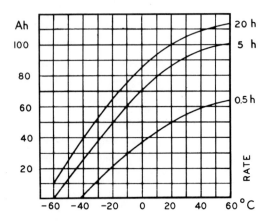

FIG. 36. Effect of temperature on capacity for gelled electrolyte maintenance-free batteries [154].

calcium wires interlocked, by hot-pressing, with polymeric monofilaments
of suitable thermoplastic materials such as polypropylene [155-158].
Pocket-type separators are often used around the shedding prone positive
plates. The cases and vents are made of plastic materials (see Sec. 4.1.7),
and venting is atmospheric (no internal pressures are allowed).

The electrolyte solution is usually contained as a free liquid, and thus
the battery is not attitude-insensitive. In some instances provision is made
for excess electrolyte solution, e.g., 60%, above the plate elements, as a
reserve [159].

3.4.3. Hermetically Sealed Constructions

Hermetically sealed storage cells and batteries represent the ultimate in a
maintenance-free design. To date, this approach has been demonstrated
in a variety of alkaline systems (e.g., nickel oxide-cadmium batteries with
KOH electrolyte). The reduction to practice, though, has not been easy and
a variety of difficulties have been reported: sealing problems, negative
"fade" (e.g., passivation of the negative cadmium electrode with time),
gradual decrease in the effectiveness of the negative active material reserve which is provided in some designs in order to minimize—if not entirely suppress—hydrogen evolution during charge and overcharge, the general problem of coping with the recombination of nonstoichiometrically
evolving oxygen and hydrogen, etc.

The technical approaches necessary to achieve gas-tight operation in a
lead-acid battery are quite similar.

In addition to the problems mentioned above, there are some peculiar
to the lead-acid system: hydrogen evolution during self-discharge, particularly for systems using lead-antimony plate grids, and an electrolyte which
participates in the cell reaction, which makes it more difficult to limit the
amount of electrolyte solution used whenever this is required for the optimization of the internal gas recombination process.

A variety of existing alkaline sealed-cell designs as well as some of the
recent acidic new systems use auxiliary (third) electrodes as overcharge
recombination reactors [160] or as voltage-sensing overcharge-limiting
controls. Generally speaking, however, two approaches are most fre-

quently in use or under consideration for hermetically sealed batteries:

1. Operation based on the "oxygen cycle"; in this case hydrogen evo-
    lution is suppressed and the evolved oxygen (during charge and
    overcharge) is recombined at the always partially discharged nega-
    tive electrode [161-168]. A schematic of such a process is given
    in Fig. 37. The cycle is symbolized in the upper right-hand corner
    of the figure, while the remainder is a more detailed representa-
    tion where each of the chemical species involved appears as both a
    source and a sink at steady state [168].
2. Catalytic recombination of hydrogen and oxygen inside or outside
    the battery; in the latter case, provision is made for the return of
    the product water to the electrolyte chamber [165, 166, 169-174].

The same constructional features apply as in the case of the vented designs: plate grid alloys chosen so as to increase the hydrogen overvoltage,
immobilized electrolyte systems, venting mechanisms or rupture dia-

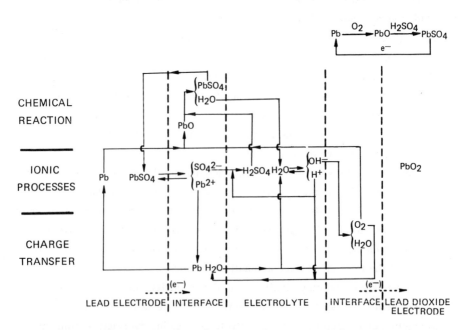

FIG. 37. Schematic of the oxygen cycle in a gas-tight maintenance-
free battery [168].

phragms, except that in this case they are activated at pressures which can be significantly higher than ambient. Indeed, some of the commercial immobilized electrolyte maintenance-free batteries equipped with check-valve vents do operate at times in a limited oxygen recombination mode.

Hermetically sealed lead-acid batteries have become practical, at increased cost, for some applications presently using other systems such as nickel oxide-cadmium. Both approaches listed above pose practical problems. The use of a catalyst is predicated on finding the right kind. Typically, in an acid medium only noble metals would be effective at acceptable recombination rates. Since the lead electrode is highly susceptible to poisoning under these conditions, with attendant self-discharge and hydrogen evolution, the catalyst beds can only be used external to the battery. The overall volume of the package thus becomes unwieldy.

The situation could change if other catalysts, e.g., phthalocyanines [165], are proven practical, but even in such a case additional internal volume would be required inside the battery in order to accommodate the recombination reactor.

The most attractive approach is to use the oxygen cycle and no third electrode. Normally, the battery requires a controlled electrolyte-electrode interface which would leave unflooded active sites available for gas recombination. Such a battery operates in an "electrolyte-starved" mode and at steady state internal pressures higher than the ambient, such as 2 or 3 atmospheres absolute. Some promising progress has been made in this area, including the feature of recombination capacity even for hydrogen [168]. The latter proceeds at significantly lower rates than for oxygen but may still be of interest since self-discharge and hydrogen evolution have not yet been completely suppressed even in the most maintenance-free type of lead-acid battery.

Gates Energy Products, Inc. [175], a subsidiary of the Gates Rubber Corp., designed a spirally wound cell which, after many years of development, has reached a high state of perfection. In Figs. 38 through 42 the principal performance characteristics of those cells are illustrated.

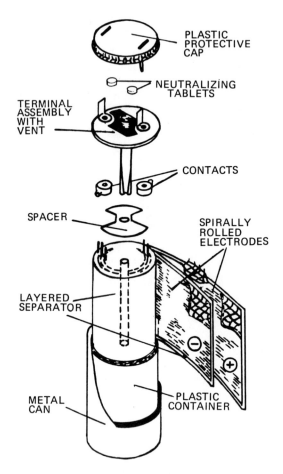

FIG. 38. Construction of D-size cell (Gates Rubber Co., 2.5 Ah at C/5 rate).

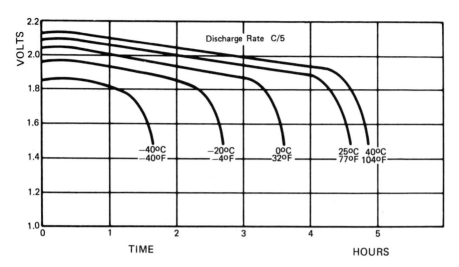

FIG. 39. Discharge time as a function of rate at various temperatures [175].

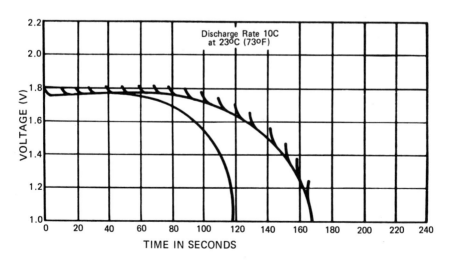

FIG. 40. High pulse current discharge [175].

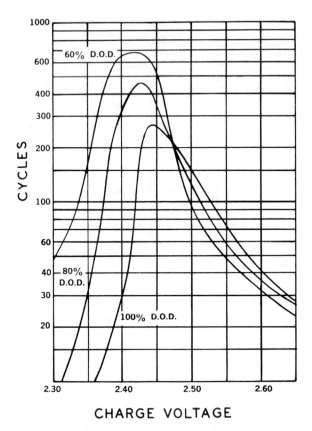

FIG. 41. The general effect of depth of discharge on cell life [175].

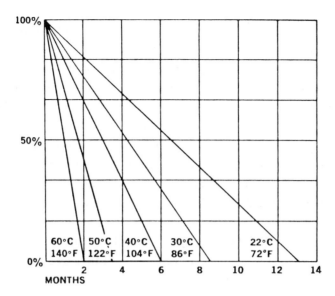

FIG. 42. Percent remaining capacity as a function of time and temperature [175].

4. TECHNOLOGY AND MANUFACTURE

4.1. Battery Components and Raw Materials

4.1.1. Lead Sponge and the Lead Oxides

The battery industry accounts for the largest consumption of lead in the world. The latest figures for the United States alone indicate a yearly consumption in excess of 600,000 short tons [176]. About 80% of it consists of recovered lead from scrap.

$\underline{\text{Pure lead}}$ has a density of 11.3 g/cm^3 at 20°C, melts at 327°C—casting range 420 to 443°C—and possesses a resistivity of approximately 2.1 × 10^{-5} Ω-cm at 18°C [177]. Small amounts of impurities will affect its electrochemical (e.g., hydrogen overvoltage) and mechanical (e.g., stress-strain relationships) characteristics appreciably. It withstands the corrosive action of up to 75 wt % cold sulfuric acid. In pasted electrochemically formed negative battery plates, it is present in a highly porous form, dubbed lead "sponge."

Lead monoxide, PbO, is present in two forms: (1) the yellow oxide or massicot, β-PbO, which has an orthorhombic crystal structure, and (2) the red oxide or litharge, α-PbO, which is stable at lower temperatures and has a pseudotetragonal and, at the same time, a rhombic lattice. α-PbO can take up more oxygen than provided for by stoichiometry, in which case its symmetry becomes increasingly rhombic [178, 179].

Red lead or minium, Pb_3O_4, was used in oxide blends having a composition selected based on processing requirements, such as paste chemistry, electrochemical formation times, etc. (see also Sec. 4.2.2). "Battery red lead" usually contained 25 wt % litharge and represented a typical oxide mix.

When the active material used to manufacture battery plates contains a higher proportion of minium, at the expense of litharge, the electrochemical formation step (Sec. 4.2.5) proceeds at a faster rate. This is due in part to the lower bulk density of minium resulting in more porous pastes, and more efficient mixing with sulfuric acid. It is also due to the fact that minium already incorporates a partially formed lead dioxide product and thus requires less current for the conversion of PbO.

Nowadays, minium does not usually figure in paste mixing formulas such as are used for SLI battery manufacture, and the leady oxides (see below) are seeing exclusive use.

Lead dioxide, PbO_2, constitutes the active material of the charged positive plate in a lead-acid battery. It is never used as a raw material in the manufacturing process (except, in a sense, as minium—see above) but prepared in situ during the electrochemical formation process.

Two forms are known[*]: the blue-gray orthorhombic α-PbO_2 [181] and the reddish-brown tetragonal β-PbO_2 [182], each possessing characteristic lattice constants [183-185]. The α form has been defined as akin to a deformed fluorite-type lattice [183], while the β form is a deformed rutile type. α-PbO_2 crystals are harder, somewhat more conductive (see below) and about two orders of magnitude larger. Their BET specific surface

---

[*]A third crystallographic form, pseudotetragonal γ-PbO_2, has also been reported [180].

area is correspondingly lower than that of β-PbO_2, namely: 0.5 versus 7.5 to 9.5 m^2/g [45,186,187].

The composition of lead dioxide in the battery plate is nonstoichiometric: $PbO_{1.80-1.98}(OH)_{0.11-0.26}$ [10,47]. The resulting free electrons in the conduction band impart a conductive, semimetallic character to the oxide. Aguf and co-workers [10] determined the following resistivities for the two oxide species, extrapolated to zero porosity: $\rho_\alpha = 10^{-3}$ and $\rho_\beta = 4 \times 10^{-3}$ Ω-cm. The actual plate resistivities will depend on the porous structure and morphology but, at any rate, the current will preferentially flow through the more conductive lead grid members ($\rho_{Pb} = 2.1 \times 10^{-5}$ Ω-cm, as noted above). The slightly lower resistivity of the α species would be expected, on structural grounds [49].

Ghosh [188] and Caulder [189] measured the active oxygen content in nonstoichiometric lead dioxide and reported differences between thermogravimetry and chemical analysis data. The incorporation of "bound water" was postulated in order to explain these differences. Work in this area is continuing.

In a lead-acid battery α-PbO_2 will form under conditions promoting highly localized pH zones, such as when the paste is highly compressed, rich in PbO, densely coated with $PbSO_4$, or acid-starved (low concentration of formation acid). Temperature, free lead content in the paste, antimony content, tetrabasic lead sulfate concentration in the paste (see Secs. 4.2.2 and 4.2.5) also play a role in establishing the level of equilibrium α-PbO_2 concentration in the plate.

Battery performance depends on the relative amounts of the two dioxide species. α-PbO_2 is reported to be beneficial in imparting higher structural strength to the plate such as evidenced by a longer cycle life at elevated temperatures. β-PbO_2 will facilitate higher discharge rates and capacities [190-191]. It is noteworthy, however, that α-PbO_2 is less stable than the β form. The electrochemically formed α-PbO_2 tends to convert upon heating in air (at approximately 300°C) to β-PbO_2 [15,192]. Quantitative conversion of the α to the β species also occurs electrochemically after several charge-discharge cycles, as determined by X-ray and neutron diffraction studies [193,194]. This, in turn, induces an increase in battery capacity after a few initial cycles (see also Sec. 2.1) and is associated with an increase in the lead dioxide crystallinity [194].

The higher stability of β-PbO$_2$ is accounted for by a standard free energy of formation difference of 0.40 kcal/mole relative to α-PbO$_2$. Wynne-Jones and co-workers [10] expressed the Nernst relationships for the two dioxide forms, at 25°C:

$$E_\alpha = 1.6971 - 0.1182 \text{ pH} + 0.0295 \log a_{SO_4^{2-}} \qquad (13)$$

$$E_\beta = 1.6871 - 0.1182 \text{ pH} + 0.0295 \log a_{SO_4^{2-}} \qquad (14)$$

which apply up to pH = 4.

Actual measurements have been reported at 25°C, where E_α = 1.698 V and E_β = 1.690 V [186, 187]. At 31.8°C and in 4.4$\underline{M}$ H$_2$SO$_4$, E_α = 1.7085 $\pm$ 0.0005 V and E_β = 1.7015 $\pm$ 0.0005 V [195].

Conversely, the oxygen overvoltage of β-PbO$_2$ is higher than for α-PbO$_2$ [50].

The electrochemical differences between the two dioxide species, which were recently confirmed also by cyclic voltammetry [52], have been suggested as the result of various "active oxygen" contents [196].

"Leady oxides," also known as gray oxides because of their gray-black color, are a fine-grained mixture of particles containing free lead, normally to the extent of 25 to 50 wt %, combined with litharge. This is the raw material for plate paste preparation. Inasmuch as minium is still used for certain specialized positive plate formulations, it is usually present at concentrations of up to 20 wt %. In the presence of moisture, the free lead in the mix oxidizes spontaneously to the red litharge form.

The properties of the leady oxides depend, as expected, on their method of manufacture (Sec. 4.2.1).

4.1.2. Lead Alloys and the Current Collection Grid

The function of a grid in the lead-acid battery plate is twofold: (1) to collect the current, and (2) to provide structural support for the active material (paste).

Some of the properties required of an adequately selected grid are (1) low materials and processing costs, (2) adequate corrosion resistance, and (3) good "bondability" of the active material by chemical and/or mechanical means.

The structural support function of a grid is particularly critical if one considers that grid "growth" and potential plate deformation are intrinsic in the functioning of a lead-acid battery plate. This is due to density changes during cycling; it represents a particularly important problem for the negative plate, which sees active material volume increases of 180% if completely discharged.

While lead is the principal material of construction of the grid, it is mechanically too weak in its pure form. Suitable alloying changes the picture significantly, as exemplified in Table 5 for some representative binary lead-antimony compositions [197].

There is a great deal of work going on in the area of new or improved alloys of lead. Suitable innovations will have to take into account the general requirements listed above and a variety of other ones as well, such as sufficiently high hydrogen and oxygen overvoltages, adequate aging charac-

TABLE 5

The Effect of Lead Alloying on

Some of Its Physical Characteristics[*]

		Antimony-Lead	
	Lead	4% Sb	8% Sb
Tensile strength, kg/cm^2	140.0	281.4[†] 816.9[‡]	325.5[†] 864.5[‡]
Hardness, BHN	3.0-6.0	8.1[†] 24.0[‡]	9.5[†] 26.3[‡]
Melting point, °C	327.4	252-299	252-271
Density, g/cm^3	11.34	11.04	10.74

[*]Based on Ref. 197.

[†]Commercially cold rolled, 95% reduction.

[‡]Annealed at 235°C, quenched and aged 1 day.

teristics including good thermal stability, adequate tensile strength, and creep resistance for pasting and subsequent handling during the various assembly and use stages, and ease of fabrication. The last aspects include such items as castability at high production rates, i.e., also good fluidity for mold filling and low as-cast porosity, adequate strength and ductility characteristics immediately after ejection from the mold, low rate of drossing and of losses of alloying elements and/or inoculants, ease of welding.

The grid alloys presently in use or under investigation usually have one major constituent in addition to lead. At present, antimony has the lion's share of the market, but calcium is making significant inroads owing to the increasing requirements for maintenance-free batteries (Sec. 3.4).

4.1.2.1. Lead-Antimony Alloys

Antimony concentrations of 2 to 12 wt % have been reported for this most widely used grid alloy, though 4 to 6 ± 0.25 wt % is the normal range in most applications. The phase diagram (Fig. 43) shows that 3.45 wt %

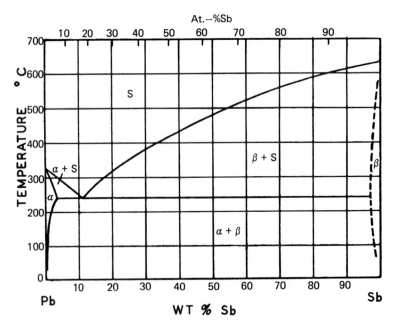

FIG. 43. Phase diagram of the lead-antimony system [198]. α, Sb in Pb; S, miscible welt. Lead—5 wt % antimony alloy.

antimony is the limit of equilibrium solid solution at the eutectic temperature of 252°C; concentrations of 2 wt % or less are also possible as a result of coring [178]. One also notes complete miscibility of lead and antimony in the liquid state and a eutectic at 11.1 wt % antimony [198].

In the finished battery grid the solid lead–antimony system consists of
the eutectic (with about 0.5 wt % solid solution of antimony in lead, at room
temperature) incorporated in a solution of antimony and lead (Fig. 44). It
should be noted that, upon solidification, there occur wide deviations from
the equilibrium diagram, particularly in the hypoeutectic portion.

Some of the reasons for the wide acceptance and use of lead–antimony
alloys in the lead–acid battery industry are as follows:

1. Superior physical properties compared to pure lead, in terms of
    tensile strength (Fig. 45), ductility, stiffness, grain strengthening
    effect [199].
2. Castability at lower temperatures, owing to a lower melting point
    than lead and a lower shrinkage tendency [200].
3. Lower thermal expansion coefficient and less of a tendency for the
    grid to "grow" during cycling.
4. More uniform corrosion mode than lead.
5. Improved adhesion of the shedding-prone positive (lead dioxide)
    active material to the grid.

Small amounts of tin, of the order of 0.15 to 0.50 wt %, are occasionally incorporated into the alloy to retain more antimony in solution and, in
so doing, to improve the fluidity and castability and reduce antimony-
induced brittleness [200].

Other additives, such as arsenic, are meant to increase the alloy's
resistance to anodic corrosion, refine the grain structure, and reduce ductility. Additions of silver are also used to increase corrosion resistance
[200].

Available impurity specifications for lead–antimony grid alloys differ
depending on the country where they are issued. An example is given in
Table 6 for the United States [201].

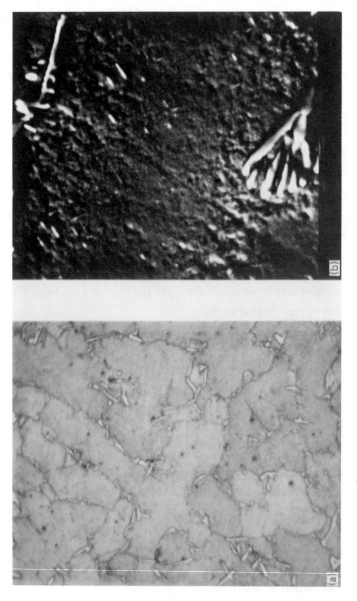

FIG. 44. Lead–5 wt % antimony alloy. (a) Microphotograph; magnification 400×; darker portions are the solid solution and lighter portions the eutectic. (b) Scanning electron picture; magnification 2400×, showing the white platelet-shaped antimony eutectic.

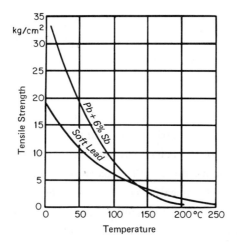

FIG. 45. Tensile strength comparison between pure lead and a typical lead-antimony alloy [198].

TABLE 6

Impurity Concentration

Specifications for Various Grid Alloys (%)

Impurity	Lead-antimony* grids [129, 201]	Lead-calcium** grids [129]
Sb		Trace
As		Trace
Bi	0.05	0.01
Cu	0.05	0.002
Fe	0.005	0.002
Pt	0.00001	0.00001
Ag		0.002
Zn	0.002	

*Sb is specified as not to exceed 8 wt % in lead-antimony positive grid alloys.

**Ca is specified not to exceed 0.085 wt % in lead-calcium positive grid alloys and 0.10 wt % in negative grids.

4.1.2.2. Electrochemical Effects of Antimony

The main disadvantage of lead-antimony alloys is their lower hydrogen
overvoltage at the negative plate, i.e., their tendency to induce a higher
rate of self-discharge in the battery as compared to pure lead or other
types of lead alloys.

The use of Sb [124] radiotracers has indicated how antimony migrates
during cycling from the positive plate grid to the positive active material
which retains most of it. Some, however, will be transported into the
electrolyte solution and through the separator (at various rates, depending
on the separator) to the negative counterelectrode, causing hydrogen evolution [202-204]. Toxic stibine (SbH_3) can also be evolved at the end of
charging and during overcharge, causing a well-known hazard when antimonial lead-acid batteries are used in enclosed environments without adequate charge control.

Regions of the grid that are not covered with active material will release significantly more antimony to the electrolyte, almost in direct proportion to the antimony content of the alloy. The antimony content of the
positive active material fluctuates during cycling and is higher during
charge than during discharge [205].

Antimony transport processes are strongly temperature-dependent.
At room temperature, 20 times as much antimony will be released as
stibine gas than at 50°C. Conversely, 10 to 20 times as much positive-
originated antimony will be found after extensive charging at the negative
plate at 50°C than at room temperature. Previously migrated antimony
present in the negative plate can again pass into solution in overcharge and
be partially released as stibine at room temperature. This phenomenon,
however, is not prevalent in the case of negative grids without active material [205].

In the positive plate of the lead-acid battery, the presence of antimony
has been reported to be beneficial for the development of significant amounts
of α-PbO_2 during formation [206] with an attendant increase in plate cycle
life. The paste texture is firmed up with a prismatic crystallization of the
lead dioxide particles [206], and the particle size of the surface crystals

is maintained small [47]. The overall result is a higher electrochemical capacity for the battery.

In the absence of antimony the amount of initial α-PbO_2 is smaller, the paste is softer, and the prismatic crystal agglomerates are replaced by a smaller and less effective nodular structure [206].

Some recent studies [48, 207] have shown that, in the case of pure lead, a maximum electrochemical capacity is reached which does not increase further after repeated cycles of charge and discharge. With the presence of antimony in the system, capacity does increase with cycling as a result of two effects:

1. Promotion of the intercrystal bonding of the two polymorphs of lead dioxide (the α and β forms)
2. Antimony's action as a nucleation catalyst for β-PbO_2 in the corrosion product attached to the grid surface.

As discussed in Secs. 2.2 and 2.3, antimony also plays an important role in the self-discharge mechanism of the positive plate.

Electron microscopy studies reported by Burbank [208] have confirmed that, for PbO_2 discharging on lead-antimony grids, the nucleation and growth of $PbSO_4$ crystals occurs at active sites and end of discharge corresponds to an exhaustion of these sites. Crystal growth itself has also been suggested by Burbank as a possible discharge rate-limiting step. Additional studies [209] also point to antimony as a nucleating agent for PbO_2 during positive plate formation. At the same time it acts as a crystal growth inhibitor.

It is noteworthy that in a pure lead system grain growth and crystallization proceed to a larger extent and result in early active material breakdown with attendant capacity reduction.

Recent investigations based on instrumental analysis techniques including radiotracers [210, 211] have shown that in the battery electrolyte Sb(III) exists both as an anion, $SbOSO_4^-$, and as a cation, SbO^+. Its reactions in aqueous sulfuric acid can be represented as follows [210]:

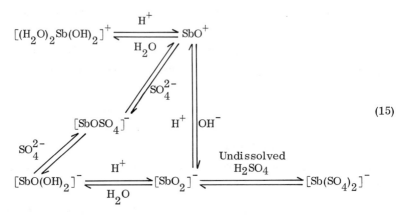

$$(15)$$

Sb(IV) is present only as a stable anion, SbO_3^-, or in the trimer form, $Sb_3O_9^-$.

Based on these investigations, and on the general background available on the subject, an antimony transference scheme has been proposed for the lead-acid battery [211] (Fig. 46). Additional considerations based on crystallographic data have recently suggested that antimony acts in the positive plate via the compound lead antimonate, $PbSb_2O_6$ [212].

4.1.2.3. Lead-Calcium Alloys

Lead-calcium alloys, first discussed in the mid-1930s [213-215], have been gaining increasing acceptance as of the late 1950s [216, 217], and are now widely considered for maintenance-free battery applications (Sec. 3.4). The main secondary constituent, calcium, is electronegative (anodic) relative to lead. It is usually present to the extent of 0.06 to 0.09 wt % (see Table 6), and its hydrogen overvoltage is similar to that of pure lead, i.e., about 0.2 V higher than that of lead-antimony alloys; hence, there are lower rates of self-discharge in the battery and it is closer to being maintenance-free.

It has also been found that positive lead-calcium grids "grow" at a slower rate than their lead-antimony counterparts in float service [218]. The positive plate growth phenomenon depends on corrosion and creep resistance. With the lead dioxide active material expanding to the extent of 40% relative to the volume of the paste from which it was initially formed, it is easy to see that the resulting built-in tension will cause more growth

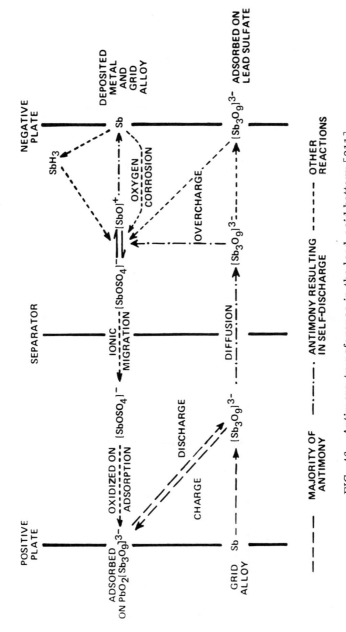

FIG. 46. Antimony transference in the lead–acid battery [211].

or deformation as the grid members become weakened by corrosion. It is
noteworthy that depth of corrosion as well as rate of growth have been
found negligibly small in float service only for alloys containing no more
than 0.1 wt % of calcium [218].

The improvement in corrosion resistance, which occurs when the grid
surface is already partially passivated with lead sulfate, is apparently due
to a significant reduction in the number of spaces between the protecting
lead sulfate crystals. Thus, fewer active lead sites are available for the
corrosion process [219].

Offsetting these advantages is the poorer active material adhesion to
the positive plate grid. In order to cope with this inherent cycle life-
shortening aspect, phosphoric acid has been used as an electrolyte solution
additive (see also Secs. 4.1.3 and 4.1.6). Special separator configurations, such as the pocket type, are also used at some additional expense
relative to the conventional ones (Sec. 4.1.5).

The electrical conductivity characteristics of lead-calcium alloys are
quite favorable. Thus, a Pb-0.1 wt % Ca alloy exhibits a conductivity 20%
higher than that of Pb-9 wt % Sb [207]. Furthermore, the alloy can be
readily cold-rolled or processed in a variety of modes other than casting.
Casting, however, requires certain precautions (see Sec. 4.2.3).

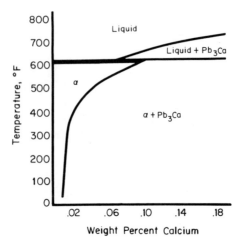

FIG. 47. The lead-calcium phase diagram [220].

Figure 47 is a phase diagram of the lead-calcium system restricted to the concentration ranges of practical interest. It indicates that below 0.1 wt %, calcium is soluble in lead at the melting point. This is a precipitation hardening type of alloy and, above the 0.01 wt % Ca levels, either heat treatment or controlled solidification, say by air cooling or a water quench, will yield favorable grain sizes and distribution of Pb_3Ca crystals in the matrix [220-221].

The Pb_3Ca crystals become larger at concentrations higher than 0.1 wt % Ca, when a separate calcium phase is also present, and act as anodic corrosion sites.

4.1.2.4. Other Lead Alloys[*]

A complete discussion of the characteristics of various types of lead alloys, including the lead-antimony and lead-calcium varieties [178], as well as a detailed survey of applications for battery use [221], have been published. The relatively intense activity in this general area is due to a variety of factors: (1) the disturbing tendency of antimony prices to fluctuate upwards; (2) the need to come up with new types of batteries such as the maintenance-free system; (3) the ever-present challenge to develop a higher specific energy or power lead-acid battery such as by the use of significantly thinner grids; and (4) the need to minimize anodic corrosion effects as much as possible.

Some of the elements investigated, primarily principal constituents in binary and ternary systems, include Li, Na, Cu, Ag, Mg, Sr, Ba, Cd, Hg, Ce, Tl, Ti, Sn, Bi, As, Se, Te, Mn, and Co. In a recent review [222] of the metallography of various binary and ternary alloys cast under air or inert atmospheres, with controlled solidification rates, the following systems were discussed in some detail: Sb-Pb, Ca-Pb, Li-Pb, Cd-Pb, Ag-Pb, As-Pb, Sn-Pb, As-Sb-Pb, Ag-Sb-Pb, Cd-Sb-Pb, and Sn-Sb-Pb.

Each of the element additives has been used in order to enhance one or more of the required grid properties. For example, tin in lead-calcium alloys (e.g., up to 1 wt % Sn + approximately 0.1 wt % Ca) is becoming in-

---

[*]See also Sec. 4.2.3, discussion on age-hardening additives.

creasingly interesting from a processing and grid property standpoint since it affects favorably the fluidity and castability.

Although antimonial alloys possess higher strength characteristics immediately following the casting operation, aged lead-calcium-tin alloys yield stronger and more ductile grids in the long run [220]. A typical composition within the concentration limits mentioned above is Pb-0.08 wt % Ca-0.5 to 1.0 wt % Sn. Increases in the tin content also increase the fully aged properties of the final product. Furthermore, adequate material compatibility is evident with some types of ternary alloys, such as Pb-0.09 wt % Ca-1.0 wt % Sn, which corrode uniformly and independently of the casting conditions used [220, 223].

Tellurium has been mentioned as having a particularly pronounced grain boundary-strengthening effect [199]. Silver, whose effectiveness in lead-antimony-silver combinations is apparently concentration-dependent, reduces the rate and extent of corrosion as compared to the lead-antimony alloy [224].

The grain boundary-strengthening effect is significant if one considers the two stages in the corrosion of lead-antimony alloys as defined by microscopy studies: uniform and nonselective attack with antimony being removed from the corrosion product, followed by a higher rate of corrosion, particularly at the grain boundaries, with antimony remaining in place [225].

Silver also appears to promote the formation of β-PbO_2 [226], which is normally present in the anodic corrosion film together with α-PbO_2, $PbSO_4$, and PbO [12, 15, 59, 227-230]. This would extend the alloy stability region to higher anode potentials.

Alloys consisting of lead with small amounts of tellurium, silver, and arsenic have been reported effective for low-gassing applications such as those involving stationary or submarine batteries [231].

Lithium has been investigated in binary alloys [232-234] or in a ternary system including tin [235]. For adequate corrosion resistance it is limited to concentrations below 0.03 wt % and tin, when present, is recommended at levels of 0.05 to 0.50 wt %. While such compositions are of interest for maintenance-free applications, owing to the low observed rates of self-discharge, reduction to practice on an industrial scale has yet to be demonstrated (see also Sec. 4.2.3).

A detailed study of the lead-antimony-cadmium system has recently been published [236] reporting, among other things, a maximum in tensile properties for an as-cast alloy at the composition: Pb-3.0 wt % Cd-2.5 wt % Sb. Cadmium is of interest, as a ternary alloy constituent, because of the need to offset the high self-discharging tendency in batteries using the binary lead-antimony system. In a different example, for the case of 4 to 5 wt % SbCd, superior charge retention and overcharge capabilities were claimed [237].

Cobalt has been reported as an effective positive grid corrosion-preventing additive, whether in the paste, the electrolyte, or the alloy itself. Typical patented compositions include 0.001 to 0.100 wt % Co, in combination with 0.05 to 0.25 wt % Ag, 1 to 12 wt % Sb, and 0.5 wt % Sn [238]. The basic concept of cobalt addition dates as far back as 1931 [239]. Investigations using radioactive ^{60}Co have shown that, during charge, cobalt in solution will precipitate at both the negative and the positive plate. At small electrolyte concentrations (<100 mg Co/liter), more cobalt will discharge at the negative, and vice versa. The antimony migration processes are apparently uninfluenced by cobalt during normal cycling but, on extended overcharge, cobalt inhibits the release of antimony from the positive plate [240]. This may provide a partial explanation of the claimed beneficial effect of this additive, such as for fast charging of traction batteries (Sec. 3.2).

Bulgarian researchers have recently reported that copper increases the tensile strength properties, as well as the fluidity, of Pb-Sb-Cu alloys up to the 0.06 wt % concentration level, which happens to correspond to the eutectic point [241]. Better corrosion resistance is also noted. Standard tests of cycle life, discharge capacity, overcharge, and self-discharge, for batteries using grids made from Pb-6.12 wt % Sb-0.069 wt % Cu, have been claimed to indicate its superiority over the binary lead-antimony system.

4.1.2.5. Dispersion-Strengthened Lead

Lead containing fine lead oxide particles (for example, 1 to 2 wt % of the oxide, with particles of the order of 2 to 100 μm in size), also known as dispersion-strengthened lead, has been receiving considerable attention owing to its greatly improved creep and tensile strength characteristics [242-247]. These properties are of practical significance for the development of high-rate thin-plate elements such as would be required for traction applications. Furthermore, the need for strengthening, albeit electro-chemically contaminating alloying agents, such as antimony, would be eliminated by the dispersion-strengthened product. [*]

The adherence of the active material to the dispersion-strengthened grid appears to be comparable to that for pure lead, or lead-calcium alloys, i.e., inferior to the lead-antimony system [247]. From a corrosion stand-point, particularly at anodic potentials, there is conflicting evidence as to the properties of dispersion-strengthened lead; this is just one of the aspects requiring further study [247,248]. Another important factor is the nature of the processing technique, e.g., metal expanding [249], stamping, weaving, forging, or powder metallurgy and the economic implications, especially since the automatic gravity casting of lead-antimony grids has reached a high degree of perfection at very low cost per manufactured grid (e.g., several U.S. cents per automotive grid) [250]. Newly developed laminar plates are shown in Fig. 48.

Satisfactory performance has been reported for cells with dispersion-strengthened grids in capacities of up to 250 Ah. However, reliable techniques for the joining of grids to the straps (bus bars), especially for the case of positive plates, are yet to be demonstrated. Fusion welding is not possible and has to be replaced, instead, by soldering or brazing with low-antimony or antimony-free lead [251].

4.1.2.6. Grids Made of Other Metals

Titanium has been considered a promising material of construction for thin lead-acid battery grids [252]. The recent issuance of patents to S. Ruben

---

[*]Dispersion-strengthened systems have been investigated which also contain additives other than PbO, such as Sn, Cu, etc. These aspects and many more have been covered in a detailed review on the subject [247].

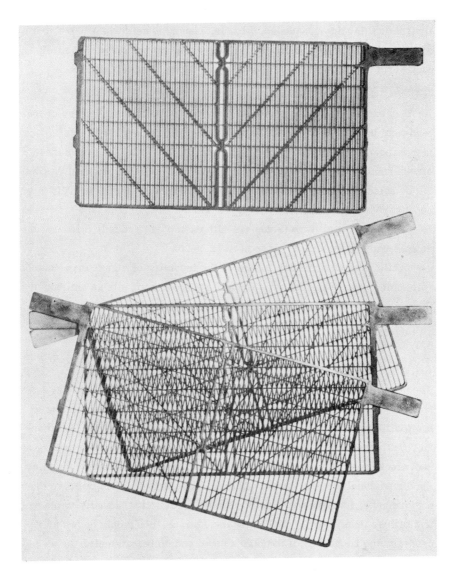

FIG. 48. Laminar plates and assembly. (Courtesy ESB, Inc.)

covering the technology of nitrided titanium grids [253-256], with or without a gold surface flash, has generated revived interest in the possibility
of developing thin-plate high-specific-energy and power lead-acid batteries.
Some work has been going on in this area, involving a variety of grid con-

figurations (tubular, expanded, fibrous, conventional design) [257], but a great deal remains to be done before the economics and short- as well as long-term performance aspects will have been adequately defined.

4.1.2.7. Lightweight Grids

Another extensively tried approach to increase the specific energy of the lead-acid battery is based on the utilization of lightweight materials, mostly polymeric, as the plate "backbone" in lieu of lead. Under these circumstances the active material, with or without a thin conductive coating or selected conductive members applied to the nonmetallic grid, must conduct at least some of the current in and out of the plate [258].

The patent literature is replete with examples, and only a few recent sample references will be given here [259-262].

While such an approach combines the incentive of high energy content with maintenance-free features (lack of such contaminants as antimony), the question of technical and economic feasibility has yet to be settled.

4.1.3. Active Material Additives

4.1.3.1. Negative Plate Additives

The negative plate of a lead-acid battery is subjected to dimensional stresses and strains during each charge-discharge cycle owing to the large difference between the densities of lead and lead sulfate. Furthermore, it constitutes the performance-limiting component of the battery at high discharge rates and low temperatures. In order to better maintain the dimensional stability of the negative active material and improve the performance, additives are included in its formulation, usually to the extent of less than 3 wt %. Without such additives, commonly known as ex-panders, the capacities of the plates could end up being smaller by as much as several hundred percent.

There are several kinds of commonly used expanders, all of which are normally present in the negative plate.

a. Lignin Derivates. Lignin derivates such as sodium lignosulfonate in quantities of less than 1 wt % (for example, 0.05 to 0.3 wt % [263]).

Lignin is a major constituent of wood, being present in amounts of 33 to 40 wt %. The other major constituent is cellulose, and lignin is sepa-

rated from it by sulfonation. The exact structure of the lignin molecule is
not known, though its basic building block is a phenyl propane derivative
[264].

The lignosulfonate molecule is thus believed to consist of randomly coiled
cross-linked polyaromatic chains.

There are numerous applications for lignosulfonates, outside the lead-
acid battery industry, which exploit their characteristics as surface-active
agents, dispersants, or binders. In the lead-acid battery case, their apparent role is to adsorb preferentially onto the lead "sponge" surface [265]
and prevent the formation of too impervious a coating of lead sulfate. A
thicker surface layer of coarse lead sulfate crystals forms during discharge,
prior to the lead passivation stage [266,267], and the generally more porous plate surface structure significantly improves the low-temperature high-
discharge rate capability of the battery. Typical sulfate crystal dimensions
precipitated in the absence and in the presence of a lignosulfonic expander
are 0.3 to 0.8 μm and 1 to 3 μm, respectively, as determined by electron
microscopy [268].

Further improvements in cold (-18°C) capacity and cycle life have also
been reported for modified lignosulfonic acid molecules, such as those
prepared by condensation with phenol resulting in an increased size expander molecule [269]. The opposite is true when certain lignosulfonic

acid salts are used, such as the calcium salt, since they induce paste thin-
ning with a subsequent "cementation" effect on the active material of the
pasted plate product in storage.

A variation of this approach, claimed to improve charge acceptance
and life by raising the hydrogen overvoltage, is based on the use of sulfonic
acid-substituted napthalene formaldehyde condensation products, in the negative paste (0.001 to 0.100 wt %) or in the electrolyte (0.0005 to 0.2000 wt
% [270].

According to recently reported microscopy studies by Simon et al.
[78], lignin acts to restrict the development of dendritic lead branching
during electrochemical formation or charge; this, in turn, ensures adequate room for lead sulfate growth during discharge and is particularly important in the case of high-rate discharges. The latter are accompanied by
the formation of much larger lead sulfate crystals constituting a steric hindrance to effective utilization of the active material. Furthermore, during
the advanced stages of charging, lignin may induce a higher rate of gas generation, assisting in the formation of an adequately porous lead "sponge"
structure.

It should, however, be kept in mind that this type of protective inhibition caused by the surface-active lignosulfonate molecule also leads to
higher—of the order of 0.2 V—charging voltage requirements. This, as
well as other types of surface effects, were recently investigated by
Yampolskaya and co-workers [271]. In general, the influence of surface-
active-type paste expanders was found to be twofold:

1. Decrease in the paste strength by an adsorption-induced weakening
   of the cohesive forces between the basic lead sulfate crystals;
2. Decrease in the concentration of basic lead sulfate in the paste.

An alternate type of surface-active expander is the class of synthetic
tannins. Their use is recommended in concentrations of 0.05 to 2.00 wt %
[272].

b. Carbon black. Carbon black, which is seldom present in amounts
larger than 0.2 wt %, has been claimed to be effective in reducing the required charging voltage and in providing for higher discharge capacities at
low temperatures. There is some debate, however, about the exact role

of carbon expanders (if any), and further studies are required on the subject. Effective or not, the addition of carbon black to the negative plate
has been reported to make it easier, in production, to differentiate by
color between the final positive and negative plate products.

c. Barium Sulfate. Barium sulfate, which is used in various proportions
(e.g., from a fraction of 1 wt % to as much as 3 wt %) has an orthorhombic
crystal structure quite similar to that of lead sulfate. Its apparent role is
to provide nucleation sites for the formation of the lead sulfate precipitate
on discharge. The net result is a plate morphology enabling higher discharge capacities at a given polarization. This is particularly evident at
the start of discharge, especially at high rates and low temperatures, and
can also be expressed as the effect of decreasing the supersaturation of the
electrolyte solution with respect to lead sulfate [78,273].

Barium sulfate has also been reported to be effective in maintaining
an adequate balance of the surface-active organic expander (e.g., lignosulfonate) within the plate structure. This is achieved by regulation of the
supply of organic expander to the active reaction sites and its desorption
in the discharged state. It has been suggested that this adsorption-
desorption mechanism would also protect the organic expander from anodic
oxidation conditions such as would be encountered at the negative plate during cell reversal [273].

Other additives have also been mentioned in the patent literature. An
example would be the use of carbohydrates (<1 wt %), sulfur (3 to 8 wt %),
and sometimes also salts of calcium, sodium, or ammonium, in conjunction with the other types of expanders [274,275].

d. Oxidation Inhibitors. Oxidation inhibitors are of importance, particularly in the manufacture of dry-charged batteries (see Sec. 4.2.5), to minimize the risk of damaging oxidation of the lead plates in the presence of
ambient moisture and oxygen. Some of these inhibitors have been found to
induce changes in the size and shape of lead sulfate crystals deposited from
the electrolyte solution during the discharge process. Thus, α-
hydroxynaphtoic acid increases the degree of dispersion of the lead sulfate
deposit, while α-nitroso-β-naphthol and α-naphthol lead to the formation
of very long acicular crystals of the sulfate [267]. Other methods of dry-

charged negative plate oxidation protection suggested by the patent literature are application of boric acid to the plates or its introduction into the
formation solution or the active material (0.5 to 10 wt %) [276], use of silicone coatings, such as methyl silicone, containing anion-active emulsify-
ing agents, such as fatty alcoholic polyglycol ethers [277,278], or use of
coatings based on about 0.01 wt % of a polymerized rosin containing at
least 10 wt % of abietic acid dimer [279]. A variation of the rosin treatment uses dimerized colophonium rosins (Poly-Pale) containing approximately 40 wt % dimerized abietic acid [280].

4.1.3.2. Positive Plate Additives

Although the use of expander-type additives is mainly related to the requirements of the negative plate structure, various attempts have also been
made to improve the positive plate. This is to be expected if one considers
that the latter often represents the performance-limiting component in a
lead-acid battery, particularly from the standpoint of cycle life.

Carbon black has been reported to be effective as an additive to the
paste used in the manufacture of positive plates [281,282]. Its main role
appears to be that of a pore structure expander with attendant improvements in the plate capacity and in the faradaic efficiency on discharge.
Optimum carbon concentrations have been recommended, typically 0.3
wt %.

Another means of increasing the plate porosity and, as a result, its
capacity, relies on the use of normally less than 1 wt % of anionic or non-
ionic foaming agents such as lauryl sulfate, a stabilizer such as sodium
carboxymethylcellulose or polyvinylalcohol, and a structure strengthener
such as water emulsions of polyvinyl chloride [283,284] or fibrous materials (e.g., Dynel). Preparation of a uniform active material slurry is
sometimes aided by sparging with carbon dioxide [285] or mixtures of carbon dioxide and other gases such as oxygen, nitrogen, air, or argon [283].

The utilization of phosphoric acid, which is discussed in greater detail
in Secs. 2.4 and 4.1.6, is based on a laminated active material structure
[286,287]. There is a denser paste layer adjacent to the grid, to offer
better protection against electrolyte penetration and self-discharge. One
or more of the active material layers contain small amounts of phosphoric

acid. The grid of such a plate is of the more shedding-prone lead-calcium alloy variety.

A different type of additive of potential interest, namely, antimony oxide, Sb_2O_3, has been the subject of a recent communication [288] describing microscopic studies of lead dioxide electrodes. It has already been established that plates made with antimonial lead grids have an active lead dioxide material consisting of prismatic crystals, while nonantimonial plates (e. g. , with lead-calcium grids) exhibit a rather indistinct globular structure [206] (see also Sec. 4.1.2). Further, it was suggested that the presence of antimony prevents the formation of excessively large lead dioxide crystals, with a resulting stronger plate [215]. It now appears that antimony oxide added to the active material can be associated with the formation of needle-shaped, acicular, crystals which serve to strengthen the plate structure. The quantity of antimony observed, for this phenomenon to occur, cannot be accounted for only in terms of transport from the grid.

A variation on this theme is the preparation of lead oxides (as by a Barton Pot-type process, see Sec. 4.2.1) predoped with antimony, and using the product to prepare plates with nonantimonial grids [289].

Boric acid has been mentioned in the preceding section as an oxidation inhibitor for dry-charged negative plates; it has also been reported as an inhibitor of detrimental surface carbonation for stored dry-charged positive plates. It is normally applied by immersion or spraying at a level of 0.3 to 1.0 wt % relative to the positive active material [290-292].

Finally, cobalt has been mentioned in a variety of patents as a positive plate corrosion-protective additive to the electrolyte or grid alloy (Sec. 4.1.2.6) as well as to the active material in the form of $CoSO_4 \cdot 7H_2O$.

4.1.3.3. Multipurpose Additives

According to some claims, both plates can benefit from the use of certain types of additives. An example in point is the addition to the battery paste of a bulking agent such as carboxymethylcellulose (e.g. , about 0.3 wt %, and/or another bulking agent such as mixed 1/8-in. and 1/4-in. Dynel fibers (also about 0.3 wt %), plus expanded pumice (e.g. , Corcel No. 46), Dicalite, or any other type of siliceous hollow globules, to the extent of about 1.0 wt % or more [293]. The negative plate also contains expanders, and the resulting composition is claimed to be particularly effective for nonantimonial grids of the Delco "wound wire" design patented for use in maintenance-free batteries (Sec. 3.4.1). In the case of antimonial positive plates, one notes that the use of Dynel fibers is not always recommended.

4.1.4. The Plates

Numerous types of negative ("sponge" lead) and positive (lead dioxide) plates in a great variety of sizes and configurations are in use, depending on the applications for which a lead-acid battery is designed. Broadly speaking, one can summarize the state-of-the-art according to the following categories:

a. Pasted Plates (Faure Plates). Pasted or Faure plates are the type used most extensively. They consist of cured active material which had previously been pasted onto a suitably selected grid. The positive plates are distinguishable by the darker active material from the light-gray negatives. In some applications, requiring greater mechanical stability and resistance to corrosion, the pasted grids are reinforced by heavier lead bars alternating with lighter grid bars.

Normally the positive plates in a lead-acid battery are made of thicker grids, owing to the intrinsically unstable positive plate structure. This, of course, is the result of grid corrosion through local galvanic action between the lead in the grid and the lead dioxide in the active mass. The specific surface area of their active material is 10 to 20 times greater than that of the negative plates (Sec. 4.1.1).

Figure 49 is a schematic representation of such a plate showing a parallel network of vertical lead strands intersected by thinner horizontal strands approximately triangular in cross section [294]. In a thick plate design the horizontal strands are shaped and emplaced differently, as indicated in the figure, in order to afford more effective retention of relatively large amounts of active material.

There are a variety of reasons for such a design:

1. To facilitate grid extraction from the casting mold;
2. To obtain high ratios of active material to grid weight;

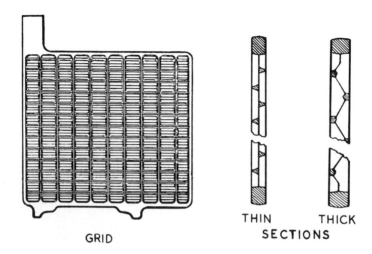

THIN   THICK

GRID   SECTIONS

FIG. 49. Cross sections of a pasted positive plate [294].

3. To achieve effective interlocking between the active material and the grid with maximum active material exposure to the reacting electrolyte.

A variation of the Faure design is the so-called "box" plate used for certain types of stationary battery negative plates (see Sec. 3.3). Figure 50 shows such an electrode, consisting of a pair of grids riveted together, with the active material sandwiched between them [295].

b. Planté Plates. For certain applications the plates consist of a lead substrate with electroformed rather than pasted active material. These are known as Planté plates and are designed in a variety of configurations: some rather "exotic" designs such as the Manchester positives (not in wide use any more), where the active material consists of electroformed spirals pressed into button-type configurations on the plate, grooved lead positive plates (Fig. 50), etc. Planté plates are normally used in conjunction with pasted negatives.

Planté plates are normally larger and heavier than the pasted variety, while possessing lower capacities. This is compensated by a higher durability which counts for their use in stationary and industrial batteries, es-

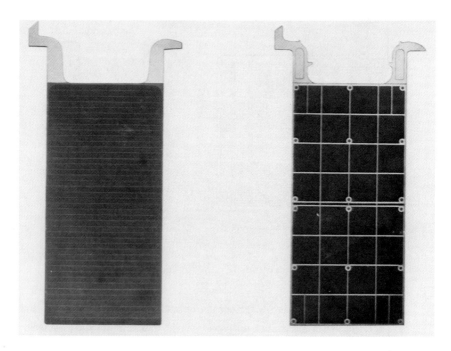

FIG. 50. Negative box plate and enlarged surface area Planté positive plate of the grooved lead type [295].

pecially in those instances where space and weight considerations are less
important.

The so-called high performance Planté type (HPP), widely used in
Europe, is thinner—6 to 8 mm instead of the standard 13 mm—and pitched
at much closer separation. As a result, the internal resistance is lowered
by a factor of 2 [95].

c. Tubular Plates. First developed in 1920, tubular plates consist of a
composite of cast vertical lead spines, forming the grid, surrounded by
active material encased in slotted tubes (Fig. 24). Again, these designs
are intended mainly for the life- and performance-limiting positive plate,
with the negative plates being of the usual pasted variety.

Three main constructional variations are in use today:

1. Resin-bonded braided glass wool tubes commonly used in the
    United States and Japan.

2. The P. G. Swedish Tudor design, consisting of braided glass wool tubes supported by an outside sleeve of perforated PVC.

3. The British Exide Ironclad, in which the tubes are integrated into a one-piece gauntlet made from woven or nonwoven terylene [also known as PzS (Panzer) cells in Germany] and considered the most economic design.

Tubular plates are intended for applications requiring long life and low maintenance. Their increased price is offset by a higher reliability for deep discharge uses (e. g., traction, signaling systems, and the like). Paste shedding and grid corrosion are minimized, the active material to grid weight ratio is superior, and nominal or close to nominal capacity is reached after only a few charge-discharge cycles, quite a departure from ordinary pasted plate behavior. Conversely, high-rate performance is inferior to that obtainable from the other plate designs.

It appears that the capacity of tubular plates is more extensively influenced by the electrolyte concentration, while less sensitive to the active material density [294]. Performance adjustments can be arrived at by suitable modifications in the tube design and cross-section shape.

Tubular plates are manufactured so as to contain mostly the larger surface area, higher capacity β-PbO_2. Typically, as the plate becomes partially discharged and lead sulfate fills the pores, the remaining active material in the interior of the tubes can no longer be reached by the reacting acid. The resulting internal pressure is contained by the tube walls and there is negligible shedding. Both elliptical and round cross-section tubes are effective in containing the internal pressure of a discharged material; square or rectangular cross-section designs provide 3 to 5% more capacity with a somewhat higher risk of active material dislodging during cycling [231]. Therefore, for deep discharge applications such as those encountered in batteries for electric traction, circular or elliptical designs are preferred.

Detailed reviews on the general subject of tubular plates have been published elsewhere [231, 296].

d. Bipolar Plates. Bipolar plates are of potential interest for applications requiring high-rate capabilities. Their design incorporates the lead and lead dioxide structure as opposite faces of the same electrode. Some work has been reported in this area [102-107], and similar structures have been used successfully in a variety of other battery types (e.g., lead/ fluoroboric acid/lead dioxide proximity fuze batteries).

Further developments are to be expected, in particular in the burgeoning traction battery field and for military applications, though the ultimate practicality of the concept, from an economic and cycle-life standpoint, has not yet been determined.

e. Lightweight plates. Lightweight plates incorporate polymeric materials in the construction of the grid (see also Sec. 4.1.2.7). They have received extensive coverage in the patent literature. Some tests of actual battery configurations were also conducted. Thus, portable 12-V, 90 Ah communication batteries have been evaluated [258]. They used polycarbonate grids with some lead spines provided for effective current collection.

An alternative approach is the molding together of a plastic grid with a lead plate lug (terminal), after which the grid is coated with conductive lead [297]. Other examples abound in the patent literature, but effective reduction to practice has yet to be demonstrated.

One should note that such lightweight designs also exhibit low, maintenance-free type rates of self-discharge (see Sec. 3.5) because the lead used for current collection purposes is usually pure and thus imparts relatively high hydrogen overvoltage characteristics to the plate.

f. Sintered Lead Plates. Similar in structure to sintered alkaline battery components, such as NiO electrodes, sintered lead plates have been reported for the first time as a practical and superior-quality product in Bulgaria [298]. It is claimed that their performance has been successfully evaluated over a test period of 3 years in light Wartburg cars, with evidence of better starting and nominal capacity characteristics than the standard batteries in use. Their manufacture is described in Sec. 4.2.4.

4.1.5. Separators

As in any other battery, in particular of the storage variety, the presence of separators between the anodes and the cathodes of each set of plates is required in order to retard as much as possible shorting through dendrites

("treeing"), "moss," and shedded active material as well as the transfer
of undesirable species such as antimony. For certain types of batteries,
the separators—envelope type—actually contribute to the retention of the
active material in good contact with the current collector grid (Sec. 3.4).
Examples of separator structures are shown in Fig. 51.

The types of separators presently in use for lead-acid batteries depend
on the applications. In most instances they are made of cellulose derivatives, of microporous polymeric materials such as PVC, polyethylene,
polyesters, rubber, or of sintered materials such as PVC or polyethylene.
Other types of materials are also used sometimes, e.g., resin-bonded
paper, latex/Kieselguhr, etc.

Many separators are provided with longitudinal ribs which effect adequate interplate spacing while allowing natural convection to minimize temperature and concentration gradients in the bulk of the electrolyte. In other
instances, e.g., the Gel/Cell line of batteries for portable appliances (Sec.
3.4) the separator is of a wraparound glass wool mat variety and spacing
is provided by additional plastic inserts of various configurations, such as
PVC-expanded screen forms. Some of the mat-type separators are also
used, at times, as electrolyte-immobilization matrices in order to make
the batteries attitude-insensitive and/or to provide for controlled electrolyte-
electrode interfacing. A great variety of separator combinations, using
several types together, is also encountered in practice depending on the
application.

With the advent of the maintenance-free battery utilizing nonantimonial
grids, i.e., more shedding-prone positive plates especially when no phosphoric acid is added to the electrolyte, microporous pocket separators
have become popular in lead-acid battery designs. They are used wrapped
around the positive plate to provide added support to the active material.

Zehender and co-workers [299] defined the effectiveness of a separator
in a battery using lead-antimony grids in terms of the extent of transport of
radioactive ^{124}Sb from the electrolyte adjacent to the positive (lead dioxide)
plate to the electrolyte adjacent to the negative (lead) plate. They used a
quantitative criterion consisting of the product of electric resistance and
antimony diffusivity. Low values (for example, 1 to 2×10^{-4} Ω-cm^3/min
for wood) are characteristic of good separators, which prevent the poison-

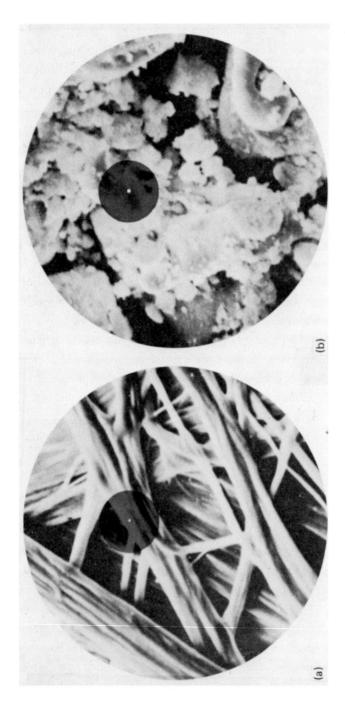

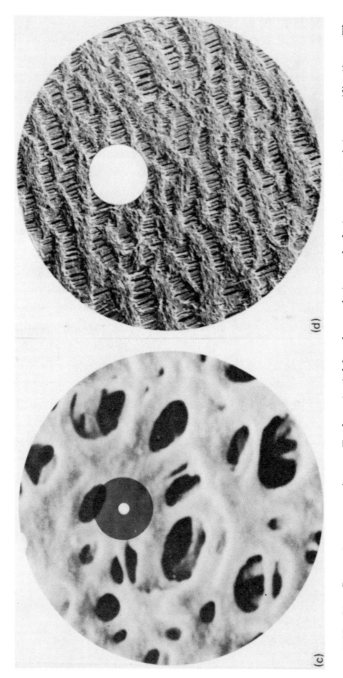

FIG. 51. Separator comparison. Each material has been photographed at an appropriate magnification. The white dot in each photograph is 5 μm in diameter. (a) Non–woven textiles (mag. 500:1). (b) Microporous rubber (mag. 1,000:1). (c) 0.45–μm membrane filter (mag. 5,000:1). (d) Celgard (mag. 32,000:1). (Courtesy Celanese Plastics Co.)

ing of the negative plates by antimony ion transport from the positive plates, and the opposite holds true for ineffective separators. This criterion can be used to differentiate quantitatively between various types of separators made of similar materials (e. g., values of 20 to 70 $\times 10^{-4}$ Ω-cm^3/min for paper-type materials, and 2 to 90 $\times 10^{-4}$ Ω-cm^3/min for plastics).

In general, good separators contain narrow and tortuous capillaries or semipermeable pore walls, which are permeable to hydrogen ions but impermeable to undesirable constituents such as antimony ions. Upper pore size limits of about 25 μm, over a range of sizes where 5 μm predominate, are acceptable. Larger sizes would accommodate particles shed primarily off the surface of the positive plate, such as during gas evolution on charge or during very high rate discharges. Gross pore clogging by lead dioxide particles would eventually produce undesirable metallic lead shorts as a result of electrochemical conversion.

In addition to antimony migration, the transport of organic negative plate expanders (such as the lignin derivatives) to the positive plate—where they are oxidized—can also be a problem depending on the type of separator used.

An effective separator must meet a variety of criteria defined by specialized measurements. Pertinent data are obtained covering such properties as permeability, porosity, pore size distribution, specific surface area, mechanical design and strength, ionic conductivity, and chemical compatibility in the electrolyte solution. Conductivity and surface area appear to be critical factors in the selection of adequate separators [300].

Details on separator evaluation and selection can be found in the literature [301].

Figures 52 and 53 indicate the effect of separator selection on discharge performance and life, respectively, for a chosen example involving SLI batteries.

4.1.6. Sulfuric Acid Electrolyte in Its Various Forms

4.1.6.1. Liquid Electrolyte Batteries

The lead-acid battery electrolyte is an aqueous solution of sulfuric acid. Based on the overall electrochemical reaction (3) (Sec. 2.1), 3.66 g of H_2SO_4/cell/Ah are theoretically required. This makes the acid storage

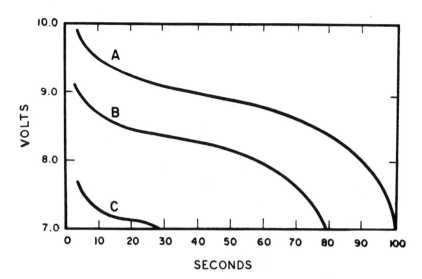

FIG. 52. Automotive batteries discharging at 40 A per positive at
-30°C. A, battery with microporous PVC separators with deep fine ribs;
B, battery with microporous PVC separators with normal depth of fine ribs;
C, battery with sintered polyethylene separators with integral wetting agent
[301].

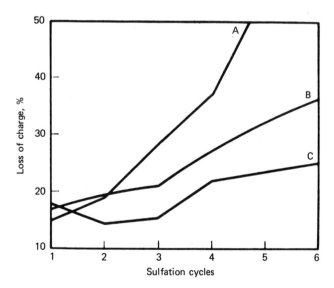

FIG. 53. Sulfation cycle test on automotive batteries. A, battery with
resin-bonded paper separators; B, battery with sintered PVC separators;
C, battery with microporous PVC separators [301].

battery different from its alkaline counterparts, where the electrolyte
serves only the function of an ion transfer medium.

Table 7 lists a variety of pertinent physicochemical data [302], and
Fig. 54 represents the variation of an important operational parameter,
namely the freezing point, at various electrolyte concentrations and temperatures.

Table 8 gives a few examples of recommended acid concentration
ranges depending on application. It is interesting to note that in no case
is it possible to reconcile any given requirement with the points of maximum conductivity or minimum freezing point on the concentration scale.

A factor to consider is the degree of purity of the electrolyte solution.
The negative lead sponge plates of a lead-acid battery will be affected by
ionic impurities depositing on them and inducing self-discharge through
local galvanic action. The result is hydrogen evolution and plate sulfation.
Low-hydrogen-overvoltage species, such as platinum, are particularly
harmful, but other elements such as iron, copper, zinc, manganese,

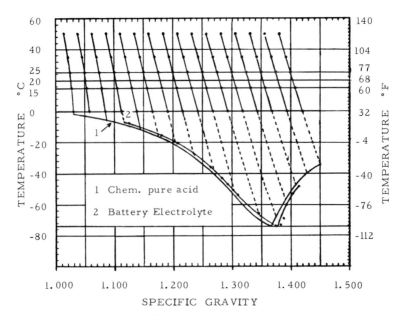

FIG. 54. Freezing point curve for sulfuric acid solutions (density =
sp. g. $T°/25°C$) [8].

TABLE 7

Physicochemical Properties of Sulfuric Acid Solutions[*]

Density 15°C (g/cm³)	Density temp. coefficient at 15°C (°C)	g of H_2SO_4 per 100 g of solution at 15°C	g of H_2SO_4 per 1 cm³ of solution at 15°C	Density (°Bé)	wt % H_2SO_4 at 15°C	Normality N	Conductivity at 18°C mhos/cm	Viscosity at 20°C cp
1.050	0.00033	7.37	0.077	6.7	7.5	1.56	0.29	1.14
1.060	0.00036	8.77	0.093	8.0				
1.070	0.00040	10.19	0.109	9.4				
1.080	0.00043	11.60	0.125	10.6				
1.090	0.00046	12.99	0.142	11.9				
1.100	0.00048	14.35	0.158	13.0	14.4	3.20	0.53	1.26
1.110	0.00051	15.71	0.175	14.2				
1.120	0.00053	17.01	0.191	15.4				
1.130	0.00055	18.31	0.207	16.5				
1.140	0.00058	19.61	0.223	17.7				
1.150	0.00060	20.91	0.239	18.8	21.0	4.91	0.67	1.58
1.160	0.00062	22.19	0.257	19.8				
1.170	0.00063	23.47	0.275	20.9				
1.180	0.00065	24.76	0.292	22.0				
1.190	0.00066	26.04	0.310	23.0				
1.200	0.00068	27.32	0.328	24.0	27.3	6.53	0.73	1.92
1.210	0.00069	28.58	0.346	25.0				
1.220	0.00070	29.84	0.364	26.0				
1.230	0.00071	31.11	0.388	26.9				
1.240	0.00072	32.28	0.400	27.9				
1.250	0.00072	33.43	0.418	28.8	33.4	8.47	0.72	2.26
1.260	0.00073	34.57	0.435	29.7				
1.270	0.00073	35.71	0.454	30.6				
1.280	0.00074	36.87	0.472	31.5				
1.290	0.00074	38.03	0.490	32.4				
1.300	0.00075	39.19	0.510	33.3	39.3	10.4	0.68	2.61
1.310	0.00075	40.35	0.529	34.2				
1.320	0.00075	41.50	0.548	35.0				
1.330	0.00076	42.66	0.567	35.8				
1.340	0.00076	43.74	0.586	36.6				
1.350	0.00077	44.82	0.605	37.4	44.8	12.3	0.62	2.93
1.360	0.00077	45.88	0.624	38.2				
1.370	0.00078	46.94	0.643	39.0				
1.380	0.00078	48.00	0.662	39.8				
1.390	0.00079	49.06	0.682	40.5				
1.400	0.00079	50.11	0.702	41.2	50.1	14.2	0.54	3.26
1.410	0.00080	51.15	0.721	42.0				
1.420	0.00080	52.15	0.740	42.7				
1.430	0.00081	53.11	0.759	43.4				
1.440	0.00081	54.07	0.779	44.1				
1.450	0.00082	55.03	0.798	44.8				
1.460	0.00083	55.97	0.817	45.4				
1.470	0.00083	56.90	0.837	46.1				
1.480	0.00084	57.83	0.856	46.8				
1.490	0.00085	58.74	0.876	47.4				

[*]Based on Ref. 302; wt % column is based on interpolations from Table 122, p. 184, in Perry's Chemical Engineers' Handbook, 3rd ed., McGraw-Hill, New York, 1950.

TABLE 8

Some Recommended Electrolyte

Concentration Ranges for Various Battery Applications

Application	Room temperature specific gravity range
Stationary	1.210-1.240
Industrial	1.260-1.280
SLI	1.260-1.280
SLI in the tropics	1.200-1.230
Aviation	1.260-1.285
Portable, maintenance-free (immobilized electrolyte)	1.275-1.285

arsenic and, as amply discussed in this chapter, antimony, also belong to the category of undesirable contaminants.

Iron will affect both the negative and the positive plates by local redox action. Organic contaminants, such as may be leached out of the separators or of the negative plate expanders, will interfere with the operation of the positive plate and cause increased polarization losses.

At high current densities and electrolyte concentrations, especially on overcharge, arsenic and antimony will release toxic gaseous by-products, namely: arsine, AsH_3, and stibine, SbH_3, respectively.

Standard specifications for battery acid are available and vary from country to country. Criteria established in the United States and Great Britain [303] are exemplified in Table 9.

4.1.6.2. Electrolyte Solution Additives

Various claims have been made relative to improvements in the performance of a battery by means of electrolyte solution additives. For example, it has been stated that by using stable surfactants of the perfluoroalkane sulfonic acid variety [304], the surface tension of the electrolyte solution will be lowered sufficiently to improve active material wetting and penetration. The result is a decrease in polarization despite the offsetting effect of surfactant-induced inhibition.

TABLE 9

Impurity Specifications for Battery Electrolyte

(in ppm referred to 100% H_2SO_4)

Impurity	American standards [175]	British standards [303]
Organic matter	To pass test defined in Federal Spec. 0-S-801-b, 4-14-65	N.A.
Fixed residue	750.00	500.0
Fe	30.00	41.0
SO_2	15.00	17.0
As	0.40	6.8
Sb	0.40	N.A.
Mn	0.07	1.4
NO_3	2.00	
Nitrogenoxides		17.0
NH_4	4.00	
Ammoniacal nitrogen		170.0
Chloride	40.00	24.0
Cu	25.00	24.0
Zn	15.00	N.A.
Se	7.00	N.A.
Ni	0.40	N.A.
Pe	To pass test defined in Federal Spec. 0-S-801-b, 4-14-65	N.A.

Cobalt salts, such as the sulfate, have been claimed as effective grid corrosion-minimizing agents especially under high charging rate conditions (for a detailed discussion of cobalt effects and applications, see Secs. 3.2 and 4.1.2.4). Cobalt has been added alone, in concentrations of the order of 0.1 wt % [239], or in conjunction with silver sulfate to minimize detrimental effects on cellulosic separators. A recommended formulation is 0.001 to 0.020 wt % $CoSo_4 \cdot 7H_2O$ + 0.05 to 0.20 wt % Ag_2SO_4 [305].

Phosphoric acid has been used as a positive active material "strengthener" in nonantimonial systems (Secs. 3.4 and 4.1.2.3), usually to the extent of 10 to 35 g/liter, or for special applications (Sec. 4.1.6.4), though not without some minor capacity reduction effects (Sec. 2.4).

4.1.6.3. Immobilized Electrolyte Batteries

The immobilization of sulfuric acid solutions has found wide application in maintenance-free and position-insensitive designs (see Sec. 3.5.1 for a complete discussion). The special case of reserve batteries to be activated by addition of water is discussed in Sec. 4.1.6.4.

4.1.6.4. Water-activated Batteries

Since the development of the dry-charged lead-acid battery (see Sec. 4.2.5), it has become of interest to design a reserve system which would only require the addition of water for activation. A variety of military and possibly also some commercial applications are visualized for such a system.

The advantages are fairly obvious: safety, since no acid handling is required; and ease in logistics and the attendant economies, since no provisions have to be made for separate storage and transport of the electrolyte solution. As water is relatively easy to come by, a water-activatable battery will essentially be equivalent to the wet (i.e., fully activated) storage battery plus having the advantage of significantly longer shelf life and reduced shipping weight.

A feasibility study on the subject was issued in 1965 [306]. Several approaches to the problem have been reported, mostly in the patent literature:

1. Concentrated acid encasement in suitably positioned plastic bags or containers. After the battery has been filled with water the acid is released by a variety of mechanical or chemical means [307–313].

2. Storage of the required amount of sulfate in partially discharged dry plates [314,315]. After addition of water, the required strength of electrolyte solution is obtained electrochemically.

3. Gelation of concentrated sulfuric acid by chemical means. The water-soluble or hydrolizable gel is appropriately positioned within

the battery and activation proceeds the moment the battery has been
filled with water [87, 88, 316-325].
4. Sorption of concentrated sulfuric acid, such as by the use of porous
containment materials. The acid is leached out after addition of
water to the battery [326-331].

Reduction to practice of some of these concepts has taken place. The
Prestolite Inc. SLI battery line, known under the trade name Oasis, is being marketed based on the sorbed acid concept [322-331]. The added water
percolates through a side-mounted porous phenolic foam block, and the resulting sulfuric acid solution fills the battery from the bottom up. The disadvantage of this approach is the larger weight and volume required and the
relatively slow activation. The process is further complicated by stratification of the resulting solution, requiring subsequent charge "boosting" of
the battery, since the first, highest-gravity fractions leached out from the
containing side block end up at the battery bottom.

The plate sulfate approach has been reduced to practice in 1971 by
E. S. B.'s Wisco Division for portable lantern applications.

Attempts have also been made to exploit the gel approach, though there
is as yet no commercial reduction to practice. Two possibilities exist:
(1) the use of hydrolizable sulfonated polyelectrolyte complexes, generally
in membrane form and disposed in parallel to the plate elements [316, 317,
319-321]; the use of water-soluble gels, particularly of the boron phosphate
variety [318, 322, 324, 325], typically consisting of approximately 2 mol %
H_3PO_4 (about 10 g/liter) and 5 mol % H_3BO_3 [87, 88]. The latter approach
appears to be more practical, and offers the advantage of completely water-
soluble concentrates. Water activation can readily be adapted to existing
battery configurations without extensive redesign problems, by positioning
most or all of the gel above the plate elements (Fig. 55). Upon activation,
uniform mixing is achieved by natural convection, and stratification problems are eliminated.

The disadvantage of the gelled acid approach is the slight reduction in
capacity, usually no more than 10%, caused by the phosphoric acid required
to prepare the gel. Conversely, when used with batteries prepared with
lead-calcium grids, such as the maintenance-free variety, the phosphoric

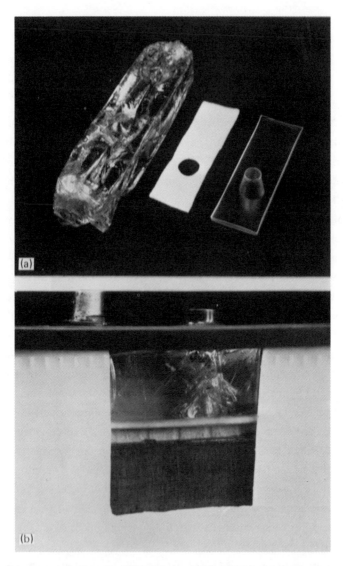

FIG. 55. Cutaway view of water-activatable battery using boron phosphate gel (b). Activation components, from left to right: bag-encased gel, supporting tray with absorbent matrix (a) [88].

acid electrolyte addition could become a useful design feature because of its beneficial effect on positive plate active material shedding.

4.1.7. The Battery Case

4.1.7.1. SLI Batteries

Until recently, vulcanized (hard) rubber was the standard material of construction for the battery case. The rubber usually contains fillers such as carbon black.

Polymers are now becoming standard materials of construction. Polypropylene [96, 97, 332] and polyethylene-polypropylene copolymers (e.g., Sears' Die Hard line built by Globe-Union), with or without glass fiber fillers (e.g., Prestolite's Oasis battery) are receiving large-scale acceptance owing to the increased weight and volume-specific energy of batteries using such thin-walled cases. Ease of fabrication, such as by injection molding of polypropylene [333-336], is another important incentive.

The bottom of the battery case is provided with plate element rests, also known as "mud" rests, in order to maintain the plates a certain distance above the case bottom and, in so doing, provide a dead volume for the accumulation of shedded active material.

The case covers are molded in one piece. They are provided with one combination vent-filling aperture per cell, plus a second opening at each end cell in order to fit the tapered terminal posts (Fig. 56). The sealing of the terminal post to the cover material is of critical importance in order to avoid its getting loosened with time, with attendant leakage of electrolyte. The integrity of the cover-to-case seal is of similar importance.

According to the type of material used, various methods of sealing are applicable, e.g., solvent sizing, epoxy adhesive, and heat sealing (such as for thermoplastic cases).

The vent cap designs are varied; they include single- and three-cell combinations [337]. Their purpose is to provide a vent path for gases while keeping the liquid in.

An alternative three-piece case design, claimed to be superior for automated assembly purposes (e.g., E.S.B.'s Star models), consists of a

FIG. 56.   One-piece battery covers [94].

straight sleeve case ultimately joined to separate cover and bottom pieces
by heat sealing [338, 339].

Other modifications are associated with the new generation of vented
maintenance-free batteries (Sec. 3.4.1). In such a case the cover only
requires a suitable common vent opening. It is provided at the top or on
the side of the case, with a protective hinged coverlet, and is usually combined with flame (explosion)-arresting porous components [340]. Filling
and terminal post openings are eliminated. Instead, the battery case itself
is provided with side terminals (Fig. 57).

The design of intercell connectors in a partitioned multicell battery
case is of considerable importance, especially since high-rate capabilities,
i.e., minimum IR losses, are becoming increasingly pertinent in modern
applications. This topic will be covered in detail in the section dealing
with the battery assembly (Sec. 4.2.6).

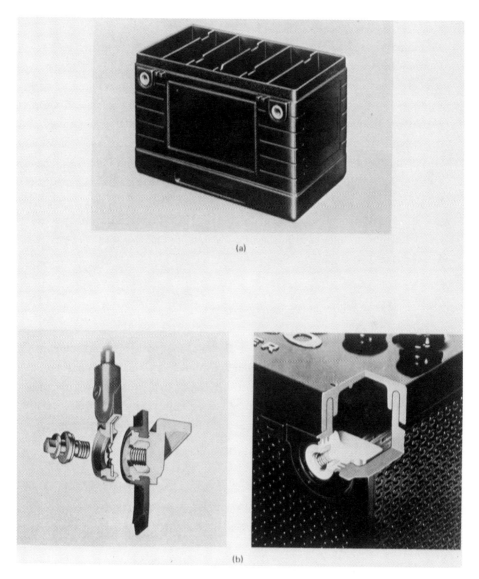

(a)

(b)

FIG. 57. Side-terminal battery container (a), with details of the terminal assembly (b) [94].

4.1.7.2. Other Types of Batteries

As might be expected, there exists considerable variety in design and selection of materials of construction for the diversified non-SLI types of
lead-acid battery applications. Stationary batteries, or single cells, will
sometimes be assembled in glass jars or ebonite containers as well as in
rubber or thermoplastic cases. Polystyrene, ABS, and at times steel,
are popular in a variety of maintenance-free portable batteries. The corresponding vents, cover and joint designs, terminals, and intercell connectors vary depending on the requirements of the product and the ingenuity of
the designer.

Examples abound in the manufacturers' literature, and some specific
cases have been discussed in the Applications sections of this chapter (Secs.
3.2 through 3.4).

4.2. The Manufacturing Process

Figure 58 is a schematic diagram of the lead-acid battery manufacturing
process detailing the two main production lines: dry- and wet-charged batteries [341]. Particulars on these two types of batteries, as well as on a
new process modification trade named Power Spin, are given in Sec. 4.2.5.3.

The various manufacturing steps are discussed in some detail in the
following sections.

4.2.1. Lead Oxide Manufacture

There are two main processes used for the manufacture of lead oxides:
ball mill and Barton pot. The relative merits of the resulting products are
still subject to widespread debate.

Ball mill oxides are produced by the grinding of lead "balls" (essentially chopped lead pigs) in rotating mills at predetermined conditions of
relative humidity, oxidizing air flow, and temperature. In some of the
more modern mills, whole pigs can be fed into the grinding equipment.
This is the lowest-temperature process for the preparation of leady oxides.

The air is normally introduced from two sources: through vents in the
lower portion of the mill casing, and through the hole in the shaft used for

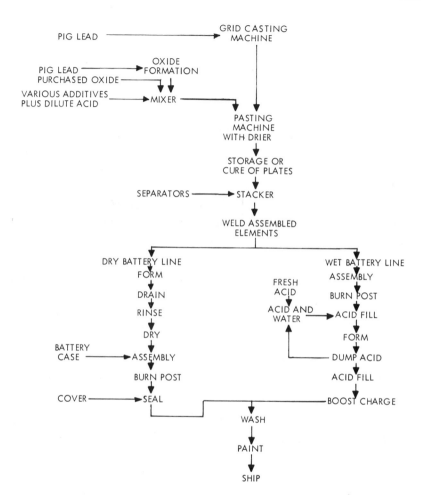

FIG. 58. Schematic diagram of the lead-acid battery manufacturing process [341].

feeding the mill with lead. The air inlets are so designed as to optimize mixing with the leady material, and water is sometimes introduced with the air for temperature control purposes.

The final product particles can be removed through openings in the cylindrical shell (e.g., Tudor-type mill) or carried out of the cone-bottomed drum by a stream of air (Hardinge-type mill). They are obtained at various states of oxidation and bulk densities, such as 10 to 90 wt % PbO equivalent (for example, 10 wt % in Chamberlain-type mills and 90 wt % in Hardinge-type mills [342] and 1.1 to 1.7 g/cm^3 [343-345], respectively.

In the presence of atmospheric oxygen and moisture, spontaneous oxidation of the free lead in the powder can occur. Maximum oxidation rates take place at water contents of 3 to 7 wt %.

Although either of the two lead monoxide polymorphs (see Sec. 4.1.1) can be partially converted to the other by ball milling, over a wide range of temperatures, the commercial product consists mostly of litharge in a virtually strain-free condition [346], though with a lattice plastically deformed by the milling action [347]. It has been suggested that an increase in the microstrain concentration, i.e., in the stored strain energy, such as by processing in vibrating mills, would make the material more reactive [346], a desirable feature.

The flat or flake-like particle product (median particle diameter: 2 to 6 μm [345]) contains, in addition to free lead, some lead particles covered with lead hydroxide or carbonate. The product color is responsible for the terminology "gray oxide."

Recent developments have been reported for mills able to produce over 200 tons/week of oxides with a PbO content that can be controlled within the range 55 to 65 wt %. Typically such equipment has drums about 2.1 m in diameter and is powered by a 200-hp motor [348].

Barton pot oxides are made by a process occurring above the melting point of lead, though below the litharge melting point. The process consists of blowing air and/or steam into a heated stirred pot reactor containing molten lead. Lead particles are produced, with a coating of lead monoxide.

Improvements in product quality have been reported, particularly in Europe, based on a modified stirring and product removal technique which takes out the fine lead oxide particles and leaves the larger ones in the reactor for further subdivision [349]. Alternatively, the Barton pot product can be subjected to a grinding process for particle size reduction.

The amount of free metallic lead can vary from 10 to 50 wt % [342], depending on the air (steam) and lead feed rates and the pot temperature. The median particle size range is 6 to 10 μm or 2 to 6 μm for the ground product [345]. Their size distribution is narrower than that of ball mill particles, and their shape is spheroidal and more regular than the jagged contours exhibited by the ball mill product. The composition includes red tetragonal and yellow orthorhombic oxides, with the former likely to be produced in quantity at the lower end of the operating temperature range (we note again that the ball mill product is mostly of the tetragonal variety).

Barton pot oxides have been reported to exhibit surface resistivities several orders of magnitude higher than the ball mill product, for example, 800 versus 20 MΩ, respectively, when compressed at 4 long tons/cm^2 [347]. This is the result of the much thicker layers of lead monoxide surrounding the Barton pot-produced particles.

The Barton pot oxide production method had been favored in the United States owing to the higher production rates possible per unit space. The ball mill variety is produced on a relatively larger scale in Europe and is now catching up in the United States as well. It has been claimed to be more reactive than Barton pot oxides and to yield basic lead sulfates at significantly higher rates during the subsequent plate manufacturing step [347] (Secs. 4.2.2 and 4.2.4). A higher plate strength is also claimed by some ball mill oxide producers as a result of a better "particle interlocking" mechanism. This would seem to be in agreement with studies showing that positive plates prepared with Barton pot oxides possessing a 50 wt % orthorhombic oxide content have less lead dioxide and smaller BET surface areas after formation than plates prepared from Hardinge ball mill oxides [350] (usually 3.5 to 4.5 versus 5 to 6.5 m^2/g). The ball mill product was also found to contain a higher concentration of the β form of the dioxide, which itself is higher in surface area than α-PbO$_2$.

It is, of course, possible to control the characteristics of the oxide by air fractionation, additional milling steps, etc., in order to achieve desired plate properties. Varying the relative amount of free lead also plays a role. Thus, positive traction battery plates may be prepared with coarser oxide particles having a higher percentage of free lead than would be acceptable for a SLI-type product.

Quality control: Size classification of either type of leady oxides is carried out in suitable screening equipment or cyclone separators. Thereafter, the principal control parameters are (1) acid absorption and (2) apparent (bulk) density.

Typically, 3.3 to 3.5$\underline{N}$ sulfuric acid (sp gr approximately 1.1) is used for the acid absorption tests.

Typical acid absorption and bulk density values encountered in practice are listed in Table 10.

In normal storage, i.e., in the presence of heat, moisture, and oxygen, product characteristics change with time. This represents a problem not only in controlling plate quality but also in achieving reproducible R, D, & E test results using various lots of oxides.

4.2.2. Lead Paste Preparation

4.2.2.1. Formulations and Reaction Products

There are various formulations for the preparation of the active material paste to be used in the manufacturing of negative and positive plates, depending on the manufacturer and the type of product to be made. They have all been arrived at with varying degrees of empiricism, and they all incorporate a measure of manufacturing trade secrets. Very careful control of the processing parameters is required, which makes this one of the more critical steps in battery manufacture. In particular, consistent characteristics are required of the leady oxide mix used for the process.

Paste preparation and quality control have been described in detail by various sources [351-352]. Basically, the process involves the mixing of a blend of lead oxides, sometimes of both the tetragonal and the ortho-

TABLE 10

Quality Control Parameters of Lead Oxides

	Acid absorption, mg/g of oxide	Bulk density (g/cm^3)
Ball mill oxide	250–300	1.1–1.7
Barton pot oxide	130–160	1.6–1.8
Ground Barton pot	180–200	1.2–1.6

rhombic variety, containing a certain fraction metallic lead (see Sec.
4.2.1) with water and a dilute solution of sulfuric acid, in that order. The
acid concentration will depend on a variety of factors: typically of the order
of sp gr 1.325, it may go down to sp gr 1.250 for higher free lead contents
in the oxide mix, or up to sp gr 1.400 for some essentially lead-free mixes,
especially when using large muller and dough-type mixers.

After the addition of water, some hydrates are reportedly formed, of
the types $2PbO \cdot H_2O$; $3PbO \cdot H_2O$; $5PbO \cdot 2H_2O$. The water content of the
formed hydrates is usually four to seven times less than the weight ratio
of water to leady oxide mix. After the addition of acid, polybasic lead sulfates and their hydrates are formed with considerable heat evolution. Primarily, the tribasic species is detected: $3PbO \cdot PbSO_4 \cdot xH_2O$, though other
compounds result as well: $PbSO_4$, $PbO \cdot PbSO_4$, $4PbO \cdot PbSO_4$.

Regardless of the method of acid addition, the paste system is alkaline
with a pH of 9 to 10, except momentarily at the points of acid contact [342].

The paste mixing process is controlled in terms of composition of the
ingredients, temperature and concentration of the acid, and duration and
mode of mixing with or without cooling. Amounts of acid ranging from 55
to 130 ml/kg of oxide have reportedly been used with various amounts of
the same type of sulfate being the end result [353, 354]. As more acid is
added, the paste stiffens. Another effect of increased amounts of acid in
the paste composition is that of higher porosity and surface area in the resulting plates [350]. This is true, however, only if the paste temperature
is not allowed to rise above normal; otherwise both porosity and surface
area actually decrease in the finished product.

Several processes occur in situ, either separately or in various combinations: recrystallization of the tribasic sulfate to the tetrabasic form,
especially after relatively long exposure to high temperatures and humidities, oxidation of the free lead, and evaporation of some of the water [354].
Various reaction product compositions have been determined, depending on
the oxide mix and the maximum reaction temperature allowed [343]. A few
examples are given in Table 11. A more detailed description of the chemical changes occurring during paste mixing and the subsequent curing process is given in Fig. 59.

TABLE 11

Paste Mixing Products for Various

Raw Material Compositions and Mixing Temperatures

(from Ref. 353)

Oxide mix composition	Maximum temperature (°C)	Mixing products
72% PbO (tetrag.) + 28% Pb	60	27% $3PbO \cdot PbSO_4$ + 45% PbO (tetrag.) + 28% Pb
80% PbO (tetrag.) + 20% Pb	85	15% $3PbO \cdot PbSO_4$ + 17% $4PbO \cdot PbSO_4$ + 48% PbO (tetrag.) + 20% Pb
70% PbO (orthorhomb.) + 30% Pb	85	12% $3PbO \cdot PbSO_4$ + 30% $4PbO \cdot PbSO_4$ + 30% PbO (orthorhomb.) + 28% Pb

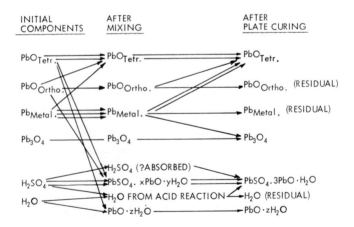

FIG. 59. Transitions of battery paste components [342].

The nonproprietary literature contains various methods and formulations for paste preparations. It is claimed, for instance, that better paste
quality control is achieved by preparing water-lead oxide mixes in such a
way that the next processing step involves the addition of relatively concentrated acids [355].

Typical plate formulations will also require various paste additives.
These are discussed in Sec. 4.1.3.

4.2.2.2. Presulfated Paste Materials

A better-defined active material morphology can be obtained by starting
with discrete crystals of lead sulfate or oxysulfates which are relatively
regular in shape and size. The geometry of such crystals should largely
determine the resulting plate pore structure, since the dissolution and recrystallization reactions which determine crystal morphology when leady
oxide pastes are cured are largely eliminated with presulfated lead oxide-
type pastes. There has also been increasing interest in this area lately
because of the advantage of paste processing by water mixing only.

A reduction to practice of this concept has been reported for the new
Bell Labs design of cylindrical stationary batteries, utilizing a positive
paste of tetrabasic lead sulfate and water (see Sec. 3.3).

4.2.2.3. The Process and Its Control

Various types of equipment are currently used in the industry for paste
preparation. They commonly include hopper assemblies provided with
stirring action and connected to the pasting equipment (the latter is described in Sec. 4.2.4). The stirring is carried out by a variety of methods:
"can" mixing where the "can" holding the paste mix rotates in one direction
while a multiblade mixing head rotates counter to it, a modification of this
principle (e.g., the Simpson mixer) where shearing and rolling effects are
achieved by a wide rimmed wheel rotating inside the paste container with
suitable baffles directing the mix under the wheel, or a "dough" mixer using sigmoid blades for the mixing action.

Normally, batches of 800 to 1,100 kg of oxide are processed at one
time with a total amount of 155 to 200 ml of solution/kg of dry oxide. The
exact mixing procedure varies depending on the type of equipment used and
each manufacturer's proprietary process. It involves adding first the

water, and then the appropriate sulfuric acid solution. Some heat removal is usually necessary by means of water-cooled jackets or by additions of excess water and evaporation by forced air correction. Mixing is continued until the temperature drops to a predetermined value, normally 40 to 50°C, and final paste consistency adjustments are sometimes made by small water additions.

Generally, the mixing process is of a batch type, though continuous mixing systems are also used. Both types of processes are highly automated, with the latter apparently producing more homogeneous pastes from the standpoint of sulfate content and distribution.

The properties of the paste are controlled in terms of several parameters:

1. The paste density expressed as "cube weight" (using a hemispherical 2-in.^{3} cup) in g/in.^{3}. Values of the order of 55 to 75 are typical. Normally the negative paste density would have higher values (at least 70 g/in.^{3}) than the positive paste (about 65 g/in.^{3}).

2. The paste consistency (usually suggestive of a stiff mortar) as determined by penetrometry. Globe (Cone) Penetrometer units (typically 34 to 40) based on an instrument developed at Globe-Union, Inc., are frequently used.

3. The pasted plate porosity, by means of a buoyancy measurement in water.

4. Various qualitative estimates (texture workability, adhesiveness). Complete details are available in the literature [351].

Quality control and the development of alternative paste formulations are often monitored in terms of mixing curves [352] plotting, for instance, weight of water/weight of oxide versus weight of acid of defined concentration/weight of oxide, with one or more of the above-mentioned properties as a parameter.

Considering the resulting product performance, one notes that the lower the paste density, the better the active material utilization, but positive plate shedding with an attendant shorter cycle life is a very real possibility. Conversely, with too dense a paste, the resulting high-strength plates will tend to buckle in service, owing to active material volumetric changes, and suffer grid breakage. The thicker grid plates used in industrial battery designs can and do use a higher density paste.

For the negative plate, the final form of the active material is a soft lead sponge. Paste density is less critical in terms of cycle-life effects.

Inasmuch as consistency is related to the rheology of the paste, it would be indicative of plate processability. A stiff paste may crack less in service, but at the same time it would be more difficult to apply during the grid pasting operation, with possible deleterious results as far as grid-to-active material adhesion is concerned. The cohesive forces which keep the active material together by itself are also included in a measurement of consistency.

The porosity of a pasted plate will play an important role in its performance, i.e., in its discharge capacity at various temperatures and rates, faradaic efficiency, and cycle life (see Sec. 2.1).

4.2.3. Grid Production

4.2.3.1. Casting

The grids are usually gravity-cast into stress-relieved cast iron molds coated with a cork-based lining bonded with materials such as carboxymethylcellulose and alkali silicates. Depending on size, grids are cast single or assembled together in panel form using standardized equipment commercially available (e.g., Wirtz casters). The porous mold coat, plus suitably designed vents, facilitate gas release from the mold; the coat also provides a measure of thermal insulation, reduces friction, and helps shake-out; its surface texture will also influence the properties of the casting, such as grain size. Mold coats are usually reapplied every 1,500 to 2,000 castings.

Ancillary equipment includes some or all of the following: a grid trimming press, the molten lead pot connected to a lead pump and a heated lead pipe feeding the material to the casting molds, and a scrap return line (conveyor) connected to the lead pot.

Pressure die casting has also been considered as a production method, but a variety of considerations mitigate against its general adoption: particularly cost or excessive grid growth in service for certain types of

alloys. Work continues in this area, with some reports of successful reductions to practice [356] using 6, 8, and 10 wt % antimony alloys. The
incentives are some structural improvements relative to the gravity cast
product, and relatively little flash, which practically eliminates secondary
operations on the grid prior to assembly.

Continuous sheet casting is another alternative, in particular because
of the increasing interest in thinner grids (less than 1.3 mm) for higher
discharge rate capability and/or higher volumetric specific energies. Here,
again, careful economic considerations are required prior to large-scale
adoption of this processing technique. The grids could be produced by
stamping or by expanding, with the latter approach being more attractive
owing to the practical elimination of scrap, i.e., of scrap reclamation
steps. Process-induced stress corrosion aspects are to be considered,
though.

a. Antimony-based Alloys. The melt is poured into the mold at 425 to
525°C with the mold heated at 135 to 180°C. Castability depends not only
on composition but also on an adequately controlled rate of heat dissipation
achieved by air and/or water cooling. The molten metal usually drops 20
to 25 cm from a ladle to the mold bottom. This drop height can also be significant for the properties of the finished product.

The lower the antimony content of the alloy, the higher the required
casting temperature. This is of interest, since at higher temperatures the
formation of oxide containing drosses may become a problem.

With the presently available automatic grid casters (Fig. 60), production rates of 11 to 12 two-grid panels per minute are typical. It has been
reported that a skilled caster operating two machines can produce 10,000
castings of 8 wt % or more antimony per shift. At lower antimony contents
(for example, 6 wt %), requiring higher operating temperatures, production rates can fall off as much as 20% [348]. This follows directly from the
casting rate-limiting step, which is the freezing time required to obtain the
proper metal grain structure. Other factors influencing the rate of production are the type of alloy, the design of the mold, and the mold coat used.

One of the fundamental problems in casting lead-antimony alloy grids
is the microporosity of the end product, which makes it more liable to

FIG. 60. Automatic grid caster.

structural failure by corrosion [357]. Elimination of any supercooling effects during casting will do away with the intricate dendritic crystallization
occurring as the melt solidifies, which is apparently responsible for the
observed microporosity. This, however, would make it that much more
difficult to avoid segregation of the lead and antimony phases in the solid,
so that the problem may be inherent in the use of a cast lead-antimony system for battery grids.

b. Calcium-based Alloys. With the increasing interest and number of
applications requiring maintenance-free batteries, the question of rapid
and economical production of lead-calcium grids has become of importance.

Gravity casting, as in the case of antimonial grids, is possible, but requires a higher temperature for proper mold filling. A variety of other processing modifications are also necessary, owing to the higher fluidity of the melt and its tendency to oxidize. The latter aspect is usually taken care of by casting under a protective reducing atmosphere.

Howard and Willihnganz [358] have indicated that rapid freezing of a lead-calcium alloy, such as in pressure casting, may result in an abnormal, porous microstructure prone to structural and corrosion failure as revealed by accelerated cycling tests. This has been confirmed in a recent study [221] which showed that, especially in the case of thinner, SLI grids, the lack of physical soundness (e.g., large voids, interconnecting pores) in certain regions of pressure cast grids is more responsible for excessive pasted plate "growth" than any metallurgical structural differences induced by the casting conditions. Nevertheless, pressure casting conditions, rather than a particular technique, are important since they control both the active precipitate distribution and the grain size in the alloy. Thus, by casting in a mold held at the relatively high temperature of 260°C (versus, say, 190°C), alloys containing up to 0.07 wt % Ca will exhibit uniform and stable corrosion properties, i.e., good resistance to excessive grid growth [221].

Gravity casting of grids made of lead-0.08 wt % calcium alloys containing 1 wt % tin has been studied from the standpoint of casting and mold temperature optimization [220]; the results are summarized in Fig. 61. Commercial casting speeds have been achieved. In general, mold design and temperature, as well as alloy composition, will determine the metallurgical characteristics of the grids, whereas ladle temperatures, casting speeds, and quench rates appear to have little if any effect [220].

c. Scrap Reclamation. An important economic aspect of the lead-acid battery manufacturing process is lead reclamation from scrapped batteries. The same also holds true for some of the alloying additives such as antimony. A detailed review on the subject has been published [200].

d. Age Hardening and Additive Effects. Further improvements in the grid manufacturing process are possible in terms of production rate, quality control, reduction of grid thickness, and costs. Another aspect of interest is the age hardening of the grid alloy under various conditions. An impor-

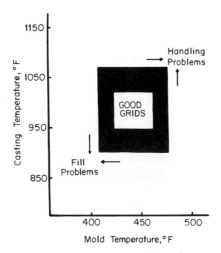

FIG. 61. Optimization of grid casting parameters for lead-0.08 wt % calcium-1.0 wt % tin alloys [220].

tant general parameter is the nature of additives to a principal alloy (see also Sec. 4.1.2) and, not surprisingly, considerable work has been reported for the lead-antimony system.

Arsenic is a popular hardening additive [359-361], by itself or in combination with tin and/or copper [359]. Hardness reaches an optimum value at more than 0.10 wt % As in Pb-7.0 wt % Sb, and at 0.15 wt % As in Pb-4.5 wt % Sb [360]. The hardening effect is interpreted as the result of an enhanced microsolubilization of Sb, particularly when undesirable microsegregation of the constituents is prevented by an adequate thermal treatment [359]. The microstructure refinement enables a more uniform corrosion over the grid surface and, as a result, increases the corrosion resistance; furthermore, castability is enhanced and grid life expectancy appears to be increased owing to α-PbO$_2$ adhering to the arsenic-antimony-lead grid surface [360].

Another type of age-hardenable alloy is the Pb-Cd-Sb system. A recent study [362] indicated that alloys containing 2 to 3 wt % CdSb, with Cd and Sb in a 1:1 atomic ratio exhibited the highest tensile strength after 6 months aging.

Torralba and co-workers investigated the effect of temperature and quenching, with or without plastic deformation, on age hardening of lead-antimony alloys [363-366] in terms of such factors as antimony content, time, vacancy diffusion mechanisms, and retention of quenched-in vacancies by a low-temperature treatment.

Another additive of interest is lithium [225-228] (see Sec. 4.1.2), and its castability has been reported as being midway between commercial antimony-lead and calcium-lead alloys. To improve its processability relative to the antimony alloy, a casting method other than by gravity may have to be used.

4.2.3.2. Other Methods

Powder metallurgy techniques have recently been given increased attention as a practical alternative to high-production-rate casting. The use of dispersion-strengthened materials (see Sec. 4.1.2.5) makes it possible to roll thin sheets of a product nearly as strong as antimonial lead. However, problems such as anisotropy in the product properties, i.e., greater strength and corrosion resistance in the rolling direction, and the economies of the process are still to be considered before commercial production is possible.

4.2.4. Plate Manufacture

4.2.4.1. Faure Plates

a. Pasting. The paste is applied to the grid under pressure, on a continuous basis, with some excess paste being removed. Various types of equipment are in use. Commonly they include the paste preparation assemblies provided with stirring action (Sec. 4.2.2), discharging through rolls and suitable wipers onto a horizontally moving conveyor (e.g., Winkel belt paster) feeding the pasted plates to a flash drier. The short-duration, high-temperature, flash-drying process, for example, to 275 to 325°C for 20 to 30 sec, only removes surface stickiness and is necessary for subsequent stack curing of the plates (see below).

Alternatives involve vertical grid positioning, such as in the Lund paster, or fixed-orifice paste-dispensing equipment, such as in the Donath

or Globe-Union pasters. Of particular significance are the new generation
of fixed-orifice pasters (Globe-Union) that provide not only high accuracy
of weight and thickness control (such as plate weights of the order of
80 ± 2 g) but also shorter changeover times from one type of plate to
another, of the order of minutes rather than hours as in the past.

The operation proceeds swiftly, with customary production rates in excess of 100 plates/min/paster. In the case of a fixed-orifice paster type
of machine, variations in the paste characteristics can be critical, since
they cause uneven pressure to be applied to the grid during the pasting operation.

The surface area of the thin layer of paste, after application on the
grids, is increased considerably from its previous bulk condition. The
water content is approximately 10 to 11 wt % [367]. Sometimes the plates
are partially dried as they move through the pasting process, and this may
reduce their water content to the highly oxidation-active value of 5 to 7 wt
% (see also Sec. 4.2.1).

As previously mentioned, the paste density is one of the performance-
and life-determining factors. For instance, in the positive (lead dioxide)
plate, a higher paste density, within acceptable limits, will improve cycle
life while reducing the high-rate discharge capacity. When lead-calcium
grids are used, active material shedding and attendant cycle life reduction
are possible. An improved laminated structure consisting of a thin inner
layer of dense paste and a thicker outer layer of regular density paste has
been suggested [286, 287].

b. Curing. After completion of the pasting operation, the plates are cured
under controlled conditions of relative humidity, temperature, and duration
[368]. There is a great variety of curing, chemsetting, or seasoning processes in use today, depending on the manufacturer, but their purpose is the
same: the strengthening of the active material prior to electrochemical
formation (Sec. 4.2.5). Together with paste mixing, this is one of the most
critical processing steps for the resulting battery's performance and cycle
life.

While stack curing of pasted plates is basically a simple operation, accurate processing conditions have to be selected in order to convert most of

the free lead—for example, 25 to 30 wt % in the original mix—to litharge. Any residual lead will have to be below the concentration level at which the active material will shed during the electrochemical formation step. The conversion will be retarded in the presence of sulfate ions and at low pHs such as -0.5 [369].

The curing conditions are more critical for the preparation of positive plates possessing adequate strength and porosity characteristics. Depending on a variety of composition and process factors, curing can take as little as 20 h or as much as 4 to 5 days.

The flash-drying step, previously mentioned as following the pasting operation, has been replaced at times by a somewhat more awkward method from a materials-handling standpoint. The plates are instead "set" for several hours at 100% relative humidity. Some hydration reactions take place during this period, and the amount of free lead is reduced. Thereafter, the plates can be dried rapidly without encountering excessive shrinkage and cracking problems.

It has been suggested that an acceleration of the curing process is possible by a closer control of the curing condition [370]. The method is based on setting the plates up vertically and maintaining a temperature of $30°C$ at 100% relative humidity. At significantly higher curing temperatures, such as $100°C$, large crystals of the tetrabasic lead sulfate, $4PbO \cdot PbSO_4$, can be formed throughout the plate active material. This may be of significance for specific applications such as long-life traction batteries [369], although it must be kept in mind that in the subsequent electrochemical formation step it is more difficult to convert the tetrabasic sulfate than the tribasic species to the dioxide.

As the moisture content of the paste decreases to values below 7 to 8 wt %, the spontaneous and exothermic oxidation of free lead takes place at rates high enough to generate significant amounts of heat. This, in turn, contributes to the plate drying process. At the same time, a moisture range of 7.0 to 8.5 wt % is also adequate for maximum rates of lead oxidation in the paste [370].

Paste structures based on coarser leady oxide particles cure faster. Ritchie suggested a schematic for the combined diffusion and electrochem-

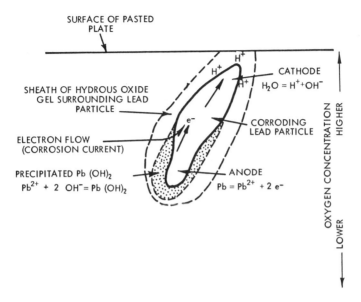

FIG. 62. Electrochemical oxidation of lead particles during plate curing [342]. Atmospheric oxygen diffuses into the paste through the sheath and depolarizes the cathode areas by combining with the hydrogen ions and electrons furnished by the lead anode (O + 2H^{++} 2e$^-$ = H$_2$O).

ical steps involved in the curing process (Fig. 62). In it one can see how a better-defined separation of anodic and cathodic areas, in larger particles, would be conducive to higher corrosion currents and a net acceleration of the overall curing process.

Figure 63 depicts typical changes in the concentrations of the various plate active material constituents during curing [371]. A schematic of the composition transitions in the battery paste during mixing and subsequent curing has been given in Fig. 59. It should be noted that if the plates are cured at higher temperatures and humidities, such as in superheated steam or at temperatures higher than 65°C, the resulting tetrabasic sulfate concentration increases in the active material.

In general, polybasic sulfates, such as 3PbO· PbSO$_4$· H$_2$O, will be formed preferentially at higher acid/oxide ratios in the paste mix. Many of the present paste manufacturing processes end up with high concentrations of the tribasic sulfate species [369]. Depending on the subsequent curing conditions, the sulfate crystals can appear globular or dendritic in

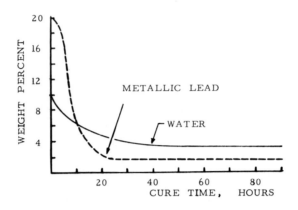

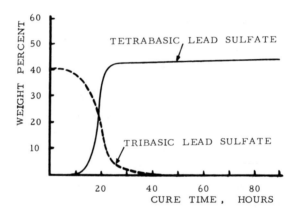

FIG. 63. Compositional parameters versus cure time [371].

shape [354]. As mentioned above, at higher curing temperatures the generally finely divided crystals of tribasic lead sulfate will recrystallize as
large coarse crystals of tetrabasic lead sulfate, with an attendant change
in plate color from yellowish-beige to dark reddish-orange.

Despite the disadvantage of slow formation in depth (complete conversion to PbO_2), the tetrabasic sulfate crystals also have some desirable
features. The electrochemical formation of their surface proceeds relatively rapidly and facilitates current distribution throughout the plate.
Furthermore, their size and high density serves to enhance the mechanical
strength of the active material.

Additional strength in the unformed plates is sometimes contributed by
carbonate films gradually forming on the surface.

4.2.4.2. Planté Plates

Planté plates are manufactured from high-purity (99.96+ wt %) rolled or cast lead sheets cut or stamped to the right size and shape. The sheets are then grooved, swaged, or machined in such a way as to produce a roughness factor of 6 to 10. The final shape of the surface depends on the design and the application, one example being the Manchester positive (see Sec. 4.1.4).

4.2.4.3. Removal of Surface Antimony

Crompton and Vitenbroek have suggested that the treatment of cast antimonial lead grids with mixtures of hydrogen peroxide and sulfuric acid at room temperature will dissolve appreciable amounts of surface antimony [372]. For example, positive 4.5 wt % antimonial grids pickled for 8 h with a 1.2 to 1.3 sp gr sulfuric acid/2 to 4 volumes* of hydrogen peroxide solution, will release in actual operation only 50% of the usual amount of antimony. In the case of 10 wt % antimonial alloys, only 80% will be released as compared to untreated grids.

The treatment can be supplemented by a 10 wt % diethylene triamine dip, to remove surface deposits.

Significant increases in cycle life have been reported as a result of this process modification, and it appears that further developments may be reported in the future along these lines.

4.2.4.4. Utilization of Additives

Various manufacturing techniques have been tried to improve the structural characteristics of the plates or to increase their faradaic efficiency and, as a result, their specific energy and power.

4.2.5. Formation

4.2.5.1. Faure Plate Formation

The pasted, flash-dried, and cured plates are electrochemically formed into lead sponge negatives and lead dioxide positives. Initially, the formation process was only referred to in the context of Planté plate manufacturing, but now the terminology is accepted in a general sense.

---

*One volume concentration H_2O_2 is that concentration (0.3 g of H_2O_2/ liter) which upon decomposition produces 1 ml of O_2 at STP/ml of peroxide solution.

The plates, alternating positives and negatives, are mounted in tanks containing dilute sulfuric acid with a specific gravity usually in the range 1.050 to 1.150. As the acid concentration increases, so does formation time. If the formation temperature is lowered, the resulting end product may have a higher low-temperature capacity, at the expense of service life. Overformation, which is harmful to the structural integrity of the active material, is particularly to be watched for in the case of the positive plates.

α-PbO$_2$ is preferentially formed in neutral or alkaline solutions and various appropriate means to prepare it have been reported [183,373], while acidic solutions will generate β-PbO$_2$ [50,374]. Indeed, the ratio of α- to β-PbO$_2$ resulting during the formation process is influenced by local pH changes brought about by diffusion processes within the plate [369]. These, in turn, depend on the types of basic sulfate present prior to formation (see Sec. 4.2.4 and Fig. 6).

The preferential formation of α- or β-PbO$_2$ will be the result not only of given concentrations of forming acid, but also of the forming current density, the forming temperature, and the paste density (with higher α/β ratios for higher paste densities) [184,375]. Electron microscopy studies have revealed that, when it forms, α-PbO$_2$ will appear preferentially at the grid-active material interface. The studies were conducted in sp gr 1.060 forming acid, using Pb-5.75 wt % Sb grids [376].

A typical process may require 24 h of high-rate formation followed by 12 h of low-rate formation [372]. Positive plates form better in more dilute acid (1.075 to 1.225 sp gr), whereas the negatives are not as sensitive to acid concentration (typical preferred specific gravity range 1.200 to 1.300).

The "one-shot" forming method, after overcoming initial difficulties such as a longer duration of formation, is now seeing wide usage. It consists of the simultaneous formation of both positives and negatives. As a rule-of-thumb, about 225% of the theoretical capacity is required and the final specific gravity of the forming acid is about 0.030 to 0.040 higher [377].

Additional improvements in the process, such as a higher concentration of starting acid and controlled heat removal, are being sought by the
industry in order to shorten the duration of the process.

Uniform product color, acceptable levels of plate gas evolution, and
constant reference voltage readings (using cadmium stick electrodes) are
indicative of end of formation. Sometimes this stage is reached only after
one or more discharges, followed by additional charging.

The formed negative plate surface normally exhibits a certain amount
of blistering, when a formation-in-a-vat-type process is used. Otherwise,
the sandwiched assemblies of negative and positive plate elements are
pressed together sufficiently tightly to prevent the occurrence of blisters.
Yarnell [378] has studied the phenomenon for negative plates made by
applying tetrabasic lead sulfate pastes with 2 wt % expander onto lead 1.05
wt % calcium grids. Under constant-current formation conditions, a solid
layer of lead sulfate forms on the electrode surface. Its subsequent reduction to metallic lead induces fissures between the surface material and the
underlying lead. Hydrogen entrapped in such fissures forms the surface
blisters.

The extent of blistering decreases with a decrease in forming acid concentration, but the resulting product has a lower starting capacity, at least
for an initial cycling period. Thus one is faced with a trade-off between the
structural tolerance for blisters and the starting capacity required.

Conversion of tetrabasic lead sulfate was mentioned in other sections
of this chapter (Secs. 3.3 and 4.2.2.2). Tetrabasic lead sulfate is the
starting paste material for the (Bell Labs.) conical-plate stationary battery.
It was reported that it can be efficiently oxidized to lead dioxide in sp gr
1.005 sulfuric acid [380], although (relative to a tribasic lead sulfate active
material) the process is likely to be slower and more difficult to carry out
to completion.

The role of expanders, relative to the structural integrity and performance of the negative plate, was discussed in Sec. 4.1.3.1. A recent microscopic study has shown that one of the expander types, the lignin derivatives, also has an effect on the formation process by reducing the size and
complexity of the dendritic lead crystals in the active material [379]. An-

other commonly used expander, barium sulfate, was shown not to have any appreciable effect on the microstructure of the formed negative plate.

A very common approach of using fillers consists of blending into the active material nonreactive diluents in fiber, particle, or other form, such as tubelets. Added benefits have been claimed for cases where the surface of the diluent has been pretreated, for better adhesion and/or cohesion with the active material. An example of such pretreatment is the coating of the particles with lignosulfonates [381].

A recent Bulgarian development [298] is the preparation of alkaline battery-type sintered plaques, only this time using lead as the starting material. This application is for the SLI battery. Ball mill leady oxides are used. They are blended with pore-forming agents in a ball mill and then pressed (optimum pressure: approximately 200-atm) onto a standard automotive grid which has been presulfated in sp gr 1.250 sulfuric acid to impart to it better adhesion characteristics.

The finished plates are also treated with sulfuric acid, by spraying, and sintered at 180 to $210°C$ for 15 to 20 min, after which they are electrochemically formed. A typical negative plate paste composition is 10 wt % ammonium sulfate (both a sulfating and a pore-forming agent) and carboxymethylcellulose (usually reported in the patent literature as a bulking and surface-active agent; in this case it is considered a plasticizer, poreforming, and surface-active agent), plus an expander mix (0.5 wt % $BaSO_4$, 0.4 wt % sawdust, 0.3 wt % lignosulfonic acid).

As mentioned in Sec. 4.1.4, superior automotive service characteristics have been reported for batteries using this type of sintered lead plate.

4.2.5.2. Structural Investigations

Microscopic investigations have shown that the formation of both the negative and the positive plates begins, as expected, at the grid-active material interface [191]. This is different from discharge, which begins on the surface and penetrates into the bulk of the active material. By means of electron microscopic investigations, the initial formation steps of the positive plate has been described as mainly a conversion of $3PbO \cdot PbSO_4 \cdot H_2O$ and the orthorhombic PbO to $PbSO_4$, with the tetragonal PbO species being less readily converted [376]. The resulting $PbSO_4$ and PbO are then converted electrochemically to PbO_2. Interior and surface reactions are essentially the same except that, as we have seen, $\alpha\text{-}PbO_2$ will form preferentially at the grid interface.

Table 12 illustrates the number of different species likely to be available for conversion in the active material of the unformed plate. This should provide some idea of the complexity of the reactions and morphological changes involved in the formation process. Furthermore, some carbonate species can be present, such as $2PbCO_3 \cdot Pb(OH)_2$ [382, 383].

Some additional details on the formation process have been suggested by Pavlov and co-workers [229, 384, 385] from studies of pure lead grid electrodes in appropriately dilute ($1\underline{N}$) sulfuric acid solutions. Under galvanostatic oxidation conditions, lead/lead sulfate converts to lead/lead hydroxide at -400 to -500 mV versus Hg/Hg_2SO_4. This is the result of a displacement of the lead sulfate zone away from the grid, which is associated with unequal Pb^{2+} and SO_4^{2-} ion fluxes within the intercrystalline lead sulfate

TABLE 12

Properties of Compounds Likely to Be Present
in Unformed Active Material of the Lead-Acid Cell

Compound	Molecular weight	Density	Gram molecular volume (cm^3)	Moles of PbO_2 formed per mole of compound	ΔV^* (cm^3)	Percent ΔV per Pb atom
Pb	207.21	11.341	18.28	1	6.56	35.87
$PbSO_4$	303.27	6.323	47.96	1	-23.12	-48.21
PbO (orthorhombic)	223.21	9.642	23.15	1	1.69	7.30
PbO (tetragonal)	223.21	9.355	23.88	1	0.96	4.02
Pb_3O_4	685.63	8.925	76.79	3	-0.76	-2.97
$PbO \cdot PbSO_4$	526.48	7.02	75.0	2	-12.66	-33.77
$3PbO \cdot PbSO_4 \cdot H_2O$	990.92	6.5[†]	152.	4	-13.	-34.
$4PbO \cdot PbSO_4$	1196.11	8.15	146.76	5	-4.51	-15.38
$\beta\text{-}PbO_2$	239.21	9.63	24.84			

[*]Volume change upon conversion to $\beta\text{-}PbO_2$ per gram-atom of Pb in the compound.
[†]A very approximate value obtained from the material synthesized for this study.

pores. Conservation of electroneutrality within the lead sulfate precipitation zone requires alkalinization, hence the conversion to lead hydroxide. Under potentiostatic oxidation conditions three potential ranges were defined by Pavlov et al. relative to Hg/Hg_2SO_4: the lead sulfate region, from about -950 to about -300 mV; the lead monoxide region, from about -300 to about +900 mV, where both $PbSO_4$ and tetragonal PbO are formed in significant amounts, as well as some $PbO \cdot PbSO_4$, $3PbO \cdot PbSO_4 \cdot H_2O$, $5PbO \cdot 2H_2O$ or orthorhombic PbO, and $\alpha\text{-}PbO_2$; and the lead dioxide region, above +900 mV, corresponding to a rapid increase of $\alpha\text{-}PbO_2$ content in the anodic layer, and formation of $\beta\text{-}PbO_2$ at potentials more positive than +1,000 mV as a result of the oxidation of bivalent lead ions at the $\alpha\text{-}PbO_2$/solution interface.

Pavlov suggests that the composition changes on oxidation are the result of ionic conductivity modifications within the intercrystalline lead sulfate region. As an example, tetragonal PbO begins to be formed when the charge-carrying species at the lead/anodic layer interface is O^{2-}.

Additional studies, conducted in multiplate SLI cells of 14 Ah capacity [386, 387], revealed that the formation processes in the positive plate can be described as occurring in two stages (Fig. 64). The process depends principally on current density, solution pH, and the $PbSO_4$/PbO ratio in the paste at the reaction layer. These parameters vary during formation and, with them, the relative amounts of α- and $\beta\text{-}PbO_2$ being generated.

The proposed one-stage negative formation process (Fig. 65) is also governed by the same parameters.

Pierson [354] reported on the stage-by-stage transformations in a cured positive plate subjected to a formation rate of 33 mA/cm^2, based on microscopy observations. Some representative pictures are assembled in Fig. 66. Before the start of formation (Fig. 66a), one notices particles of lead within a matrix of lead monoxide, tribasic lead sulfate, and dendritic tetrabasic sulfate; during the initial process stage (Fig. 66b), formation spreads outwardly from the dendrites and continues until the conversion is completed (Fig. 66c). By appropriate modifications of the process, both the positive crystal structure or the negative lead sponge can be produced in shapes varying from fine and nodular to coarse and dendritic.

Because of the previously discussed desirable features of the tetrabasic sulfate crystals, a correspondingly regulated recrystallization of the

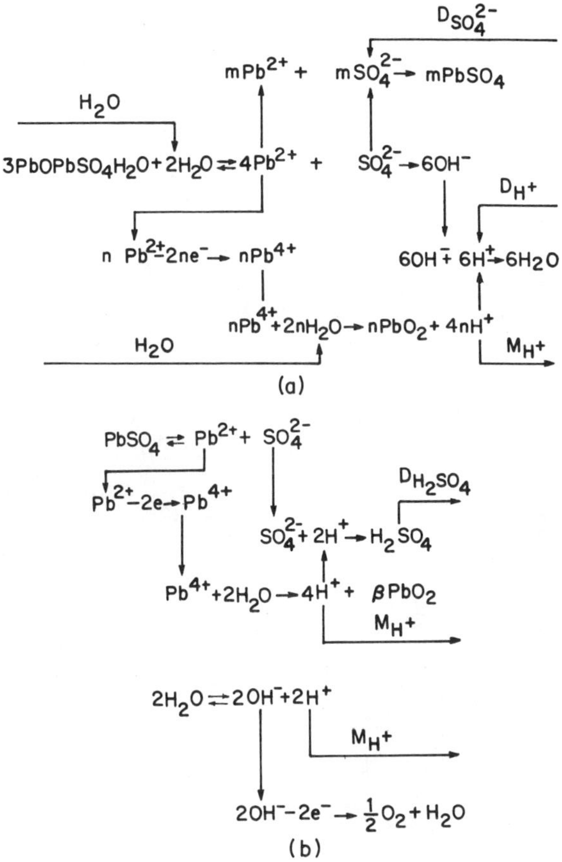

$$D_{SO_4^{2-}}$$

$$mPb^{2+} + \quad mSO_4^{2-} \rightarrow mPbSO_4$$

$$H_2O$$

$$3PbOPbSO_4H_2O + 2H_2O \rightleftharpoons 4Pb^{2+} + \quad SO_4^{2-} \rightarrow 6OH^-$$

$$D_{H^+}$$

$$n\ Pb^{2+} \underline{2ne^-} \rightarrow nPb^{4+} \qquad 6OH \mp 6H^+ \rightarrow 6H_2O$$

$$nPb^{4+} + 2nH_2O \rightarrow nPbO_2 + 4nH^+$$

$$H_2O \qquad\qquad M_{H^+}$$

(a)

$$PbSO_4 \rightleftharpoons Pb^{2+} + SO_4^{2-}$$

$$Pb^{2+} - 2e \rightarrow Pb^{4+} \qquad\qquad D_{H_2SO_4}$$

$$SO_4^{2-} + 2H^+ \rightarrow H_2SO_4$$

$$Pb^{4+} + 2H_2O \rightarrow 4H^+ + \beta PbO_2$$

$$M_{H^+}$$

$$2H_2O \rightleftharpoons 2OH^- + 2H^+$$

$$M_{H^+}$$

$$2OH^- - 2e^- \rightarrow \tfrac{1}{2}O_2 + H_2O$$

(b)

FIG. 64. Formation process schematic for the positive plate [387].
(a) First stage. (b) Second stage.

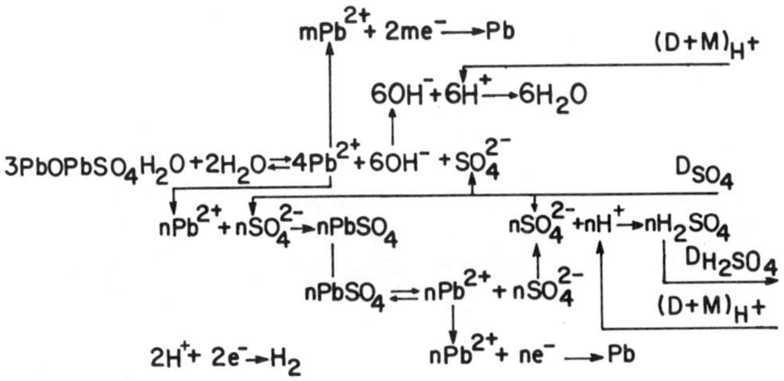

$$mPb^{2+} + 2me^- \rightarrow Pb \qquad\qquad (D+M)_{H^+}$$

$$6OH^- + 6H^+ \rightarrow 6H_2O$$

$$3PbOPbSO_4H_2O + 2H_2O \rightleftharpoons 4Pb^{2+} + 6OH^- + SO_4^{2-} \qquad D_{SO_4}$$

$$nPb^{2+} + nSO_4^{2-} \rightarrow nPbSO_4 \qquad nSO_4^{2-} + nH^+ \rightarrow nH_2SO_4$$

$$D_{H_2SO_4}$$

$$nPbSO_4 \rightleftharpoons nPb^{2+} + nSO_4^{2-} \qquad (D+M)_{H^+}$$

$$2H^+ + 2e^- \rightarrow H_2 \qquad nPb^{2+} + ne^- \rightarrow Pb$$

FIG. 65. Formation process schematic for the negative plate [387].

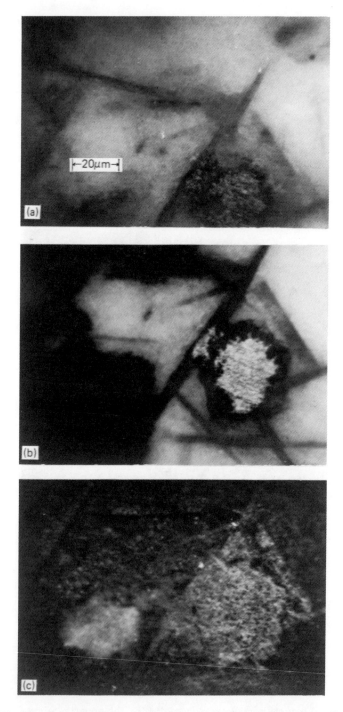

FIG. 66. Several stages in the formation of a positive plate (formation rate: 33 mA/cm^2) [355]. (a) Before the start of formation. (b) Start of formation. (c) Completion of formation.

positive plate active material may actually contribute to an improved faradaic efficiency and strength. Similarly, a more complete conversion of the
centers of the crystals to lead dioxide, preferably the β polymorph, has
been suggested as beneficial since it could yield a cellular structure possessing a dense, hard, conductive surface layer surrounding a high-capacity
finely dispersed bulk portion [371].

Electrochemical investigations have shown that all three basic sulfates
are converted to β-PbO_2 during an anodic oxidation process [388, 389].
This also holds true for other constituents of the paste: $PbSO_4$, Pb_3O_4,
$2PbCO_3 \cdot Pb(OH)_2$, and orthorhombic PbO [382, 383]. On the other hand,
the tetragonal form of PbO appears to oxidize to a lead dioxide polymorph
different from the known species, exhibiting the stoichiometry of $PbO_{1.91}$,
and retaining the diffraction pattern of the original PbO [390].

Elsewhere in this chapter (Sec. 4.1.1), it was pointed out that the addition of "red lead," Pb_3O_4, to the paste mix accelerates the formation rate.
This could be due to a more readily reactive lead oxide (PbO) constituent
which converts into hydrated basic lead sulfate nucleation sites facilitating
the start of the formation process [391]. Burbank and Ritchie [392] also
reported a distinct Pb_3O_4 effect, both for pure and for antimonial lead
grids, in terms of larger PbO_2 crystallites resulting from the formation
process. This suggested to them the formation of a stronger submicro-
scopic crystal network and, as a result, a more durable plate.

At the negative plate, the formation process is associated with a disso-
lution-recrystallization mechanism. The basic sulfate of the active material
is forced into solution by the action of the electric current, after which the
reduced lead reprecipitates in the form of acicular and, at times, platelet-
shaped crystals. This mechanism is inherently different from the one prevailing during subsequent charge-discharge cycles [79] (see also Sec. 2.1.2).

4.2.5.3. Wet- and Dry-charged Batteries

The next processing step is drying, or assembly of the plate elements while
still moist. In the first case the plates are rinsed in water until most of the
acid has been eliminated. They are subsequently dried and assembled into
so-called dry-charged batteries. These are sealed and stored until ready
for activation, i.e., for filling with the required concentration of sulfuric
acid (see Sec. 4.1.6). An inert drying atmosphere and adequate control of

the entire process are required in order to avoid rapid oxidation of the negative plates.

A great variety of methods and specialized equipment (e.g., the Tiegel
oven) are used whereby the plates are dried separately or together in their
forming racks or in special racks; other drying processes involve complete
assemblies of negative and positive plates, complete plate elements with
the separators in place, or even complete batteries using forced convection
of drying gas [393]. This is particularly pertinent in the case of a fully
assembled (E.S.B., Inc.) Star-type three-piece battery (Sec. 4.1.7) prior
to sealing the top and bottom to the main sleeve [394].

The drying process can be carried out in vacuum or in a neutral or reducing atmosphere; superheated steam drying is also used sometimes, as
well as other specialized techniques such as wet-plate immersion in water-
immiscible liquids, i.e., kerosene or other hydrocarbons heated above the
boiling point of water.

In the presence of potential oxidizing conditions, such as atmospheric
drying, oxidation inhibitors are used for the negative plates (Sec. 4.1.3.1)
and, in some procedures, carbonation inhibitors for the positive plates
(Sec. 4.1.3.2).

Normally, the maximum allowable drying temperatures are approximately $80\,^{\circ}C$ for the positive plates or complete element assemblies.

Negative plates under inert atmospheres can be subjected to higher temperatures (for example, $180\,^{\circ}C$), but require cooling prior to exposure to
the atmosphere.

Detailed descriptions of the dry-charging process are available in the
literature [11, 395-397].

In the alternative processing case, the so-called wet batteries are
assembled following the formation process, filled with the appropriate electrolyte concentration, and stored for shipment.

A new manufacturing method trade-named Power Spin (Globe-Union,
Inc.), has recently been described [62, 398]. It results in a battery that
can be handled in inventory as either a wet- or dry-charged product. The
method consists of a formation and draining step involving some processing
modifications that ensure a low rate of positive plate self-discharge in storage. The battery is subsequently centrifuged at 15 to 25 g forces, and hermetically sealed (to avoid negative plate oxidation in storage) with a finite

amount of moisture still retained in the plate elements. Manufacturing cost reductions and superior activation characteristics—generally not requiring charge boosting—have been claimed for this new development. Figure 67 illustrates the process as compared to the sequence used for a wet product, and Fig. 68 shows the equipment used.

4.2.5.4. Planté Plate Formation

In order to achieve complete formation in a reasonable time, forming agents are added to the electrolyte. These attack the lead surface chemically and thus enhance the rate of the electrolytic formation process. Examples of such agents are nitric acid and nitrates, chlorates, perchlorates, permanganates, etc. (for example, 13 g of $NaClO_4$/liter of H_2SO_4 of sp gr 1.06 to 1.08).

At the same time, the forming agents inhibit the generation of a protective covering of lead dioxide which would prevent completion of the electrolytic process, i.e., of the lead sulfation followed by conversion to lead dioxide.

Negative Plantés can also be formed by electrolytic reversal and subsequent conversion of the lead dioxide to lead sponge.

Electrochemical formation times of the order of 25 to 50 h at 1 to 5 mA/cm^2 are typical. At the end of the Planté formation process, conditions and, at times, special posttreatment steps, are selected in such a way as to completely consume the forming agents or eliminate remaining traces thereof. That is necessary in order to avoid structural deterioration of the plate after it is put into service, e.g., lead dissolution as $PbCl_2$. An example of such a postformation treatment sequence would be the dipping of the plate in water or dilute sulfuric acid, followed by immersion in ordinary battery acid. One or more discharge-recharge cycles are also customary as a final step of the posttreatment process.

4.2.6. Battery Assembly

The formed or unformed positive and negative plates are stacked as fully assembled elements, with the appropriate separators sandwiched in between with the ribbed sides facing the positive plates. The stacking is carried out manually or automatically by means of specialized equipment such as the Reed stacker (Fig. 69).

The next assembly step involves welding the plate lugs to connecting straps or bus bars of lead. The straps are cast on the multiplate elements

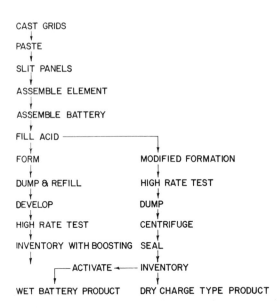

```
CAST GRIDS
   ↓
PASTE
   ↓
SLIT PANELS
   ↓
ASSEMBLE ELEMENT
   ↓
ASSEMBLE BATTERY
   ↓
FILL ACID ─────────────────┐
   ↓                       ↓
FORM                    MODIFIED FORMATION
   ↓                       ↓
DUMP & REFILL           HIGH RATE TEST
   ↓                       ↓
DEVELOP                 DUMP
   ↓                       ↓
HIGH RATE TEST          CENTRIFUGE
   ↓                       ↓
INVENTORY WITH BOOSTING SEAL
   ↓                       ↓
       ┌─ ACTIVATE ◄── INVENTORY
       ↓                   ↓
WET BATTERY PRODUCT    DRY CHARGE TYPE PRODUCT
```

FIG. 67. Power Spin (Globe-Union, Inc.) compared with wet battery process.

FIG. 68. Centrifuging installation used in the Power Spin (Globe-Union, Inc.) Process.

FIG. 69. Automatic plate stacker.

automatically in some of the more advanced processes, e. g. , Globe-Union's COS (cast-on-strap) method [399-402] (Fig. 70), or a similar in situ cast-on-strap process by General Motors. The latter involves precoating the appropriate plate portion with a lower-melting-point alloy, particularly to make it adaptable to lead-calcium systems [403]. The terminal lug welding process is known in the trade as "burning."

Various shapes and sizes of lead parts are required in the overall operation. For ease of production these are normally cast automatically with a minimum of manual labor.

FIG. 70. Equipment for automatic element interconnection by the COS technique (Globe-Union, Inc.).

The design of the intercell connectors in a lead-acid battery is of considerable importance in light of the increasing requirements for high power delivered at subzero temperatures. The material of construction is lead-antimony, or pure lead for maintenance-free operation or for minimum IR losses. There are three basic types used in SLI battery construction (Fig. 71):

   a. Postlink connectors, also known as multicover or conventional connectors; they consist of a lead link welded to the terminals of the joined elements.
   b. Up-and-over connectors, also known as over-partition connectors; they connect the elements directly over the partition, and the resulting shorter conduction path reduces the IR losses.

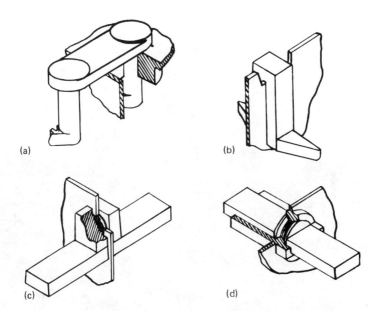

(a)   (b)

(c)   (d)

FIG. 71. The three basic types of intercell connectors in a lead-acid battery [339]: (a) Multicover or conventional connectors; (b) Over-partition or up-and-over connectors; through-partition connectors—(c) HV (Globe-Union, Inc.), and (d) Star (E. S. B. , Inc.).

c. Through-partition connectors, which reduce the IR losses to a minimum by means of a resistance-welded link made through the intercell partition, under the electrolyte level. Two types are particularly known in the United States: the trade-named HV variety, developed by Globe-Union, Inc. [404-410], and E. S. B.'s Star design,
consisting of a flanged completely straight-through configuration
and claimed to represent a slight improvement even over the HV
connector [338,339]. An approach similar to the Star design has
been patented by Lucas (Britain) [411,412].

Figure 72 illustrates the HV through-partition welding equipment, and
Fig. 73 compares the IR losses of the three types of connectors discussed
above.

Detailed discussions on the intercell connector technology are available in the literature [338,339,413,414].

FIG. 72. Equipment for through-partition HV-type connection (Globe-Union, Inc.).

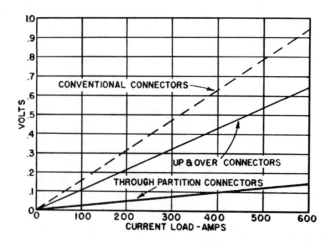

FIG. 73. A comparison of IR losses with the three basic types of intercell connectors [413].

144

4.3. Operation, Testing, and Service

4.3.1. Charging Methods

Storage battery upkeep, both from the standpoint of costs and easy maintenance, is predicated on adequate charging operations. Furthermore,
battery cycle life and performance depends to a great extent on the mode
of charging. Special problems are encountered in the case of traction applications, since high rates of charging (quick charging) are sometimes a required feature, and deep discharge cycling is typical.

Fundamentally, gas evolution during charging should be kept at a minimum and charge acceptance at a maximum. Some gassing is beneficial
since it induces convection currents and minimizes harmful stratification
of the electrolyte solution. However, beyond an acceptable rate of gassing,
significant shedding of active material will occur prematurely.

Charge acceptance, usually expressed as a percentage, is the ratio of
the capacity actually used for the active material conversion to the total
capacity supplied for charging during the same time interval. For instance,

$$K_+ = \left[1 - \frac{v_O}{209 \ C}\right] 100 \tag{16}$$

$$K_- = \left[1 - \frac{v_H}{418 \ C}\right] 100 \tag{17}$$

where

K_+, K_- = charge acceptances of positive and negative plate, respectively, %;

v_O, v_H = volume evolved of oxygen and hydrogen, respectively, cm^3
at STP;

C = charging capacity, Ah.

Peters and co-workers [415] investigated the charge acceptance of
Pb-12 wt % Sb-0.15 wt % As grid plates used in 6-V 12-Ah motorcycle batteries. Their experiments were conducted at various charging rates and
temperatures and yielded a family of charge acceptance data ranging from
70 to 100% for the negative plates and 34 to 96% for the positives. What is
significant in their findings is that, whereas the negative charge acceptance
stays at 100% for a finite time period and then decreases to zero, the posi-

tive charge acceptance passes through a minimum and a maximum prior to
tapering off to a value above zero at the end-of-charge point. This positive
plate behavior is apparently associated with the formation and subsequent
decomposition of persulfuric acid ($H_2S_2O_8$), with the charge acceptance
curve maxima coinciding with maxima in the measured persulfuric acid
concentrations.

In normal practice as applied, say, to automotive batteries, controlled
potential—i.e. tapered current—charging modes are the rule.

Unlimited or limited current modes can be used, with the latter almost
universally accepted since this reduces the charger size (peak power requirement) and the heat developed during charge, although at the expense
of an increased charging time.

In SLI systems, voltage regulation varies depending on the manufacturer, with the regulation limits established as 14.75 ± 1.25 V in some
cases or as tight as 14.2 ± 0.5 V in others. Battery life, particularly
from an overcharge standpoint, is a direct result of the quality of voltage
regulation. With the introduction of the alternator-rectifier charging system, replacing the dc generator, charging efficiency has been improved
and charging is maintained even during engine idling. Figure 74 exhibits
the differences in charging currents, as a function of engine rotational
speed, for both types of charging.

In the case of deep discharges, such as for industrial and traction batteries, the rule of thumb determines that 16 h is required for complete recharging with additional "equalizing" as needed, either by raising the end-
of-charge voltage or by maintaining the battery connected to the constant
potential charger. Additional rules of thumb provide for a charge voltage
of 2.30 V when using lead-antimony grids, at 24°C, with 0.007 V less per
each additional degree of cell temperature. The voltage setting for lead-
calcium grids is 0.15 to 0.20 V higher.

A complete recharge in less than 5 h was not considered possible without an adverse effect on cycle life, and the charge current was not supposed
to exceed the value provided by the so-called ampere-hour rule if an acceptable level of gassing and temperature rise were to be maintained. The latter aspect has been shown not to hold true in many instances, and some per-

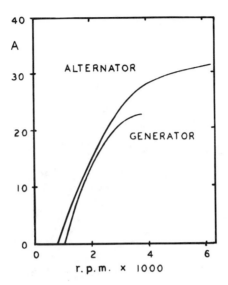

FIG. 74. Variation of charging currents for an alternator and generator with engine speed.

tinent results have recently been reported for lead-calcium batteries [416],
Fig. 75 showing significant deviation from the rule except for cells which
had been 50% discharged. We note that in any antimony-free system the
gassing voltage remains invariant with age since there is no agent present
that can lower the hydrogen overvoltage of the lead sponge negatives.

Many other charging variations are under investigation or in actual
practice, e.g., the Spegel charger (Legg Industries, Ltd., a member of
the Chloride Electric Group in England), which cuts the charging current
off at, say, 2.3 V/cell and turns it on again after the voltage has dropped
to a predetermined lower limit. The sequence is repeated until the time
on charge equals the time on stand, at which point the battery is fully
charged. Variations on this general theme involve the use of two decreasing charging current steps, each for a predetermined time, followed by a
time-controlled third and lower charging current step [417].

The completely automatic nature of such types of charging systems
represents a significant advance over the more conventional timers normally used with modified constant-potential chargers. However, since the
battery terminal voltage is a function of both temperature and age in anti-

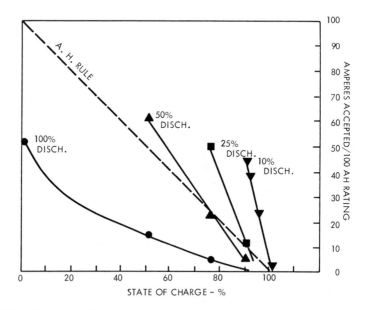

FIG. 75. Deviations from the ampere–hour rule [416]. Lead–calcium system, 1.210 sp gr electrolyte, discharge at 8-h rate, recharge to 2.17 V/ cell.

monial batteries, the same kind of voltage-related control limitations apply as in the case of simpler systems. Some attention is presently being directed to methods which use changes in the electrical impedance of a battery during the charging process to assess its instantaneous current acceptance. A charging system based on this principle would avoid many of the limitations normally associated with modified constant potential- or gas pressure-controlled devices. It is likely that more will be heard on this subject in the near future.

A study of potentiostatically controlled charging of the negative plate has recently been published [418]. It has shown that, under such conditions, the current decays exponentially according to a defined relationship:

$$i = i_K + (i_H - i_K)\exp\left(-i_H \frac{t}{Q}\right) \tag{18}$$

where

  i = current density;

i_K = integration constant;

i_H = defined by: $\varphi s \ell / R$ (φ = polarization, s = specific surface area per unit electrode volume, ℓ = electrode thickness, R = resistance component of the polarization), i_H is the exchange current density for lead sulfate charging at time zero;

t = time of charge;

Q = lead sulfate content in a discharged electrode, expressed as capacity density.

The relationship holds true for thin electrodes, for which

$$\ell \ll \left(\frac{R}{s\rho}\right)^{0.5} \tag{19}$$

where

ℓ = electrode thickness

ρ = effective resistivity

Equation (18) is related to an empirical current decay concept first suggested by Woodbridge [419].

An empirical exponential relationship of the form:

$$i = KQ_C^P \tag{20}$$

has been suggested based on available experimental data [420], where Q_C is the state of charge (measured in coulombs) and K and P are constants. P has been shown to be an increasing function of the discharge rate.

Much remains to be done in the area of lead-acid battery charge optimization, both from the standpoint of theoretical understanding and in its reduction to practice. An important area requiring development efforts is the well-known problem of low charge acceptance at low temperatures. To date, the only proposed solutions involve battery warmers!

Other specific aspects of interest are discussed in what follows.

4.3.1.1. Alternating Current Ripple Effects

An ac ripple is normally associated with the dc charging component. This has a detrimental effect on the corrosion rate of anodically polarized lead, whether pure or alloyed with antimony [421]. The corrosion mechanism depends on the depolarizing effect of the ac ripple and, as could be expected, the corrosion rate will depend on both the proportion of ac and the density of dc current actually used during the charging process.

4.3.1.2. Rapid Charging

There is an increasing interest in the possibility of devising high-rate
charging techniques by suitable modifications of the battery and/or the
charge control method. Impetus in this direction has lately been provided
by the increased applicability of the lead-acid battery to a variety of traction systems [422].

Various schemes have been worked on: charge control by hydrogen
concentration monitoring using a thermal conductivity probe [423,424],
also known as the Cyclocat system in Europe, though the reliability of this
approach has not yet been proved to every user's satisfaction [424], especially when using aged batteries; others use pressure [425] or flow [426]
sensing of the gases evolved during charge and overcharge. A resistance-
free voltage sensing method is suggested in [427]. Here, too, additional
development work is needed.

Another widely publicized approach for rapid charging is based on the
utilization of high-current pulses interrupted by discharge steps sometimes
as short as a few microseconds [428-433]. This or similar techniques,
also described in rather picturesque terms as "electrochemical burping,"
have generally proved to be disappointing in the case of lead-acid batteries
owing to excessive internal heating effects [434,435].

4.3.2. Performance Testing

The object of performance testing, to adequately define a given lead-acid
battery type in line with its required duty cycles, is under constant ap-
praisal and revision. That has been particularly true for establishing
standards of cycle life and discharge capacity at various rates and temperatures.

It is noteworthy that there are as yet no industry wide accepted methods for accelerated life testing, and considerable controversy still exists
regarding the exact significance of even a conventional and presently accepted type of life evaluation, normally requiring periods as long as 3 to
4 months.

Appendixes I through III list examples of presently accepted capacity,
cycle life, and vibration tests for SLI batteries. These have been estab-

lished and are regularly updated as a result of coordinated efforts in the United States between Battery Council International and the Society of Automotive Engineers. The International Electrotechnical Commission (Geneva, Switzerland) coordinates similar standards worldwide, but especially for the European market [436].

In the case of dry-charged batteries, adherence to a given set of activation requirements (e.g., specific gravity of electrolyte after activation, before and after a charge boost) is also monitored.

One notes that the engine starting ability of a battery is now determined by measuring the current that can be supplied for 30 sec with a minimum of 7.2 V for a standard 6-cell battery at $0°F$ (cold cranking rating). This current requirement is primarily a function of engine displacement, and 7.2 V is considered a cranking minimum—starter motors in American cars draw up to 280 A at $-23°C$ and 160 to 190 A at $27°C$. Under former rating systems, the battery discharge rate had no relationship to the amperes required for starting the automobile engine.

The 20-h capacity (ampere-hour rating) has been used as a criterion of reserve power available over a period of time at relatively low rates, such as 3 to 4 A. Today's automobile requirements are significantly higher, though, and the new rating (reserve capacity rating) expresses the number of minutes a battery can deliver 25 A and maintain a minimum of 10.2 V at $27°C$. This is meant to define the ability of a battery to operate the ignition, lights at night, windshield wipers, and defroster in the event of battery recharge system failure.

There is a great variety of other types of performance tests, particularly in the case of other than SLI systems. Standardization is not yet established for such instances (e.g., portable maintenance-free batteries).

4.3.3. Product Quality Control and Field Service

4.3.3.1. Quality Control
As for the manufacturing process, there are many procedural regulations, some of them subtle, some quite radical, depending on the producer.

a. Wet-charged Batteries. An obvious check, made on the shipping line, involves the specific gravity of the electrolyte solution. Hydrometers are

TABLE 13

Recommended Electrolyte Specific Gravity
as a Function of Climate and State of Charge [94]

State of charge[*]	Cold and temperate climates	Tropical climates
Fully charged	1.265	1.225
75% charged	1.225	1.185
50% charged	1.190	1.150
25% charged	1.155	1.115
Discharged	1.120	1.080

[*]Determined by discharge at the reserve
capacity rate.

used for this purpose. The relationship of specific gravity to the battery
state of charge as a function of the climate at the location of ultimate use
is exemplified in Table 13. The check ensures adequate state of charge
and acceptable cell-to-cell variations in this respect. Adequacy of cover
and vent seals and electrolyte level is also checked at this point.

Statistically selected batteries are given a variety of performance
tests (see Sec. 4.3.2) after having been allowed to stand a predetermined
amount of time, e.g., 10 to 60 days. High-rate (5 to 15°C) discharges
(lasting several seconds) are also customary as a quick in-line check of
mechanical failures such as imperfect internal connections, broken terminals, etc.

In storage, a boosting charge—taking care not to overcharge to any
extent—is recommended whenever the electrolyte specific gravity falls
more than a given increment, such as 0.040 corrected to 26.7°C. In warm
weather this may be required every 30 days. A typical SLI charging rate
will be 1 A per positive plate in a cell element.

b. Dry-charged Batteries. As indicated in Sec. 4.3.2, dry-charged batteries also have to be checked for their activation characteristics. In the
case of process modifications (e.g., the recently developed Power Spin
technique, Sec. 4.2.5.3), a "dry open circuit" voltage test is customary
for detection of excessive intercell leakage currents.

In storage boosting is required. After activation the usual dry-charged product, with the possible exception of some categories such as the Power Spin battery, will require a boosting charge.

c. Maintenance-free Batteries. As the name implies, maintenance-free batteries self-discharge at a very slow rate. Typical capacity loss values for some of the lead-calcium gelled electrolyte systems (Sec. 3.4.1.1) are of the order of several percent per month at the most.

These types of batteries are mostly characterized in terms of performance-test results.

4.3.3.2. Field Service

The established storage battery manufacturers are involved in extensive statistical sampling combined with tear-down and other diagnostic analyses of batteries returned from the field. Such activities also involve testing of new product or process developments in controlled field sample lots.

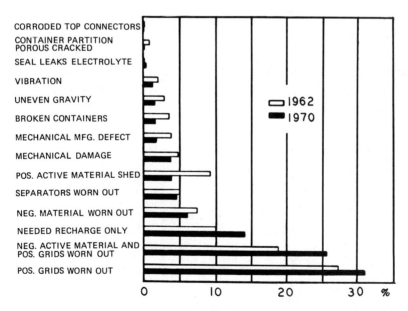

FIG. 76. Failure modes of batteries removed from service—1962 versus 1970. From BCI Technical Committee Report 1970, Chart I, per Ref. 437.

The results of failure analysis surveys conducted on a statistical basis
are also available from compilations such as those of Battery Council International, listing type and frequency of failure of SLI batteries. Figure 76
compares the years 1962 and 1970 in chart form [437]. It indicates positive grid corrosion, with and without worn-out negative active material, as
the main mode of failure. Here one should recall that this failure mechanism is aggravated by overcharging effects in a system where charge is
potential-controlled (see Sec. 4.3.1) and where antimony, if present in the
grid alloy, will progressively contaminate the negative plate and reduce its
hydrogen overvoltage.

Appendix IV lists various types of failure and some of their possible
reasons, based on accumulated field and laboratory tests.

5. THE FUTURE OF LEAD-ACID BATTERIES

The lead-acid battery is firmly established as the starting-lighting-ignition
source of the automobile, and Fig. 77 gives an idea of the increasing magnitude of its use, based on U.S. data. Similar developments are occurring
elsewhere, particularly in markets seeing a significant increase in the use
and production of automobiles. Thus, it has been estimated that the recent
rate of market growth in France and Italy has been 6 to 9% per year, less

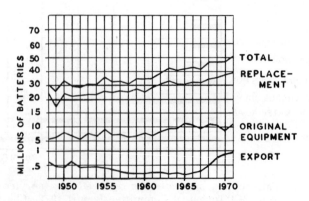

FIG. 77. U.S. shipments of SLI batteries [176].

than 6% in England, but higher than 9% in Germany, and 13% per year or
higher in Japan [95, 437].

There is an increasing emphasis on various forms of on-the-road and
off-the-road electric traction. Considerations of economics and reliability
are maintaining the lead-acid battery as the only large-scale source of
electric power. As an example, an estimated 42,000 golf carts were sold
in the United States during the 12 months ending June 1970, 95% of them
electrically powered, putting some 241,000 batteries to this end use.

Significant market expansion is also occurring in a variety of other
applications requiring the use of maintenance-free portable batteries or of
stationary emergency power systems.

Despite new technological competition from other areas, e.g., high-
energy storage batteries using sodium and sulfur, or such systems as
nickel oxide-iron, nickel oxide-zinc, rechargeable air cells, lithium-metal
sulfide, chlorine hydrate-zinc, fuel cells, etc., which may become attractive in the future for specialized applications (military use, certain on-the-
road traction systems, or load-leveling installations for power stations)
the lead-acid battery stands a good chance of holding its own on a large
scale and for an indefinite future. While lead-acid batteries will remain a
relatively low specific energy power source, they have numerous advantages to make them practical: low cost of the constituents and manufactur-
ability at tremendously high production rates, high reliability over many
hundreds of charge-discharge cycles (as an example, the U.S. market has
for some time seen SLI batteries guaranteed for 5 years; in Britain 7-year
traction batteries are standard equipment), performance flexibility over a
wide range of discharge rates and temperature, and good storage characteristics in a dry-charged state.

It was estimated that 36% of lead production was used during 1970 for
the manufacture of batteries. By 1980 projections call for a consumption
figure of 48% [438]. Further substantiation of the strong future of the lead-
acid battery can be obtained by consulting the statistical data and projections for the United States sponsored by Battery Council International [439,
440].

ACKNOWLEDGMENTS

The author would like to express his appreciation to Globe-Union, Inc.,
for permission to publish this monograph. Special thanks are due to B.
Kolbeck, who organized the manuscript typing effort and contributed a
major share thereof, and E. Ronowski who assisted in this project.

Editor's Note:

Dr. Weissman wrote this chapter in 1973/74 and it has been slightly revised and updated in the final processing of the manuscript. However, the
changes were very few. Essentially all the technical progress of these
three years has been described with foresight by the author and this is a
testimony to his familiarity with the subject. As a man directing the research organization of one of the largest and advanced lead battery manufacturers, he was obviously the right man to write this chapter. However,
it shows also that changes in the technology of such a highly developed system, as the lead-acid battery represents, are happening very slowly—
rather over decades than within years.

In the meantime a scientific treatise of the lead-acid battery by Prof.
H. Bode, in preparation for many years, has been published in The Electrochemical Society Series (see Bibliography). I participated in the English
translation of the German manuscript. This treatise provides a welcome
source for the detailed basic information which could not be included in this
book for the sake of brevity. Such data would have distracted the readers
who are more interested in new technology and battery applications.

K. V. Kordesch

APPENDIX A: STORAGE BATTERY SPECIFICATIONS[*]

Vehicular, Ignition, Lighting and Starting Types

Scope and Classification

This specification covers storage batteries of the lead-acid type as constructed in the wet-charged or dry-charged condition for starting, lighting and ignition service on passenger automobiles, commercial vehicles and other diversified off-the-highway applications.

Two Types of Specifications and Ratings

There are many specifications and performance standards used to evaluate batteries and measure battery power.

All fit into two classifications . . . those which have been established to provide the market with meaningful information and those more technically oriented tests used by automotive and battery engineers to evaluate material variations, battery design and manufacturing methods.

Pretesting Conditioning Procedure

Unless specified otherwise as for dry charge activation tests, all industry sanctioned and approved standards are established on new, fully charged batteries. Therefore, the following conditioning procedure has been established as a requirement prior to laboratory testing.

The batteries shall be charged at a rate equal to 1% of the $0°F$ $(-17.8°C)$ cranking performance test until all cells are gassing freely and the charge voltage and specific gravity of electrolyte are constant over three successive readings taken at one hour intervals.

During the period of charging, the electrolyte temperatures shall be maintained between $60°F$ and $110°F$ $(15.6°C$ and $43.3°C)$.

Measurements of Battery Performance

To provide meaningful measurements of battery performance to the end user, battery engineers supported by BCI (Battery Council International)

---

[*]Extracted from Ref. 176.

and SAE (Society of Automotive Engineers) established two methods of expressing the power characteristics of a battery—power for engine cranking for starting and, secondly, power for ignition, lighting and other accessories required in emergencies.

Cranking Performance

The primary function of the battery is to provide power to crank the engine during starting. This requirement involves a large discharge in amperes over a short span of time. Therefore, the Cranking Performance rating standard is defined as: The discharge load in amperes which a battery at $0°F$ (-17.8°C) can deliver for 30 seconds and maintain a voltage of 1.2 volts per cell or higher.

Cranking Performance Test Procedure

1. Charge battery according to the methods given under "Pretesting Conditioning Procedure."

2. Place the battery in an air ambient held at $0°F$ to $\pm2°F$ (17.8°C $\pm$ 1.1°C) until the temperature of a center cell reaches the rating temperature $\pm1°F$ ($\pm0.6°C$).

3. With the battery at the rating temperature, discharge the battery at the rating current. The rating current shall be held constant $\pm1\%$ throughout the discharge. Measure battery terminal voltage under load at the end of 30 seconds, except for Diesel batteries use voltage at 90 seconds. The terminal voltage for Diesel batteries at 90 seconds shall be equivalent to 1.0 volts per cell or greater. The terminal voltage for other batteries at 30 seconds shall be equivalent to 1.2 volts per cell or greater.

Reserve Capacity

A battery must provide emergency power for ignition, lights, etc., in the event of failure in the vehicle's battery recharging system. This requirement involves a discharge at normal temperature. The Reserve Capacity rating is defined as: The number of minutes a new fully charged battery

at 80°F (26.7°C) can be discharged at 25 amperes and maintain a voltage
of 1.75 volts per cell or higher.

Reserve Capacity Test Procedure

1. Charge battery according to the methods given under "Pretesting
   Conditioning Procedure."

2. The fully charged battery at a temperature of 80°F ± 10°F (26.7°C
   ± 5.6°C) is discharged at 25 A ± 0.25 amperes to a terminal voltage equivalent to 1.75 volts per cell.[*] The reserve capacity is defined as the time of discharge in minutes. All results shall be corrected to 80°F (26.7°C) standard. This shall be accomplished by
   maintaining the battery electrolyte temperature at end of discharge
   at 80°F ± 1°F (26.7°C ± 0.6°C) or by applying the following correction factor.

$$M_c = M_r \left[1 - 0.01(T_f - 80) \right]$$

where

M_c = corrected minutes

M_r = minutes run

T_f = temp. °F, at end of discharge[+]

0.01 = temperature correction factor

Test Sequence

1. Reserve capacity
2. Cranking performance test
3. Cranking performance test
4. Reserve capacity[‡]
5. Reserve capacity[‡]

---

[*]Record temperature of a center cell at the time the end voltage of
1.75 volts per cell is reached.

[+]Results are not considered valid if the final temperature is above
90°F (32.2°C) or below 70°F (21.1°C).

[‡]Not required if the ratings are met on first Reserve Capacity and
Cranking Performance Test. The battery will be considered passing if it
passes the Reserve capacity and Cranking performance test within the five
(5) cycles.

Compliance Standards

When evaluated in accordance with accepted statistical sampling procedures, 90% of the batteries produced should meet the reserve capacity and cranking performance ratings.

Test Procedures and Standards for Evaluation of Battery
Design and Manufacturing Methods:

In addition to the foregoing rating standards covering those battery perform-ance characteristics important to the user—cranking performance and re-serve capacity—the battery industry has endorsed additional test procedures technically important to automotive engineers and to battery manufacturers in maintaining good control of manufacturing methods.

Cold Activation Test of Dry Charge Batteries

This test is made to measure the charge retained in the plates, effective-ness of the processing, wetting characteristics of plates and separators and the behavior of the dry charged battery when activated and placed in service under cold weather conditions.

Test Procedure

1. Place battery and electrolyte $1.265 \pm .005$ sp gr at $80°F$ $(26.7°C)$ in cold box at $30°F \pm 2°F$ $(-1.1°C \pm 1.1°C)$ for at least 18 hours prior to test and hold until both battery and acid are at $30°F \pm 2°F$ $(-1.1°C \pm 1.1°C)$.
2. Remove from cold box and immediately fill the battery with the cold electrolyte.
3. Allow to stand 20 minutes after filling the last cell, then record specific gravity and temperature of the electrolyte.
4. Discharge the battery at 75% of the $0°F$ $(-17.8°C)$ cranking per-formance rating. Note and record the terminal voltage at 15 sec-onds.

The battery is considered to meet the specification requirement if the terminal voltage at 15 seconds is equivalent to 1.2 volts per cell or greater.

Charge Rate Acceptance

This test is designed to determine the ability of a new, previously untested wet battery to accept a charge from a voltage regulated charging system when the battery is at 30°F (-1.1°C) and partially discharged.

Test Procedure

1. Charge battery according to method given under "Pretesting Conditioning Procedure."

2. With the battery at 80° ± 10°F (26.7° ± 5.6°C) discharge at 25 amperes for 80% of thc rated reserve capacity in minutes. Place immediately in cold box until electrolyte of an intermediate cell reaches 30°F ± 2°F (-1.1° ± 1.1°C). With battery in the 30° ± 2°F (-1.1° ± 1.1°C) ambient, charge at a constant potential of 2.4 volts per cell and record amperes charge rate at 10 minutes. This rate shall be taken as the charge rate acceptance and the minimum requirement is 2% of the 0°F (-17.8°C) cranking performance test rating.

Overcharge Life Test

This test was developed when batteries were moved under the hood and placed next to the engine where they were subjected to high engine temperatures. It is primarily a test of positive grid alloy and grid design.

Test Procedure

1. 495 amp-hr[*] of continuous charge is given at 4.5 amp[*] with batteries in a water bath controlled at 100°F ± 5°F (37.8°C ± 2.8°C). Water shall be added daily during charge to restore electrolyte to normal level.

2. The battery is given a 48-h stand on open circuit in the water bath.

3. A 150-amp discharge check rate is given at the battery temperature obtained in (2) to an end voltage equivalent to 1.20 per cell or a minimum discharge of 30-sec whichever occurs first.

4. Repeat (1) to (3).

---

[*]For 12-v batteries rated at more than 180 minutes reserve capacity and all 6-v batteries, use two times the ampere rates and two times the ampere-hour values specified.

The overcharge life test shall be terminated when the battery fails to meet (3). The number of life units (weeks) on test shall be computed to the last week in which the minimum time of 30 seconds has been equaled or exceeded; that is, the week in which the battery failed shall not be counted in reporting the life units.

Vibration Test

The vibration test is a means of evaluating the structural integrity of the complete battery; i.e., the ability of the battery to deliver dependable power while the vehicle is operated over rough roads or other severe terrain.

Test Procedure

A fully charged battery shall be vibrated for a minimum of 2 hours at the acceleration force and frequency specified. At the start of the vibration test, the battery temperature shall be $80°F \pm 10°F$ ($26.7°C \pm 5.6°C$) and shall be discharged at approximately 1% of the cold cranking rate during the vibration test. It must maintain a steady voltage and current, and at the end of this test, there shall be no internal damage and no adhesion or cohesion failure of the sealing compound.

Hot and Cold Cycle Test

The purpose of this test is to determine the integrity of the seal binding the cover over each cell to the battery container under both hot and cold temperatures.

Test Procedure

The battery must pass ten (10) cycles of 16 hours at $0°F \pm 1°F$ ($-17.8°C \pm 0.6°C$) and 8 hours at $70°F$-$90°F$ ($21.1°C$-$32.2°C$) with no signs of cracking or loss of adhesion becoming apparent. At the 3rd, 6th, and 10th cycle, air pressure of one pound per sq. inch will be applied to each cell individually, immediately after removing from the cold chamber. This pressure must be maintained for 15 seconds. In addition to maintaining the air pressure throughout the test, there shall be no evidence of apparent cohesive cracks.

Test Sequence-Evaluation of Battery Design and Manufacturing Methods

Cold Activation[*]	1	1	1
Charge Rate Acceptance	2		
Reserve Capacity	3		
Cranking Performance	4		
Cranking Performance[†]	5		
Reserve Capacity[†]	6		
Reserve Capacity[†]	7		
Overcharge Life	8		
Vibration		2	
Hot and Cold Cycle			2

[*]To be run only if batteries are received for testing in charged and dry condition. If wet charged, bring to full charge as outlined in Pretesting and Conditioning Procedure.

[†]Not required if the ratings are met on the first Reserve capacity and Cold performance test.

Electric Vehicle Battery Test

This test is designed to simulate demands imposed on batteries engaged in supplying power to electric vehicles.

Batteries fully charged and with the electrolyte temperature at 80° ± 5°F (26.7°C ± 2.8°C) are discharged at the constant rate specified for the type battery being tested to a terminal voltage equivalent to 1.75 volts per cell. The discharge time in minutes is the battery capacity. The full charge electrolyte specific gravity is to be the same as specified by the battery manufacturer. Golf car batteries shall be tested as indicated above at a rate of 75 ± 1 ampere.

APPENDIX B: SAE STANDARD TEST PROCEDURE FOR
                    STORAGE BATTERIES J537G[*] (Revised Feb. 1972)

Applications

This SAE Standard applies to lead-acid types of storage batteries used in
motor vehicles, motorboats, tractors, and automotive industrial appliances
equipped with regulated charging systems.

ELECTRICAL TESTS

Testing Procedure

Capacity ratings are to be determined by the procedures outlined under
Sampling, Conditioning, and Sequence of Tests. Battery classifications,
dimensions, and ratings are given in Table 1.

Sampling

Five batteries shall be selected as a sample for tests. The age of the sample from date of manufacture shall be not less than 10 days nor more than
60 days.

Conditioning

Charging

The battery shall be charged at a rate equal to 1/100 of the $0°F$ Cold Cranking discharge rate until all cells are gassing freely and the charge voltage
and specific gravity of electrolyte are constant over three successive readings taken at 1 hour intervals.

---

[*] Extracted from Ref. 176.

Electrolyte Temperature

During the period of charging the electrolyte temperature shall be maintained between 60° and 110°F.

Specific Gravity

The fully charged initial specific gravity at the end of the first conditioning charge shall be adjusted to 1.265 ± 0.005 corrected to 80°F. For dry charged batteries, specific gravity at the end of the first conditioning charge shall be 1.265 ± 0.005.

Sequence of Tests

1. Charge battery according to methods given under Conditioning, and repeat this before each discharge.
2. Perform Charge Rate Acceptance Test.
3. Conduct the Reserve Capacity Test at 25 amp at 80°F.
4. Conduct the Cold Cranking Test at 0°F at the discharge rate specified in Table 1 for 0°F test.
5. Conduct the Cold Cranking Test at -20°F at the discharge rate specified in Table 1 for -20°F test.
6. Repeat step 3. [*]
7. Repeat step 6. [*]
8. Perform life tests.

The highest Reserve Capacity test value obtained for each battery in test events 3, 6, and 7 shall be used to determine an average value for the five sample batteries, which value will be considered to represent the performance of the battery type.

---

[*]To determine compliance with ratings, steps 6 and 7 are not required if the Reserve Capacity Rate is made in step 3.

Activation Test of Dry Charged Battery

Dry charged batteries when activated according to manufacturer's instructions shall meet the minimum standards for wet batteries.

Charge Rate Acceptance Test

This test is to reveal the ability of a new, previously untested wet battery (or activated dry charged battery) to accept a charge under conditions existing in the voltage-regulated electrical system of a vehicle at 30°F with the battery in a partially discharged condition.

1. Charge battery according to method given under Conditioning.
2. Discharge battery at 25 ± 0.25 amp from a starting electrolyte temperature of $80° \pm 5°F$ for a time equal to 0.8 times the Reserve Capacity rating shown in Table 1.
3. Place battery in cold box until electrolyte of a center cell reaches $30° \pm 2°F$.
4. With battery in $30° \pm 2°F$ ambient, charge at a constant-potential equivalent to 2.4 V per cell and record charge rate in amperes at 10 minutes. This rate in amperes shall be taken as the Charge Rate Acceptance.

Note: Minimum requirement in Table 1 is 2% of the 0°F Cold Cranking Rate.

Reserve Capacity Rating

The fully charged battery at a temperature of $80° \pm 5°F$ is discharged at 25 ± 0.25 amp to a terminal voltage equivalent to 1.75 V per cell.[*] The Reserve Capacity is defined as the time of discharge in minutes. All results shall be correct to 80°F standard. This shall be accomplished by

---

[*]Record temperature of a center cell at the time the end voltage equivalent to 1.75 V per cell is reached.

maintaining the battery electrolyte temperature at the end of discharge at $80° \pm 1°F$ or by applying the following correction factor:

$$M_c = M_r \left[1 - 0.01(T_f - 80) \right]$$

where:

M_c = corrected minutes

M_r = minutes run

T_t = temperature at end of discharge, F^*

0.01 = temperature correction factor

Cold Cranking Test (30 Sec Test)

The following test is a measure of the cranking capability of a battery at the rating temperature.[†]

1. Place the battery in an air ambient held at the rating temperature[†] $\pm 2°F$ until the temperature of a center cell reaches the rating temperature $\pm 1°F$.
2. With the battery at the rating temperature,[†] discharge the battery at the rating current[‡] shown in Table 1 for 30 sec. The rating current shall be held constant $\pm 1\%$ throughout the discharge. Measure battery terminal voltage under load at the end of 30 sec.
3. The acceptance criterion for this test is that the battery terminal voltage at the end of 30 sec shall be equivalent to 1.2 V per cell or greater.

Cold Cranking (90 Sec. Test for CIM Starting)

The following test is a measure of the cranking capability of a battery at the rating temperature.

---

[*]Test is not valid if T is outside temperature of $70°-90°F$.

[†]Rating temperature for purpose of this test is either $0°$ or $-20°F$ as shown in Table 1.

[‡]When a battery is built with double insulation for heavy duty service, see footnote c of Table 1.

1. Place the battery at the rating temperature $\pm 2°$F until the temperature of a center cell reaches the rating temperature $\pm 1°$F.

2. With the battery in an air ambient at the rating temperature, discharge the battery at the rating current shown in Table 1 for 90 sec. The rating current shall be held constant $\pm 1\%$ throughout the discharge. Measure battery terminal voltage under load at the end of 90 sec.

3. The acceptance criterion for this test is that the battery terminal voltage at the end of 90 sec shall be equivalent to 1.0 V per cell or greater.

Life Test

See SAE J240.

Overcharge Life Test[*]

This test was designed to provide more overcharge than that covered in the Cycling Life Test. A charge 495 amp-hr is given, followed by a discharge capacity check, thereby representing one week's testing on a continuous basis or one complete life unit. The procedure for a life unit is as follows:

1. 495 amp-hr of continuous charge is given at 4.5 amp, with batteries in a water bath as specified under Test Equipment and Specifications.

2. The battery is given a 48 hr stand on open circuit in the water bath.

3. A 150 amp discharge check rate is given at the battery temperature obtained in step 2 to an end voltage equivalent to 1.20 V per cell or a minimum discharge time of 30 sec whichever occurs first.

4. Repeat steps 1-3 without separate discharge.

---

[*]For 12 V batteries at more than 180 minutes Reserve Capacity and all 6 V batteries, use two times the ampere rates and two times the ampere-hour values specified for the Overcharge Life Test.

5. The overcharge life test shall be considered completed when the battery fails to meet step 3. The battery shall meet the minimum life unit (weeks) specified in Table 1. The number of life units (weeks) on test shall be computed to the last week in which the minimum time of 30 sec has been equaled or exceeded; that is, the week in which the battery failed shall not be counted in reporting the life units.

6. Water shall be added daily during charge to restore electrolyte level to normal.

VIBRATION TEST

This vibration test is to determine the ability of a battery to withstand vibration forces without suffering mechanical damage, loss of capacity or electrolyte or without developing internal or external leaks.

Procedure

1. Four batteries shall be selected as a sample for this test.
2. Bring batteries to full charge and pressure test each cell with 2 psi air pressure for 5 sec. Maximum pressure loss shall be 0.1 psi on a closed system which has the pressure supply blocked off.
3. Place one or more of the batteries at $80° \pm 5°F$ on vibration machines recommended for this test. The battery plates shall be oriented parallel to the axis of the rotating shaft of the machine.
4. The batteries shall be firmly held down by a hold-down means bearing on the battery in similar fashion as to that encountered in vehicle applications.
5. The electrolyte shall be at the level recommended by the manufacturer.
6. The batteries shall be vibrated for a minimum of 2 hours at an acceleration of 5 g and a frequency of 30-55 cps. Each 2 hours of vibration shall represent one unit of vibration.
7. During vibration, there shall be no electrolyte losses.

8. After vibration, each cell must maintain 2 psi of air pressure for 5 sec with a maximum loss of 0.1 psi of pressure as tested in item 2.

9. Immediately discharge the battery at 80° ± 5°F at the Cold Cranking rate for 0°F shown in Table 1. The 80°F battery must meet the minimum voltage at 30 sec.

10. The battery will be rated at the number of units it can survive while meeting the requirements of the preceding paragraph and on external and internal examination shall have no mechanical defects or leaks.

TEST EQUIPMENT AND SPECIFICATIONS

Tank

The tank shall be of sufficient size and design to permit a minimum of 1 in. battery clearance on all sides including the bottom, and a battery immersion depth of 6 in. Two battery strip supports are to be used. The width of the contact surface between the battery and each battery support shall be no greater than 1.5 in. Means shall be provided to thermostatically control the water temperature at 100° ± 5°F.

Vibration Machine

U.S. Army ordnance battery vibration machine as shown on drawing No. D7070340, or L.A.B. Reliability vibration test machine type ARV-30x40-400, or exact equivalent.

Double Insulated Batteries

Double insulation is defined for the purpose of this SAE Standard as the use of a retaining sheet of porous or perforated material between the positive plate and the customary single separator.

Location of Battery Parts

The location and polarity of the terminal posts and the position of handles, when used, shall be as shown in Table 2.

Type Designations and Markings

Type letters, numbers, or symbols, which shall enable the user to determine ratings from the manufacturer's catalogs, shall be stamped or molded on the case or cell connectors or on a name plate permanently attached to the end or side.

Terminal Posts

Polarity shall be plainly marked as follows: The positive post shall be identified by Pos, P, or + on the terminal or on the cover near the terminal. When taper posts are used for terminals of lead-acid storage batteries, the dimensions shall be as follows:

Small diameter of negative post: 0.625 in.
Small diameter of positive post: 0.688 in.
Minimum length of taper: 0.625 in.
Taper per foot, included angle: 1.333 in.

LIFE TEST FOR AUTOMOTIVE STORAGE BATTERIES—SAE J240

SAE Recommended Practice

Report of Electrical Equipment Committee approved May 1971.

Scope

This life test simulates automotive service where the battery operates in a voltage regulated charging system. This test subjects the battery to charge and discharge cycles resulting in failure modes comparable to those encountered in automotive service.

1. The battery is tested in a water bath maintained at $105° \pm 5°$ F.

2. The test cycle is performed as follows:

    Discharge: 2 min at 25 A

    Charge: (a) maximum voltage—14.8

                (b) maximum rate—25 A

                (c) time—10 min

3. The battery is continuously cycled for 100 h (for example, Monday noon until 4:00 p.m. Friday).

    A switching delay of not more than 10 s is permitted from termination of charge to start of discharge, and termination of discharge to start of charge.

4. The battery is given a 60 h stand on open circuit in the water bath.

5. Discharge battery at a rate equal to its $0°$ F cold cranking discharge rate in amperes, at the battery temperature obtained in paragraph 4, to an end voltage equivalent to 1.20 V per cell, or a minimum discharge time of 30 s, whichever occurs first.

6. Replace battery on the cycling test without a separate recharge. Start on the "charge" portion of the cycle.

7. The life test shall be considered completed when the battery fails to maintain 1.2 V per cell for a minimum of 30 s on the manual discharge (paragraph 5).

    The point of failure shall be determined by plotting the 30 s discharge voltage values.

8. Water should be added as required during the cycling portion of the test.

Note: This test only applies to 12 V batteries of 180 min or less Reserve Capacity.

SAE Technical Board Rules and Regulations

All technical reports, including standards approved and practices recommended, are advisory only. Their use by anyone engaged in industry or trade is entirely voluntary. There is no agreement to adhere to any SAE Standard or SAE Recommended Practice, and no commitment to conform or be guided by any technical report.

In formulating and approving technical reports, the Technical Board, its Councils and Committees will not investigate or consider patents which may apply to the subject matter. Prospective users of the report are responsible for protecting themselves against liability for infringement of patents.

Author's Note:

It is to be noted that specified cycle no. minima are still in the process of being established for various sizes of batteries. The number of cycles to failure can easily be of the order of 5,000 to 10,000, for a variety of SLI batteries in the 50- to 90-Ah capacity range.

APPENDIX C: OTHER CYCLING TESTS STILL IN USE

1. Old SAE Cycle Life Test (J537F)

The Cycling Life Test is a test principally to evaluate battery internal components. It does not attempt to simulate car service conditions which are extremely variable. The batteries shall be placed in a water bath such as described under Overcharge Life Test—Test Equipment and Specifications, spaced so the control of the water bath temperature at 105° ± 5°F will allow the electrolyte temperature, at no time, to exceed 115°F during the life cycles. Batteries shall be discharged for 1 hr at approximately 20 amp for a total of 20 amp-hr and recharged at the rate of approximately 5 amp for 5 hr for a total of 25 amp-hr.

The total time for one complete cycle is 6 hr, which permits 4 cycles per day or approximately 27 cycles a week. The average temperature of the electrolyte of the center cell shall be maintained at 110° ± 5°F during the life cycles. In order to determine the condition of the battery throughout the duration of this life test, one complete capacity discharge test is made each week at 20 amp. The length of time in minutes required for the battery voltage to drop to a final terminal voltage equivalent to 1.75 V per cell, while discharging at 20 amp, is recorded. When the running time (to 1.75 V per cell) of the battery on a complete discharge cycle, at the 20 amp

rate, drops below 70% of the time in minutes in Table 1 for Reserve
Capacity, the life test of the battery shall be considered completed. The
point of failure shall be determined by plotting the minutes run.

A complete capacity discharge cycle should not be made after the battery has stood on open circuit without a freshening charge of 1 hr at 10 amp
for every 24 hr on open circuit.

Water should be added as required, except that no water should be
added just before the complete capacity discharge cycle.

Performance standard: (Specified minimum) 250 cycles for 60-Ah
battery.

2. Edison Cycle Life Test

Edison Cycle Life Test for Lead-Acid SLI Batteries

Temperature:	$38°$ to $43°C$ in air environment
Cycle time:	10 min; 1,000 cycles/week
Charge	6 min, 27 sec at 5 to 6 A ⎫
Discharge	10 sec at 150 A ⎬ 10 min
Rest	3 min, 23 sec ⎭
Capacity test:	150 A to 7.2 V (1.2 V/cell)
Specified minimums:	
Capacity	>0.5 min to 7.2 V
Life	10 weeks (10,000 cycles) for 60 Ah battery

APPENDIX D: INTERNAL EXAMINATION OF CELLS[*]

The repair of used defective batteries is rarely justified. The high costs
of parts and labor as compared to the cost of replacement with a new battery make any projected repairs uneconomical. Batteries having one-piece

[*]Extracted from Ref. 94.

covers either heat sealed or sealed to the containers with permanent ce-
ments cannot in any case be opened for repair and reused.

The following discussion is presented for its informational value, and
is illustrative of some of the points covered in Part XVI, "Factors Affect-
ing Battery Life."

Inspection for Open Circuits

The condition of an incomplete circuit or a failure due to an "open" circuit
in a battery is infrequent but is a condition that could exist either on a new
battery or at some time later in the life of the battery. It would be evidenced by the lack of a voltage reading across the terminals of the battery.
In a new battery it would most likely be a result of a mechanical or manufacturing defect in the battery, such as a connection not made or a broken
lead part. After the battery has been in use this type of failure might be
caused by excessive corrosion of a poorly burned connection, a broken part
resulting from vibration or misuse or a "fused" (melted) lead connection
caused by a dead short or excessively high load across the battery terminals for too long a period of time.

Inspection for Short Circuits

This condition may result from material falling from the plates and being
deposited in sufficient quantity to short-circuit the plates at the bottom or
edges. Or it may be the result of lead growing ("treeing") from plate to
plate through a hole or split in a separator or a rough edge on a plate may
have cut through the separator. Lead may have run down during the lead
burning of either the plates to the post straps or the connectors to the posts.
Such a "run-down" may not short-circuit the element until later when considerable wear has occurred. One plate or several plates in the element
may have "buckled," causing excessive wear and failure of separators resulting in a short circuit of the element.

Examination of Positive Plates Finely Divided Shedding

Finely divided, uniform shedding over the entire surface of the positive
plates is characteristic of a battery after long service.

Grid Oxidation

Long continued over-charging is always accompanied by oxidation of the
positive grids, which reduces the metallic cross-section of the grid wires
and weakens the plate so that it is easily broken under slight pressure or
vibration. Over-charging can be corrected only by lower adjustment of the
voltage regulator to meet driving conditions.

Chunky Shedding

This may be caused by too high a charging rate on a sulfated positive plate.
It also might be caused by the battery freezing while discharged. Too high
a charging rate would probably be accompanied by oxidation of the plate
frame. Chunky shedding may also be induced by vibration due to loose
mounting of the battery in the carrier.

Cracked Outside Grid Frames

This indicates severe expansion due to the positive plates having been in a
partly discharged condition and having been permitted to stand until considerable hardening action has taken place. This condition is aggravated by
over-charging after plates have become sulfated.

Hard Plates

When the material in the plate is hard when scratched, like the surface of
an unglazed tile, it is indicative of heavy sulfation. It may be corrected by
a long, slow charge, from 60 to 100 hours at half the normal rate, unless
the sulfate has become too extensive and coarsely crystalline.

Buckled Plates

Buckling of positive plates may be caused by plates standing in an under-charged state for a considerable period of time, or may be the result of excessive charging, especially after plates have previously become densely sulfated by standing in an undercharged state.

Discoloration

Low electrolyte level will sometimes result in the formation of a distinct area or zone of whitish sulfate near the tops of the positive plates which were exposed to the air, while the area covered by electrolyte will remain brown in color. The activity of such a sulfated area is permanently impaired even though electrolyte level is restored.

Discoloration of plates with white lead sulfate may also appear in elements which have stood for considerable time in a discharged condition in electrolyte of very low specific gravity.

Examination of Negative Plates, Glossy Negative Material

Fully charged sponge lead negative material normally has a slate gray color and glosses to a metallic sheen when rubbed smooth with the back of the thumbnail.

Sandy Negative Material

This may result from operation of the battery when the element is shorted, or may be caused by high–gravity acid aggravated by high temperatures. If these conditions of high gravity and high temperature prevail for some time, the negative sponge lead may become soft and mushy.

If the material appears to be sandy when rubbed with the back of the thumbnail, it may be due only to a discharged condition. This is due to the fact that very often when a negative plate is only partially charged, the material may have an appearance similar to a sandy negative. If, after charging for at least 24 hours and until the gravity stops rising (it may take two or three days), the sandy feeling of the material is not removed, the negative plates may be considered worthless.

Loss of Active Material

Loss of negative active material from the grids may be the result of high-specific-gravity acid, of charging when the element is short-circuited, or of excessive temperature due to continuous charging at an excessive rate.

Discoloration

A white discoloration may be due to the accumulation of white lead sulfate, resulting from low electrolyte level, or standing in a semi-charged condition. Dark discolorations may be due to small deposits of metallic impurities such as antimony or copper, or may be due to the battery having been accidentally charged in reverse for a prolonged period.

Examination of Separators: Normal Condition

Separators which are firm but flexible, not black in color and without splits or holes, are considered normal.

Soft or Extremely Brittle

This condition is caused by prolonged exposure to high temperature, above $110°F$ ($43.3°C$), high-specific-gravity acid (above 1.300) resulting from water loss, or a combination of both of these conditions. These conditions usually turn the separator very dark, almost black in color. Microporous rubber of plastic separators are usually not so affected.

Worn Separators

Worn separators with ribs almost destroyed and the back web perhaps perforated are characteristic of separators subjected to excessive pressure
caused by buckled plates or excessively expanded negative material or excessive vibration. Separator deterioration resulting from excessive temperature or high-specific-gravity acid, will make the separator much
darker in color than when the cause is pressure from plates.

Pitted Separators

Pitting of separators results from pressure due to contact with loosened
active material from positive or negative plates. When caused by positive
material, the pitting starts on the ribbed side of the separator and may
show dark-colored, carbonized spots. When due to negative material, the
pitting starts on the side of the separator in contact with the negative plates
and is often characterized by a "worm-holed" appearance due to movement
of loose particles of active material.

Fringed at Bottom

This may be caused by excessive wear due to buckling of positive plates,
or may result from oxidation of the separators by the shedded positive
active material.

Notched at Bottom

Notches at points where separators rest on the bridges indicate either loose
installation of the battery or excessive looseness of the element in the cell
compartment allowing separators to chafe on the bridges.

If the battery was loose in the car, the outside of the container may
show abrasion where it came in contact with the battery-carrier supports.
If the container does not show such abrasion, it is possible that the element

was assembled too loosely in the cell. Excessive and prolonged vibration can also cause this condition in correctly assembled and installed batteries.

Examination of Containers: Abrasion

If container abrasion is noted, it is a sure sign that the battery has not been securely held in position. It is then probable that the resulting severe vibration has seriously damaged the elements and possibly caused cracks in the container. This condition may also have caused plates to wear deep notches in the element rests or the rests to have worn deep notches in the bottom of the separators.

Partition Failure Tests

If the battery is giving trouble and requires repeated recharging, even after adjustment of the charging system, and the fully charged voltage of cells while on charge is uniform, test for cracked or porous partitions.

If two adjacent cells have gravity readings considerably lower than the others, it is a fairly good indication that there is electrical leakage between the two cells. The positives of one cell, and the negatives of the next, discharge against each other.

Container Leaks

External leaks or seepage can often be detected by inspection of the box walls. If in doubt about a leak on the bottom of the box, wash the battery all over, dry it and set it on a clean dry piece of paper on a clean, dry board, where a leak will produce a wet spot.

Distortion and Bulging

Holddowns which are too tight may pull containers out of shape. High temperatures experienced in under-the-hood locations will aggravate distortion and may permit bulging of any long, unsupported cell walls by the constant pressure of the weight of liquid in the cells. The materials of which containers are made become softer when heated to high temperatures, and may distort under a steady pressure, even though it be small.

BIBLIOGRAPHY

1. F. E. Kretzschmar, Betriebseigenschaften und Krankheiten des Blei-Akkumulators, 4th ed., Verlag Technik, Berlin, 1950.

2. C. Drotschmann, Bleiakkumulatoren, Verlag Chemie, Weinheim, 1951.

3. G. W. Vinal, Storage Batteries, 4th ed., Wiley, New York, 1955.

4. H. G. Brown, The Lead Storage Battery, John Sherratt & Son, Altrincham, England, 1959.

5. M. A. Dasoyan and I. A. Aguf, The Lead Accumulator, Asia Publishing House, London, 1968.

6. W. Garten, Bleiakkumulatoren, 9th ed., VDI Verlag, Düsseldorf, 1968; 10th edition, VARTA Batterie A.G., Hannover, 1974.

7. M. Pöhler, Behandlung und Wartung von Bleibatterien für Elektrofahrzeuge, VDI Verlag, Düsseldorf, 1969.

8. E. Witte, Blei- und Stahlakkumulatoren, Otto Krausskopf Verlag, Mainz, 1969.

9. N. E. Hehner, Storage Battery Manufacturing Manual, Independent Battery Manufacturers Association, Largo, Fla., 1970.

10. W. Hofmann, Lead and Lead Alloys, Springer-Verlag, New York, 1970.

11. J. Burbank, A. C. Simon, and E. Willihnganz, The Lead-Acid Cell, in Advances in Electrochemistry and Electrochemical Engineering, Vol. 8 (C. W. Tobias, ed.), Wiley, New York, 1971, pp. 158-251.

12. H. Bode, Lead Acid Batteries, The Electrochemical Society Series, Wiley-Interscience, John Wiley & Sons, Inc., New York, 1977.

REFERENCES

1. G. Planté, Compt. Rend., 50, 640 (1860).

2. J. V. Tierney, Jr., and C. K. Morehouse, in Proceedings of the Second International Conference on Lead, Arnhem, Pergamon Press, New York, pp. 235-241.

3. K. J. Euler and A. Fleischer, in Performance Forecast of Selected Static Energy Conversion Devices (G. W. Sherman and L. Devol, eds.),

29th Meeting of AGARD Propulsion and Energetics Panel, Liège, Belgium, 1967, pp. 3-56.

4. J. H. Gladstone and A. Tribe, Nature, 25, 221, 461; 26, 251, 342, 602; 27, 583 (1882-1883).

5. I. A. Aguf, Sov. Electrochem., 4, 1022 (1968).

6. H. Bode, Elektrotechn. Z., 18, 857 (1966).

7. M. A. Dasoyan and I. A. Aguf, The Lead Accumulator, Asia Publishing House, London, 1968, p. 6.

8. G. W. Vinal, Storage Batteries, 4th ed., Wiley, New York, 1955.

9. T. Chiku and K. Nakajima, J. Electrochem. Soc. 118, 1395 (1971).

10. J. Burbank, A. C. Simon, and E. Willihnganz, in Advances in Electrochemistry and Electrochemical Engineering, Vol. 8 (C. W. Tobias, ed.), Wiley-Interscience, New York, 1971, pp. 157-251.

11. N. E. Hehner, Storage Battery Manufacturing Manual, Independent Battery Manufacturers Association, Largo, Fla., 1970.

12. P. Ruetschi and R. T. Angstadt, J. Electrochem. Soc., 111, 1323 (1964).

13. H. Bode and E. Voss, Electrochim. Acta, 1, 318 (1959).

14. S. C. Barnes and R. T. Mathieson, in Batteries 2 (D. H. Collins, ed.), Pergamon Press, New York, 1965, pp. 41-54.

15. J. Burbank, J. Electrochem. Soc., 106, 369 (1959).

16. D. Berndt and E. Voss, in Batteries 2 (D. H. Collins, ed.), Pergamon Press, New York, 1965, pp. 17-27.

17. A. C. Simon, Electrochem. Technol., 3, 307 (1965).

18. R. P. Vasileva and E. A. Mendzheritskii, Sov. Electrochem., 8, 507 (1972).

19. D. Berndt, Chem. Ing. Tech., 38, 627 (1966).

20. D. Berndt, in Power Sources 2 (D. H. Collins, ed.), Pergamon Press, New York, 1970, pp. 17-31.

21. H. Haebler, H. Panesar, and E. Voss, Electrochim. Acta, 15, 1421 (1970).

22. J. P. Carr, N. A. Hampson, and R. Taylor, J. Electroanal. Chem., 33, 109 (1971).

23. M. F. Skalozubov, J. Appl. Chem. USSR, 35, 1736 (1962).

24. M. Maja, Electrochim. Metal, 3, 63 (1968).

25. M. I. Gillibrand and G. R. Lomax, Electrochim. Acta, 8, 693 (1963).

26. M. I. Gillibrand and G. R. Lomax, Electrochim. Acta, 11, 281 (1966).

27. W. Peukert, Elektrotech Z., 18, 287 (1897).

28. P. E. Baikie, M. I. Gillibrand, and K. Peters, Electrochim. Acta, 17, 839 (1972).

29. H. Bode and J. Euler, Electrochim. Acta, 11, 1211, 1221 (1966).

30. H. Bode, J. Euler, E. Rieder, and H. Schmitt, Electrochim. Acta, 11, 1231 (1966).

31. K. J. Euler, Chem. Ing. Tech., 38, 631 (1966).

32. K. J. Euler, Electrochim. Acta, 13, 2245 (1968).

33. H. Bode, H. Panesar, and E. Voss, Naturwissenschaften, 55, 541 (1968).

34. H. Leibssle, H. Reber, W. Herrmann, and E. Zehender, Bosch Tech. Ber., 2, 159 (1968).

35. H. Bode, H. Panesar, and E. Voss, Chem. Ing. Tech., 41, 878 (1969).

36. T. Chiku and K. Nakajima, J. Electrochem. Soc., 116, 1407 (1969).

37. T. Chiku, J. Electrochem. Soc., 115, 982 (1968).

38. W. Stein, Naturwissenschaften, 45, 459 (1958).

39. W. Stein, Ph.D. thesis, Technische Hochschule, Aachen, W. Germany (1959).

40. H. Lehning, ETZ-A, 93, 62 (1972).

41. J. S. Dunning, D. N. Bennion, and J. Newman, J. Electrochem. Soc., 118, 1251 (1971).

42. J. S. Newman and C. W. Tobias, J. Electrochem. Soc., 109, 1183 (1962).

43. E. A. Grens, II, and C. W. Tobias, Electrochim. Acta, 10, 761 (1965).

44. R. C. Alkire, E. A. Grens, II, and C. W. Tobias, J. Electrochem. Soc., 116, 1328 (1969).

45. E. Voss and J. Freundlich, in Batteries (D. H. Collins, ed.), Pergamon Press, New York, 1963, pp. 73-87.

46. P. Ness, Electrochim. Acta, 12, 161 (1967).

47. J. Burbank, NRL Reports 6859 (1969) and 7256 (1971), U.S. Naval Research Laboratory, Washington, D.C.

48. E. Willihnganz, Power Sources 5, 9th International Symposium, Brighton 1974 (D. H. Collins, ed.), Academic Press, New York, 1975, pp. 43-53.

49. J. P. Carr and N. A. Hampson, Chem. Rev., 72, 679 (1972).

50. P. Ruetschi, R. T. Angstadt, and B. D. Cahan, J. Electrochem. Soc., 106, 547 (1959).

51. P. Ruetschi and R. T. Angstadt, J. Electrochem. Soc., 105, 555 (1958).

52. H. S. Panesar, in Power Sources 3 (D. H. Collins, ed.), Oriel Press, Newcastle upon Tyne, 1971, pp. 79-89.

53. K. Ekler, Can. J. Chem., 42, 1355 (1964).

54. E. Tarter and K. Ekler, Can. J. Chem., 47, 2191 (1969).

55. B. N. Kabanov, E. S. Weisberg, I. L. Romanova, and E. V. Krivolapova, Electrochim. Acta, 9, 1197 (1964).

56. J. J. Lander, J. Electrochem. Soc., 98, 213 (1951).

57. J. J. Lander, J. Electrochem. Soc., 98, 220 (1951).

58. J. J. Lander, J. Electrochem. Soc., 99, 339 (1952).

59. J. J. Lander, J. Electrochem. Soc., 103, 1 (1956).

60. J. J. Lander, J. Electrochem. Soc., 105, 289 (1958).

61. P. Ruetschi, J. Electrochem. Soc., 120, 331 (1973).

62. V. M. Halsall and R. R. Wiethaup, Society of Automotive Engineers, Automotive Engineering Congress, Detroit, 1972, paper No. 720041.

63. A. C. Simon, C. P. Wales, and S. M. Caulder, J. Electrochem. Soc., 117, 987 (1970).

64. A. C. Simon, and S. M. Caulder, J. Electrochem. Soc., 118, 659 (1971).

65. K. J. Euler, Electrochim. Acta, 13, 1533 (1968).

66. D. Simonsson, J. Electrochem. Soc., 120, 151 (1973).

67. E. Voss and G. Huster, in Performance Forecast of Selected Static Energy Conversion Devices (G. W. Sherman and L. Devol, eds.), 29th Meeting of AGARD Propulsion and Energetics Panel, Liège, Belgium, 1967, pp. 57-77.

68. N. A. Balashova, B. N. Kabanov, and L. D. Kovba, J. Appl. Chem. (USSR), 37, 909 (1964).

69. J. Burbank, in Batteries (D. H. Collins, ed.) Pergamon Press, New York, 1963, pp. 43-60.

70. A. C. Simon, in Batteries 2 (D. H. Collins, ed.), Pergamon Press, New York, 1965, pp. 63-80.

71. A. Ragheb, W. Machu, and W. H. Boctor, Werkstoffe u. Korrosion, 16, 676 (1965).

72. S. M. Caulder, A. C. Simon, and J. T. Stemmle, Fall Meeting of the Electrochemical Society, Miami Beach, Fla., 1972.

73. G. Archdale and J. A. Harrison, J. Electroanal. Chem., 34, 21 (1972).

74. G. Archdale and J. A. Harrison, J. Electroanal. Chem., 39, 357 (1972).

75. G. Archdale and J. A. Harrison, 8th International Symposium held at Brighton, 1972, paper No. 22.

76. H. S. Panesar and V. Portscher, Metalloberfläche Angew. Elektrochem., 26, 252 (1972).

77. L. S. Sergeeva and I. A. Selitskii, Russian J. Phys. Chem., 39, 107, (1965).

78. A. C. Simon, S. M. Caulder, P. Gurlusky, and J. R. Pierson, Fall Meeting of the Electrochemical Society, Miami Beach, Fla., 1972.

79. A. C. Simon, in Power Sources 2 (D. H. Collins, ed.), Pergamon Press, New York, 1970, pp. 33-53.

80. D. Pavlov and R. Popova, Electrochim. Acta, 15, 1483 (1970).

81. M. Kugel, German Pat. 480, 149 (1929).

82. M. Kugel, U. S. Pat. 1,748,485 (1930).

83. D. Evers, H. Gumprecht, and M. Rasche, U. S. Pat. 3,011,007 (1961).

84. S. Tudor, A. Weisstuch, and S. H. Davang, Electrochem. Technol., 3, 90 (1965).

85. S. Tudor, A. Weisstuch, and S. H. Davang, Electrochem. Technol., 4, 406 (1966).

86. S. Tudor, A. Weisstuch, and S. H. Davang, Electrochem. Technol., 5, 21 (1967).

87. R. F. Amlie, E. Y. Weissman, C. K. Morehouse, and N. M. Qureshi, J. Electrochem. Soc., 119, 568 (1972).

88. R. F. Amlie, E. Y. Weissman, and N. M. Qureshi, Proceedings of the 25th Power Sources Symposium, Atlantic City, N. J., May 23-25, 1972, pp. 67-71.

89. L. V. Vanyukova, M. M. Isaeva, and B. N. Kabanov, Dokl. Akad. Nauk SSSR, 143 (2), 32 (English translation) (1962).

90. E. Voss, Proceedings of the Second International Symposium on Batteries, Bournemouth, England, 1960, paper No. 16.

91. H. Bode and E. Voss, Electrochim. Acta, 6, 11 (1962).

92. F. Huber and M. S. A. El-Meligy, Z. Anorg. Allg. Chem., 367, 154 (1969).

93. R. F. Amlie and T. A. Berger, J. Electroanal. Chem., 36, 427 (1972).

94. Battery Service Manual, 7th ed., Battery Council International, Burlingame, Calif., 1972. Revised in 1976, Battery Council International, Chicago, Illinois, 1976.

95. M. Barak, Symposium on Lead-Acid Batteries, Indian Lead Zinc Information Centre, New Delhi, 1971, pp. 1-25.

96. V. M. Halsall and D. Orlando, The Institution of Mechanical Engineers, Symposium on Auxiliary Power Service and Equipment for Road Vehicles, Cranfield, England, 1969.

97. V. M. Halsall, Association of American Battery, Quebec City, 1969.

98. B. S. Herder, Proceedings of the Second International Conference on Lead, Arnhem, Pergamon Press, New York, 1967, pp. 171-178.

99. H. Jacob, Proceedings of the Second International Conference on Lead, Arnhem, Pergamon Press, New York, 1967, pp. 183-188.

100. Australian Lead Development Association, The Electric Industrial Truck, Melbourne, 1972.

101. I. S. Payne, Proceedings of the Second International Conference on Lead, Arnhem, Pergamon Press, New York, 1967, pp. 179-182.

102. P. Eisler, Brit. Pat. 956, 553 (1964).

103. P. Eisler, Brit. Pat. 1,092,271 (1967).

104. P. Eisler, Brit. Pat. 1,092,272 (1967).

105. R. D. Nelson and R. E. Biddick, AD 686,115, U.S. Dept. of Commerce, Office Tech. Serv., Washington, D.C., 1968.

106. R. E. Biddick and R. D. Nelson, Intersociety Energy Conversion Engineering Conference Record, Boulder, Colo., 1968, pp. 47-51.

107. R. D. Thornton and W. W. Carson, Intersociety Energy Conversion Engineering Conference Record, Las Vegas, Nev., 1970, paper No. 709, 199, pp. 16-27.

108. L. G. Reed, U.S. Pat. 2,511,943 (1950).

109. R. R. Aronson, U.S. Pat. 3,518,127 (1970).

110. J. C. Duddy, U.S. Pat. 3,453,145 (1969).

111. J. C. Duddy and J. B. Ockerman, Proceedings of the 2nd International Electric Vehicle Symposium, Atlantic City, N.J., 1971, Electric Vehicle Council, New York, Publication No. 72-1000, pp. 516-530.

112. A. R. Cook, Proceedings of the Symposium on Batteries for Traction and Propulsion, Columbus Section of the Electrochemical Society, (1972), pp. 65-101.

113. B. Agruss, J. Electrochem. Soc., 117, 1204, 1207 (1970).

114. B. Agruss, J. Electrochem. Soc., 118, 375 (1971).

115. K. V. Kordesch, Electrochemical Society Spring Meeting, Los Angeles, Extended Abstracts No. 273, pp. 658-663 (1970).

116. K. V. Kordesch, J. Electrochem. Soc., 118, 812 (1971).

117. K. V. Kordesch, Intersociety Energy Conversion Engineering Conference, Boston, Paper No. 719015, 1971, Proceedings, p. 103, publ. by Society of Automotive Engineers, New York.

118. D. F. Taylor and E. G. Siwek, Society of Automotive Engineers, International Automotive Engineering Congress, Detroit, 1973, paper No. 730252.

119. H. P. Silverman, N. R. Garner, R. R. Sayano, R. E. Biddick, and E. T. Seo, 7th Intersociety Energy Conversion Engineering Conference, San Diego, Calif., 1972, paper No. 729026, publ. by American Chemical Society, Washington, D.C.

120. P. D. Gibbons, Lead Development Association, London, 1965.

121. P. G. S. Chick, Lead Development Association, London, 1965.

122. E. R. Sanderson, Power Works Eng., 59, 6 (1964).

123. A. Prescott, The Consulting Engineer, 65 (May 1967).

124. P. J. Edwards and B. W. Baxter, 4th International Conference on Lead, Hamburg, 1971.

125. Å. Ljungblom, Erricson Rev. (Sweden), 45 (4), 142 (1968).

126. D. E. Koontz, D. O. Feder, L. D. Babusci, and H. J. Luer, The Bell System Tech. J., 49, 1253 (1970).

127. D. O. Feder, Proceedings of the 6th Advances in Battery Technology Symp., So. Calif.-Nev. Section, The Electrochemical Society (1970), pp. 7-45.

128. Australian Lead Development Association, Lead Batteries for Auxiliary Power Supply, Melbourne, 1972.

129. Interim Federal Specification W-B-00134-B (GSA-FSS), 1970, Battery, Storage (Lead-Acid, Industrial Floating Service).

130. G. Stover, Chem. Engineering, 79 (14), 126 (1972).

131. S. Sada and S. Haraguchi, Rev. Electro. Comm. Lab. (Japan), 19, 99 (1971).

132. P. C. Milner, The Bell System Tech. J., 49, 1321 (1970).

133. Globe-Union, Inc., Battery Division, Milwaukee, Catalog 64-1.

134. L. D. Babusci, B. A. Cretella, D. O. Feder, and D. E. Koontz, U.S. Pat. 3,434,883 (1969).

135. L. D. Babusci, B. A. Cretella, D. O. Feder, and D. E. Koontz, U.S. Pat. 3,532,545 (1970).

136. A. G. Cannone, D. O. Feder, and R. V. Biagetti, The Bell System Tech. J., 49, 1321 (1970).

137. F. A. Berberick, H. E. Durr, A. A. Haller, and E. J. Biron, The Western Electric Engineer, 17 (1), 18 (1973).

138. R. V. Biagetti and M. C. Weeks, The Bell System Tech. J., 49, 1305 (1970).

139. As reported by D. Gribble of the Lead Development Association, London, in The Battery Man, pp. 4 et seq. (Dec. 1971).

140. E. T. DeBlock and J. R. Thomas, Electronics World, 75 (6), 32 (1966).

141. J. A. Orsino and H. E. Jensen, Proceedings of the 21st Annual Power Sources Conference, Atlantic City, N.J., 1967, pp. 60-64.

142. J. R. Thomas and D. R. Wolter, Proceedings of the 21st Annual Power Sources Conference, Atlantic City, N.J., 1967, pp. 64-68.

143. J. P. Malloy, Proceedings of the 21st Annual Power Sources Conference, Atlantic City, N.J., 1967, pp. 68-73.

144. K. Eberts, Elektrotechnik Z., 21, 297 (1969).

145. O. C. Wagner, AD695626, U.S. Dept. Comm., Office Tech. Serv., Washington, D.C., 1969.

146. K. Eberts, in Power Sources 2 (D. H. Collins, ed.), Pergamon Press, New York, 1970, pp. 69-92.

147. A. I. Harrison and K. Peters, in Power Sources 3 (D. H. Collins, ed.), Oriel Press, Newcastle upon Tyne, 1971, pp. 211-225.

148. K. Eberts and O. Jache, Proceedings of the Second International Conference on Lead, Arnhem, Pergamon Press, New York, 1967, pp. 171-178.

149. A. Sabatino, Battery Council International Meeting, Denver, Colo., 1972.

150. S. Bastacky, U.S. Pat. 3,711,332 (1973).

151. J. R. Smyth, J. P. Malloy, and D. T. Ferrell, Jr., Proceedings of the Second International Conference on Lead, Arnhem, Pergamon Press, New York, 1967, pp. 193-197.

152. S. Hattori, Proceedings of the Third International Conference on Lead, Venice, Pergamon Press, New York, 1970, pp. 155-165.

153. D. H. McClelland and J. L. Devitt, Ger. Offen. 2,137,908 (1972).

154. Globe-Union, Inc., Battery Division, Milwaukee, Gel/Cell Catalog.

155. E. G. Wheadon and N. L. Willmann, U.S. Pat. 3,556,854 (1971).

156. E. G. Wheadon and C. P. McCartney, U.S. Pat. 3,652,336 (1972).

157. E. G. Wheadon and N. L. Willmann, U.S. Pat. 3,679,789 (1972).

158. E. G. Wheadon and N. L. Willmann, U.S. Pat. 3,690,950 (1972).

159. As reported by a variety of sources in Battery Research New Batteries in Japan, Vol. 2, U.S. Branch of the Electrochemical Society of Japan, 1972, pp. 5-6.

160. P. Ruetschi and J. B. Ockerman, Electrochem. Technol., 4, 383 (1966).

161. Accumulatoren-Fabrik A. G. Hagen, Ger. Pat. 1,782,752 (1959).

162. E.S.B., Inc., Brit. Pat. 1,032,852 (1966).

163. S. Hills and D. L. K. Chu, J. Electrochem. Soc., 116, 1155 (1969).

164. H. Reber, U.S. Pat. 3,457,112 (1969) and 3,462,303 (1969). Bosch Techn. Berichte, 3 (3), 3 (1970).

165. B. K. Mahato and E. C. Laird, Power Sources 5, 9th International Symposium, Brighton, 1974 (D. H. Collins, ed.), Academic Press, N.Y., 1975, pp. 23-41.

166. K. V. Kordesch, U.S. Pat. 3,258,360 (1966).

167. K. Terae, Ger. Offen. 1,941,997 (1970).

168. B. K. Mahato, E. Y. Weissman, and E. C. Laird, J. Electrochem. Soc., 121, 13 (1974).

169. As reported by a variety of sources in Battery Research and New Batteries in Japan, Vol. 1, U.S. Branch of the Electrochemical Society of Japan, 1971, pp. 27-31.

170. L. Kreidl, U.S. Pat. 3,630,778 (1971).

171. M. Fukuda, T. Miura, and K. Takahashi, U.S. Pat. 3,598,653 (1971).

172. S. Sekido, M. Yamashita, and M. Matsumoto, U.S. Pat. 3,622,398 (1971).

173. G. Sassmanshausen and N. Lahme, U.S. Pat. 3,701,691 (1972).

174. J. I. Dyson and E. Sundberg, 8th International Symposium, Brighton, 1972, paper No. 23.

175. Battery Application Manual, Gates Energy Products, Inc., Denver, Colo.

176. Battery Council International, The Storage Battery Manufacturing Industry, 1971-72 Yearbook, Burlingame, Calif. Yearly additions and revisions: Battery Council International, Chicago, Ill., 1976.

177. Lead Industries Association, Lead—Publication MET-2/Lea, 1969.

178. W. Hofmann, Lead and Lead Alloys, 2nd ed., p. 71, Springer, New York, 1970.

179. J. L. Orsini, Proceedings of the 3rd International Conference on Lead, Venice, Pergamon Press, New York, 1970, p. 204.

180. G. Perrault and J. Brenet, Compt. Rend., 250, 325 (1960).

181. A. I. Zalarsky, Y. A. Kondrashov, and S. S. Tolkachev, Dokl. Akad Nauk SSSR, 75, 559 (1950).

182. U. B. Thomas, Trans. Electrochem. Soc., 94, 42 (1948).

183. H. Bode and E. Voss, Z. Elektrochem., 60, 1053 (1956).

184. H. R. Thirsk, Batterien, 14, 111 (1960) and 113 (1961).

185. V. H. Dodson, J. Electrochem. Soc., 108, 401 (1961).

186. P. Ruetschi, J. Sklarchuk, and R. T. Angstadt, Electrochim. Acta, 8, 333 (1963).

187. P. Ruetschi, J. Sklarchuk, and R. T. Angstadt, in Batteries (D. H. Collins, ed.), Pergamon Press, New York, 1963, pp. 89-103.

188. S. Ghosh, J. Electrochem. Soc. Japan, 34, 38 (1966).

189. S. M. Caulder, private communication (1971).

190. C. Drotschmann, Batterien, 17, 472 (1963).

191. N. E. Bagshaw and K. P. Wilson, Electrochim. Acta, 10, 867 (1965).

192. J. Bousquet, J. M. Blanchard, and J. C. Rémy, Bull. Soc. Chim. France, 3206 (1968).

193. S. Ikari and S. Yoshizawa, J. Electrochem. Soc. Japan, 27, E247 (1959).

194. D. Kordes, Chem. Ing. Tech., 38, 638 (1966).

195. P. Ruetschi and B. D. Cahan, J. Electrochem. Soc., 105, 369 (1958).

196. S. Ikari, S. Yoshizawa and S. Okada, J. Electrochem. Soc. Japan, 27, E223 (1959).

197. R. Smith, Lead and Its Alloys, in Metals, Vol. 2, Design Engineering Series, Morgan-Grampian Ltd., London, 1970.

198. J. Verney, Metall., 23, 836 (1969).

199. T. F. Yelsukova, M. A. Bolshanina, T. M. Cherkasova, and K. F. Titova, Phys. Metals Metallogr. (USSR), 26 (1), 78 (Aug. 1969).

200. P. Costa and A. Mambelli, 37th International Foundry Congress, 1970, Brighton.

201. Interim Federal Specification W-B-00133 B (GSA-FSS), 1970, Battery, Storage (Lead-Acid, Industrial Portable Service).

202. W. Herrmann and G. Proepstl, Z. Elektrochem. (Ber. Bunsen.), 61, 1154 (1957).

203. W. Herrmann, W. Ilge, and G. H. Proepstl, Second U.N. Intl. Conf. on the Peaceful Uses of Atomic En., A/Conf. 15/P/988, Geneva, 1958.

204. W. Herrmann, ETZ-B, 16, 643 (1964).

205. H. Leibssle and E. Zehender, Bosch Tech. Berichte, 3, 163 (1970).

206. A. C. Simon and S. M. Caulder, Power Sources 5, 9th International Symposium, Brighton, 1974 (D. H. Collins, ed.), Academic Press, N.Y., 1975, pp. 109-122.

207. J. Burbank, in Power Sources 3 (D. H. Collins, ed.), Oriel Press, Newcastle upon Tyne, 1971, pp. 13-34.

208. J. Burbank, J. Electrochem. Soc., 118, 525 (1971); 111, 1112 (1964).

209. E. J. Ritchie and J. Burbank, J. Electrochem. Soc., 117, 299 (1970).

210. J. L. Dawson, J. Wilkinson, and M. I. Gillibrand, J. Inorg. Nucl. Chem., 32, 501 (1970).

211. J. L. Dawson, M. I. Gillibrand, and J. Wilkinson, in Power Sources 3 (D. H. Collins, ed.), Oriel Press, Newcastle upon Tyne, 1971, pp. 1-11.

212. D. E. Swets, J. Electrochem. Soc., 120, 925 (1973).

213. H. E. Haring and U. B. Thomas, Trans. Electrochem. Soc., 68, 293 (1935).

214. E. F. Schumacher and G. S. Phipps, Trans. Electrochem. Soc., 68, 309 (1935).

215. F. V. Goler, Die Giesserei, 25, 242 (1938).

216. H. E. Jensen, Proceedings of the 11th Power Sources Conference, Atlantic City, N.J., 1957, pp. 73-77.

217. E. Willihnganz, Proceedings of the 13th Power Sources Conference, Atlantic City, N.J., 1959, pp. 73-75.

218. U. B. Thomas, F. T. Forster, and H. E. Haring, Trans. Electrochem. Soc., 92, 313 (1947).

219. A. A. Abdul Azim and K. M. El-Sobki, Corros. Sci., 12, 371 (1972).

220. T. W. Caldwell and U. S. Sokolov, Power Sources 5, 9th International Symposium., Brighton, 1974 (D. H. Collins, ed.), Academic Press, New York, 1975, pp. 73-95.

221. U. Heubner and H. Sandig, 4th International Conference on Lead, Hamburg, 1971.

222. G. W. Mao, J. G. Larson, and P. Rao, Metallography, 1, 399, (1969).

223. J. A. Young, J. B. Barclay, Battery Council International Meeting, San Francisco, 1973.

224. G. W. Mao and P. Rao, Brit. Corros. J., 6, 122 (1971).

225. A. C. Simon, J. Electrochem. Soc., 114, 1 (1967).

226. O. Hyvarinen and M. H. Tikkanen, Acta Polytech. Scand., Chem-Met. Series (Finland), 89, 5 (1969).

227. J. Burbank, J. Electrochem. Soc., 104, 693 (1957).

228. I. I. Astahov, E. S. Vajsberg, and B. N. Kabanov, Dokl. Akad. Nauk SSSR, 154, 1414 (1964).

229. D. Pavlov, C. N. Poulieff, E. Klaja, and N. Iordanov, J. Electrochem. Soc., 116, 316 (1969).

230. J. A. von Fraunhofer, Anti-Corrosion Methods and Materials, 15, (11), 9 (1968).

231. E. Sundberg, Proceedings of the Second International Conference on Lead, Arnhem, Pergamon Press, New York, 1967, pp. 227-233.

232. G. W. Mao, T. L. Wilson, and T. G. Larson, J. Electrochem. Soc., 117, 1323 (1970).

233. T. L. Wilson and G. W. Mao, Metall. Trans., 1, 2631 (1970).

234. G. W. Mao, T. L. Oswald, and B. J. Sobczak, in Power Sources 3 (D. H. Collins, ed.), Oriel Press, Newcastle upon Tyne, 1971, pp. 61-78.

235. G. W. Mao, U. S. Pat. 3,647,545 (1972).

236. P. Rao and G. W. Mao, J. Inst. Metals, 100, 13 (1972).

237. N. E. Bagshaw, Proceedings of the Third International Conference on Lead, Venice, Pergamon Press, New York, 1970, pp. 209-219.

238. H. E. Zahn, U. S. Pat. 2,841,491 (1958).

239. F. E. E. Boos and R. N. Chamberlain, U. S. Pat. 1,826,724 (1931).

240. E. Zehender and W. Herrmann, Bosch Tech. Berichte, 1, 126 (1965).

241. A. Kirow, T. Rogatschev, and D. Denew, Metalloberfläche Angew. Elektrochemie, 26, 234 (1972).

242. D. J. I. Evans, H. R. Huffman, and J. P. Warner, Materials in Design Engineering, 59, 105 (Apr. 1964).

243. D. H. Roberts and N. A. Ratcliff, Metallurgia, 70, 223 (1964).

244. J. A. Lund, E. G. v. Tiesenhausen, and D. Tromans, Proceedings of the Second International Conference on Lead, Arnhem, Pergamon Press, New York, 1965, pp. 7-16.

245. A. Lloyd and E. R. Newson, Proceedings of the Second International Conference on Lead, Arnhem, Pergamon Press, New York, 1965, pp. 31-34.

246. N. E. Bagshaw and T. A. Hughes, in Batteries 2 (D. H. Collins, ed.), Pergamon Press, New York, 1965, pp. 1-15.

247. M. Torralba and J. L. Ruiz, Rev. Metal, CENIM, 7, 41 (1971).

248. S. Feliu, J. A. Gonzalez, and J. J. Royuela, Chem. Age India, 19, 281 (1968).

249. J. A. Evans and N. Shanks, U. S. Pat. 3,621,701 (1971).

250. M. N. Parthasarati and N. Srinivasan, Symposium on Lead-Acid Batteries, Indian Lead Zinc Information Centre, New Delhi, 1971, pp. 57-75.

251. T. L. Bird, I. Dugdale, and G. G. Graver, Proceedings of the Third International Conference on Lead, Venice, Pergamon Press, New York, 1970, pp. 221-224.

252. J. B. Cotton and I. Dugdale, in Batteries (D. H. Collins, ed.), Pergamon Press, New York, 1963, pp. 297-307.

253. S. Ruben, U. S. Pat. 3,486,940 (1969).

254. S. Ruben, U. S. Pat. 3,499,795 (1969).

255. S. Ruben, U. S. Pat. 3,576,674 (1969).

256. S. Ruben, U. S. Pat. 3,615,831 (1970).

257. P. Faber, 8th International Symposium, Brighton, 1972, paper No. 24.

258. R. Thomas, Proceedings of the Third International Conference on Lead, Venice, Pergamon Press, New York, 1970, pp. 169-184.

259. N. L. Willmann and J. L. Helms, U. S. Pat. 3,516,863 (1970).

260. N. L. Willmann and E. G. Wheadon, U. S. Pat. 3,516,864 (1970).

261. N. L. Willmann and E. G. Wheadon, U. S. Pat. 3,621,543 (1971).

262. W. R. Scholle, U. S. Pat. 3,738,871 (1973).

263. C. O. Schilling, U. S. Pat. 3,446,670 (1969).

264. Lignosol Chemicals Co., division of Dryden Chemicals Ltd., Quebec, Catalog 1.

265. T. F. Sharpe, Electrochim. Acta, 14, 635 (1969).

266. E. G. Yampolskaya, M. I. Ershora, I. I. Astakhov, and B. N. Kabanov, Sov. Electrochem., 2, 1211 (1966).

267. I. A. Aguf, M. A. Dasoyan, L. A. Ivanenko, E. V. Parshikova, and K. M. Soloveva, Sov. Electrochem., 6, 1591 (1970).

268. E. G. Yampolskaya, M. I. Ershora, V. V. Surikov, I. I. Astakhov, and B. N. Kabanov. Sov. Electrochem., 8, 1209 (1972).

269. E. G. Yampolskaya, I. A. Smirnova, M. I. Ershora, S. A. Sapotniskii, and L. I. Kryukova, Sov. Electrochem., 8, 1289 (1972).

270. S. Ikari, U. S. Pat. 3,481,785 (1969).

271. E. G. Yampolskaya, M. I. Ershora, I. I. Astakhov, G. L. Teplitskaya, and B. N. Kabanov, Sov. Electrochem., 7, 562 (1971).

272. V. Fiedler, V. Malikova, and H. Weber, Brit. Pat. 1,176,330 (1970).

273. Y. B. Kasparov, E. G. Yampolskaya, and B. N. Kabanov, J. Appl. Chem. USSR, 37, 1922 (1964).

274. J. L. Limbert, H. G. Proctor, and D. T. Poe, U. S. Pat. 3,523,041 (1970).

275. J. L. Limbert, H. G. Proctor, and D. T. Poe, U. S. Pat. 3,615,788 (1971).

276. H. Haebler, U. S. Pat. 3,486,941 (1969).

277. O. Jache, U. S. Pat. 3,658,594 (1972).

278. O. Jache, U. S. Pat. 3,708,338 (1973).

279. G. S. Lello and P. V. Lowe, U. S. Pat. 3,518,120 (1970).

280. G. S. Lello and P. V. Lowe, Ger. Offen. 1,817,147 (1969).

281. C. Drotschmann, Batterien, 17, 560 (1964).

282. S. Palanichamy et al., Batterien, 20, 947 (1966).

283. A. Sabatino and E. J. Jackson, U. S. Pat. 3,100,162 (1963).

284. E. J. Jackson and A. Sabatino, U. S. Pat. 3,496,020 (1970).

285. E. J. Jackson and A. Sabatino, U. S. Pat, 3,502,505 (1970).

286. S. Tudor, A. Weisstuch, and S. H. Davang, U. S. Pat, 3,447,969 (1969).

287. S. Tudor, A. Weisstuch, and S. H. Davang, Proceedings of the 20th Annual Power Sources Conference, Atlantic City, N. J., 1966, pp. 128-130.

288. A. C Simon, S. M. Caulder, and E. J. Ritchie, J. Electrochem. Soc., 117, 1264 (1970).

289. C. J. Venuto, U. S. Pat. 3,723,182 (1971).

290. H. Haebler, U. S. Pat. 3,459,595 (1969).

291. Varta A. G. (Germany), French Pat. 1,525,757 (1968).

292. H. Haebler, U. S. Pat. 3,459,595 (1969).

293. R. N. Snyder and E. G. Wheadon, U. S. Pat. 3,702,265 (1972).

294. M. Devadoss and J. Vedamuthu, Proceedings of the Symposium on Lead-Acid Batteries, Indian Lead Zinc Information Centre, Calcutta, 1968, pp. 29-35.

295. Sonnenschein GmbH (Germany), Catalog No. 1510e on Stationary Batteries.

296. P. J. Moll, Batterien, 15, 150, 166 (1961).

297. F. R. Holloway, U. S. Pat. 3,607,412 (1971).

298. D. M. Petrov and Y. K. Manikatov, Poroshk. Metall. (English translation), 4 (100), Plenum Publishing Corp. (1971), pp. 336-339.

299. E. Zehender, W. Herrmann, and H. Leibssle, Electrochim. Acta, 9, 55 (1964).

300. S. Palanichamy, M. Devasahayan, S. Sampath, and H. V. K. Udupa, Proceedings of the Symposium on Lead-Acid Batteries, Indian Lead Zinc Information Centre, Calcutta, 1968, pp. 36-40.

301. R. G. Robinson and R. L. Walker, in Batteries (D. H. Collins, ed.), Pergamon Press, New York, 1963, pp. 15-41.

302. C. Drotschmann, Batterien, 19, 761, 772, 798, 815 (1965).

303. British Standards Institution, Specification BS 3031: 1972, London.

304. H. Brown, U. S. Pat. 2,857,295 (1958).

305. J. F. Macholl and K. Patterson, U. S. Pat. 2,841,632 (1958).

306. G. L. Simmons, R. W. Adler, W. E. Elliott, and W. L. Towle, AD678594, U. S. Dept. Comm., Office Tech. Serv., Washington, D. C., 1965.

307. T. Yeoman, U. S. Pat. 2,773,927 (1956).

308. A. Sam, U. S. Pat, 3,304,202 (1963).

309. J. T. Redmon, U. S. Pat. 3,525,639 (1970).

310. W. S. Maxel and G. M. Ginnow, U. S. Pat. 3,589,947 (1971).

311. A. Sabatino, U. S. Pat. 3,643,834 (1972).

312. F. P. Daniel, U. S. Pat. 3,647,555 (1972).

313. C. K. Morehouse and R. F. Amlie, U. S. Pat. 3,649,363 (1972).

314. J. C. Duddy, U. S. Pat. 3,607,408 (1971).

315. N. J. Cortese and J. E. Bell, U. S. Pat. 3,733,220 (1973).

316. H. H. Roth, U. S. Pat. 2,596,046 (1952).

317. J. T. Rivers, U. S. Pat. 2,684, 950 (1954).

318. M. H. Little, U. S. Pat. 3,530,002 (1970) [Also: Brit. Pat. 1,183,283 (1970); Fr. Pat. 1,570,708 (1969); Ger. Offen. 2,032,404 (1971)].

319. D. L. Douglas, R. E. Biddick, and J. B. Ockerman, In Power Sources 2 (D. H. Collins, ed.), Pergamon Press, New York, 1970, pp. 93-101.

320. D. L. Douglas and H. J. Banas, U. S. Pat. 3,556,850 (1971).

321. D. L. Douglas and H. J. Banas, U. S. Pat. 3,556,851 (1971).

322. R. F. Amlie, U. S. Pat. 3,556,860 (1971) [Also: Arg. Pat. 171, 487 (1969)].

323. H. Lauck, U. S. Pat. 3,586,539 (1971) [Also: Fr. Demande 2,016,326 (1970)].

324. F. L. Marsh, U. S. Pat. 3,639,175 (1972).

325. F. Marsh and K. L. Thompson, U. S. Pat. 3,663,304 (1972).

326. R. G. Robinson, Brit. Pat. 785,848 (1957).

327. K. Parker and J. L. Brosilow, U. S. Pat. 3,408,233 (1968).

328. J. P. Badger and H. A. Bernholtz, U. S. Pat. 3,540,939 (1970).

329. H. A. Bernholtz and J. P. Badger, U. S. Pat. 3,591,422 (1971).

330. J. P. Badger, Battery Council International Meeting, Phoenix, Ariz. (1971).

331. J. P. Badger, Society of Automotive Engineers, Automotive Engineering Congress, Detroit, 1972, paper No. 720040.

332. V. M. Halsall, Society of Automotive Engineers, Mid-Year Meeting, Detroit, 1968, paper No. 680389.

333. V. M. Halsall and P. A. Cosme, U. S. Pat. 3,673,302 (1972).

334. R. M. Fiandt, U. S. Pat. 3,388,007 (1968).

335. V. M. Halsall, U. S. Pat. 3,416,969 (1968).

336. V. M. Halsall and P. A. Cosme, U. S. Pat. 3,509,603 (1970).

337. R. E. Hennen, U. S. Pat. 3,597,280 (1971).

338. H. C. Burns, Battery Council International Meeting, San Francisco, 1973.

339. J. W. Brodhacker, H. C. Burns, and A. L. Fox, Society of Automotive Engineers, Automotive Engineering Congress, Detroit, 1972, paper No. 720042.

340. R. E. Hennen and V. M. Halsall, U. S. Pat. 3,772,088 (1973).

341. E. G. Fochtman and W. R. Haas, Battery Council International
      Meeting, Denver, Colo., 1972, pp. 92-108.

342. E. J. Ritchie, Final Report, Projects LE-82 and LE-84, Inter-
      national Lead Zinc Research Organization, Inc., New York, 1971.

343. F. Scholl, Z. Anal. Chem., 245, 49 (1969).

344. F. Scholl, Bosch Tech. Berichte, 3 (3), 89 (1970).

345. V. M. Halsall and J. R. Pierson, Battery Council International
      Meeting, Denver, Colo., 1972.

346. D. Lewis, D. O. Northwood, and R. C. Reeve, J. Appl. Cryst., 2,
      156 (1969).

347. K. Appelt, Electrochim. Acta, 13, 1727 (1968).

348. W. B. Williams, Proceedings of the Second International Conference
      on Lead, Arnhem, Pergamon Press, New York, 1967, pp. 211-214.

349. M. Brachet, Proceedings of the Third International Conference on
      Lead, Venice, Pergamon Press, New York, 1970, pp. 201-204.

350. E. Skoluda, J. Kwasnik, K. Nowak, and J. Kranska, Electrochim.
      Acta, 17, 1353 (1972).

351. J. F. Dittman and H. R. Harner, Bulletin No. 7, Eagle-Picher Co.,
      Cincinnati, Ohio, 1956.

352. J. F. Schaefer and H. R. Karas, Spring Meeting of the Association
      of American Battery Manufacturers, Miami Beach, Fla., 1959.

353. H. Bode and E. Voss, Electrochim. Acta, 1, 318 (1959).

354. J. R. Pierson, Electrochem. Technol., 5, 323 (1967).

355. T. Ishikawa, H. Tagawa, and Y. Nakamura, J. Electrochem, Soc.
      Japan, 29, E85 (1961).

356. Pressure Diecasting of Battery Plates, Machinery and Production
      Engineering, Vol. 121, p. 600 (Oct. 25, 1972), Machinery Publ. Co.,
      Ltd., Brighton, England.

357. A. C. Simon, NRL Report 6723, U.S. Naval Research Laboratory,
      Washington, D.C., 1968.

358. A. M. Howard and E. Willihnganz, Electrochem. Technology, 6,
      370 (1968).

359. L. Arduini and L. Baroni, La Metall. Italiana, 60, 437 (1968).

360. G. W. Mao and T. G. Larson, Metallurgia, 236 (Dec. 1968).

361. S. Nishikawa and K. Tsumuraya, Seisan Kenkyu, 21, 596 (1969).

362. M. M. Tilman, Bureau of Mines Report No. RI 7453, 1970.

363. M. Torralba, J. J. Regidor, and J. M. Sistiaga, Z. Metallkde., 59,
      184 (1968).

364. M. Aballe and M. Torralba, Rev. Metal, CENIM, 5, 385 (1969).

365. M. Torralba, J. J. Regidor, and J. M. Sistiaga, Rev. Metal, CENIM, 5, 706 (1969).

366. J. J. Regidor, M. Aballe, M. Torralba, and J. M. Sistiaga, Fourth International Conference on Lead, Hamburg, 1971.

367. C. Drotschmann, Batterien, 15, 267 (1962).

368. R. H. Greenburg, F. B. Finan, and B. Agruss, J. Electrochem. Soc., 98, 474 (1951).

369. J. Armstrong, I. Dugdale, and W. J. McCusker, in Power Sources (D. H. Collins, ed.), Pergamon Press, New York, 1967, pp. 163-177.

370. M. E. D. Humphreys, R. Taylor, and S. C. Parnes, in Power Sources 2 (D. H. Collins, ed.), Pergamon Press, New York, 1970, pp. 55-67.

371. J. R. Pierson, in Power Sources 2 (D. H. Collins, ed.), Pergamon Press, New York, 1970, pp. 103-119.

372. T. R. Crompton and G. Vitenbroek, J. Electrochem. Soc., 119, 655 (1972).

373. R. T. Angstadt, C. J. Venuto, and P. Ruetschi, J. Electrochem. Soc., 109, 177 (1962).

374. J. P. Carr, N. A. Hampson, and R. Taylor, Ber. Bunsen Ges., 74, 557 (1970).

375. I. G. Kiseleva and B. N. Kabanov, Dokl. Akad. Nauk SSSR, 122, 1042 (1958).

376. W. O. Butler, C. J. Venuto, and D. V. Wisler, J. Electrochem. Soc., 117, 1339 (1970).

377. P. Andresen, Batterien, 17, 489 (1963).

378. C. F. Yarnell, J. Electrochem. Soc., 119, 19 (1972).

379. J. R. Pierson, P. Gurlusky, A. C. Simon, and S. M. Caulder, J. Electrochem. Soc., 117, 1463 (1970).

380. C. F. Yarnell and M. C. Weeks, Fall Meeting of the Electrochemical Society, Miami Beach, Fla., 1972.

381. A. W. Sohn and A. Erasmus, U. S. Pat. 3,480,478 (1969).

382. A. W. Sohn and A. Erasmus, NRL Report 6450, U. S. Naval Research Laboratory, Washington, D. C., 1966.

383. A. W. Sohn and A. Erasmus, in Power Sources (D. H. Collins, ed.), Pergamon Press, New York, 1967, pp. 147-161.

384. D. Pavlov, Electrochim. Acta, 13, 2051 (1968).

385. D. Pavlov and N. Iordanov, J. Electrochem. Soc., 117, 1103 (1970).

386. D. Pavlov, G. Papazov, and V. Iliev, J. Electrochem. Soc., 119, 8 (1972).

387. D. Pavlov, in Proceedings of the Symposium on Batteries for Traction and Propulsion (R. L. Kerr, J. McCallum, and D. E. Semones,

eds.), Electrochemical Society, Columbus Section, 1972, pp. 135-149.

388. J. Burbank, NRL Report 6345, U. S. Naval Research Laboratory, Washington, D. C., 1965.

389. J. Burbank, J. Electrochem. Soc., 113, 10 (1966).

390. J. Burbank, NRL Report 6613, U. S. Naval Research Laboratory, Washington, D. C., 1967.

391. F. A. Schneider, Batterien, 14, 64, 83 (1960).

392. J. Burbank and E. J. Ritchie, J. Electrochem. Soc., 116, 125 (1969).

393. H. Haebler and H. Hense, U. S. Pat. 3,556,852 (1971).

394. F. J. Port, U. S. Pat. 3,499,228 (1970).

395. C. Drotschmann, Batterien, 16, 317, 328, 362, 376, 392 (1962); 413, 422 (1963).

396. B. K. Balu, J. Electrochem. Soc. India, 3, 97 (1970).

397. E. G. Tiegel, Proceedings of the Third International Conference on Lead, Venice, Pergamon Press, 1970, pp. 191-198.

398. V. M. Halsall and E. N. Mrotek, U. S. Pat. 3,652,341 (1972).

399. A. Sabatino and P. V. Lowe, U. S. Pat. 3,087,005 (1963).

400. A. Sabatino and P. V. Lowe, U. S. Pat. 3,229,339 (1966).

401. A. Sabatino and P. V. Lowe, U. S. Pat. 3,238,579 (1966).

402. R. A. Buttke, U. S. Pat. 3,200,450 (1965).

403. R. E. Mix, U. S. Pat. 3,652,337 (1972).

404. C. A. Rigsby, U. S. Pat. 3,703,589 (1972).

405. C. A. Rigsby, U. S. Pat. Re. 25,054 (1961).

406. Globe-Union, Inc., Brit. Pat. 1,000,611 (1965).

407. A. Sabatino and D. Orlando, U. S. Pat. 3,313,658 (1967).

408. A. Sabatino and D. Orlando, U. S. Pat. 3,503,056 (1967).

409. A. Sabatino, U. S. Pat. 3,476,611 (1969).

410. W. C. Kirchberger and D. Orlando, U. S. Pat. 3,427,424 (1969).

411. S. C. Barnes, J. H. Harris, and K. S. Owen, U. S. Pat. 3,515,597 (1970).

412. B. R. Allen, U. S. Pat. 3,723,699 (1973).

413. A. Sabatino, Fall Meeting of the Association of American Battery Manufacturers, Chicago, 1963.

414. D. Orlando and T. Oswald, Automotive Engineering Congress, Detroit, 1966, SAE paper No. 660028.

415. K. Peters, A. I. Harrison, and W. H. Durant, in Power Sources 2 (D. H. Collins, ed.), Pergamon Press, New York, 1970, pp. 1-16.

416. E. Willihnganz, Fall Meeting of The Electrochemical Society, Cleveland, Ohio, 1971.

417. P. Chatenay, International Union of Producers and Distributors of Electrical Energy, Electric Vehicle Study Days, Brussels, 1972, paper No. 2.3.8.

418. V. S. Yanchenko, G. P. Mazina, A. E. Valeitenok, and I. A. Selitskii, Sov. Electrochem., 7, 848 (1971).

419. J. L. Woodbridge, Elect. Eng., 54, 516 (1935).

420. W. E. Rippel, Proceedings of the 2nd International Elect. Vehicle Symposium, Atlantic City, N.J., Electric Vehicle Council, New York, Publ. No. 72-1000, 1971, pp. 247-274.

421. R. Vijayavalli, P. V. Rao, and H. V. K. Udupa, Proceedings of The Symposium on Lead-Acid Batteries, Indian Lead Zinc Information Centre, Calcutta, 1968, pp. 65-69.

422. M. H. Hames, K. W. Nolan, and D. M. Pope, Proceedings of the Third International Conference on Lead, Venice, Pergamon Press, New York, 1970, pp. 227-238.

423. J. Rouchet, International Union of Producers and Distributors of Electrical Energy, Electric Vehicle Study Days, Brussels, 1972, paper No. 2.3.8.

424. Lead Development Association, Users' Experiences with Gas-Controlled Charging, London, 1971.

425. J. A. Mas, Society of Automotive Engineers, Mid-Year Meeting, Detroit, 1968, paper No. 680393, SAE Journal, 77 (6), 31 (1969)

426. T. W. Moore and W. F. Sommer, AD860882, U. S. Dept. Comm., Office Tech. Serv., Washington, D. C., 1968.

427. K. V. Kordesch, J. Electrochem. Soc., 119, 1053 (1972).

428. W. B. Burkett and J. H. Bigbee, III, U. S. Pat. 3,597,673 (1971).

429. W. B. Burkett and J. H. Bigbee, III, U. S. Pat. 3,614,582 (1971).

430. W. B. Burkett and R. V. Jackson, U. S. Pat. 3,614,583 (1971).

431. W. B. Burkett and J. H. Bigbee, III, U. S. Pat. 3,614,584 (1971).

432. W. B. Burkett, Battery Council International Meeting, Phoenix, Ariz., 1971.

433. R. H. Sparks, Society of Automotive Engineers, Automotive Engineering Congress, Detroit, 1972, paper No. 720109.

434. F. Lawn, Battery Council International Meeting, Denver, Colo., 1972.

435. J. Orsino, private communication.

436. International Electrotechnical Commission, Publication 95-1, 1961, and revisions (e.g., Modifications of Publication 95-1, Technical Committee No. 21, 1965).

437. Battery Council International, Meetings: San Francisco, 1973, London, 1974, Hollywood-on-the-Sea, Fla., 1975.

438. W. Sies, Fourth International Conference on Lead, Hamburg, 1971.

439. J. V. Tierney, Jr., Battery Council International Meeting, Denver, Colo., 1972.

440. R. A. Burkard, Industrial Forecast and Market Analysis, 1976-1980, Battery Council International, Meeting in Mexico City, April 1976.

Chapter 2

THE ELECTRIC AUTOMOBILE

Karl V. Kordesch[*]

Union Carbide Corporation
Battery Products Division
Technology Laboratory
Parma, Ohio

[*]Present Affiliation: Institut für Anorganisch Chemische Technologie der Technischen Universitat Graz, Graz, Austria.

1. THE NEED FOR ELECTRIC VEHICLES

1.1. Air Pollution. A Problem of the Past?

Discussing the internal combustion engine's performance, the Surgeon General of the U.S. Public Health Service made the following statement in June 1966: "The threat to health constitutes the primary impulse for the control of air pollution in the United States."

This sounded like the well-known warning against cigarette smoking, and psychologically this statement received the same promotion treatment: against a beautiful backdrop or scene out of nature's picture book, we are reminded that an elegant big car is fashionable. The facts were more grim: in 1966 the exhausts of 90 million motor vehicles constituted 61% of the nation's air pollution [1]. Table 1 shows a breakdown of the pollution components.

TABLE 1

Total U. S. Air Pollution in 1966 (10^6 Tons/Year) [1]

Contaminant	Auto	Other	Total	Percent auto
Hydrocarbons	12	7	19	63
Carbon monoxide	66	5	71	93
Nitrogen oxides	6	7	13	46
Sulfur dioxide	1	25	26	4
Particulates	1	11	12	8
	86	55	141	61

In 1966 nearly 1 ton of pollutants was emitted for every 1,100 gallons
(2 tons) of gasoline burned (assuming 15,000 miles/car/year). Especially
severe was the situation in crowded cities, where the pollutants were in-
jected into the air at street level in rush-hour periods.

Pollution from power plants and industries is normally discharged
through stacks, less irritant to the individual. However, their general effect on human health, meteorological changes, and plant growth may be
more in the long run.

To put the numbers stated in proper perspective, it should be noted
that each of the 3 million square miles of the United States has 24 million
tons of air above it. Actually, global atmospheric processes would remedy
even a tenfold increased pollution problem if it were not for special local
conditions preventing it (e.g., stagnation above Los Angeles).

It is only fair to state also the reply of the automobile companies to the
accusation of the internal combustion as the biggest polluter. In a booklet
published by General Motors, an attempt was made to rectify "misinforma-
tion of the public" [2]. The 1966 data of the U.S. Department of Health
were based on preliminary estimates. The actual data for 1968 are given
in Fig. 1. In the new compilation, the transportation section pollutes only
42%, of which the motor vehicle contributes 39%. This data should be compared with Table 1. Furthermore, if the relative effect of pollution is
established and the equivalent health damage is considered instead of the
weight of the pollutants, the data of Fig. 2 should be evaluated. Particulate

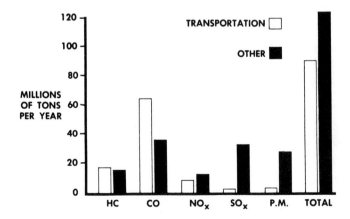

FIG. 1. U.S. air pollution (1968) on a weight basis [2].

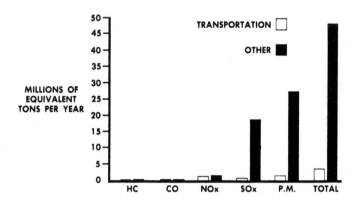

FIG. 2. U. S. air pollution (1968) on a relative effect basis [2].

matter and sulfur dioxide are then the biggest irritants. This puts the blame on the big central power stations using coal. The article does not quote the medical authorities who decided on the health effects of, e.g., CO versus SO_2, particulate matter versus hydrocarbon-caused smog. The controversy just shows how complex the area of pollution is. Besides, General Motors expected (at that time) that the cost of emission control work for 1971 would be $124 million and another $64 million for facilities, with 2,000 people employed in the task of cleaning up the automobile exhausts. Perhaps the difference between 1966 and 1968 data already reflects some success of the earlier efforts?

Two reaction products of the burned gasoline and the air are usually not considered pollutants: water and carbon dioxide. The first is probably harmless, except for condensation (fogging) in cold weather. The second is the subject of a considerable controversy [3]. Carbon dioxide progressively added to the world atmosphere at a rate of 20 to 50 ppm/decade could raise the temperature of the earth surface by $2°C$ in the middle of the twenty-first century, causing polar ice sheets to melt and flood large areas, certainly changing rainfall patterns, etc. The reasoning is based on the "greenhouse effect" of carbon dioxide: to retain the earth-heat radiation to a large percentage.

Fortunately, the reduction of the sunshine influx to the earth by the increase of particulate matter may compensate for it.

The air pollution issue is still alive today. Considerable progress has been made indeed: the emission control levels of the Clean Air Act of 1970 are to a large extent realized in modern cars. Crankcase control and exhaust control have the biggest effect (57% reduction), evaporation controls add up to 72%, and carbon monoxide (catalytic) burning results in over 80% reduction. Unleaded gas is used. Nitrogen oxide removal is the one remaining unsolved problem.

Reduction beyond the 85 to 90% level is met by strong resistance because of the increased cost of the controls and car maintenance. Emission-controlled cars have poorer driving qualities, higher gas consumption (exception: catalytic converters improve mileage because the engines can be adjusted to run efficiently).

In summary, the pollution control argument establishing the need for electric vehicles has been considerably weakened. It remains strong as an interrelated issue: vehicles with low power consumption conserve energy and pollute less.

1.2. Energy Resources

The internal combustion engine depends entirely on oil products as an energy source. Efforts to switch to steam engines or other types of heat engines operating with fuels derived from coal or atomic power have not been very successful to date, in spite of the existing technology. The depletion of petroleum supplies and the need for better energy conservation is therefore the crucial point of future development strategies.

The worldwide situation is dramatically illustrated in Fig. 3, taken from a Department of Transportation Report [4].

A more detailed picture of the U.S. petroleum supply is presented in Fig. 4, taken from a Shell Oil Company publication [5].

It is noteworthy that the oil company diagram (Fig. 4) does not show how the curve continues beyond the peak years of Fig. 3 (1990). However, to be safe, oil companies have already purchased most (~70%) of the coal reserves in the United States. It must also be realized that the estimated oil demands beyond 1973 did not follow the projection (for reasons not foreseen in 1972, e.g., inflation, causing a slowing down of car production, increases in gasoline prices, and lower population growth.)

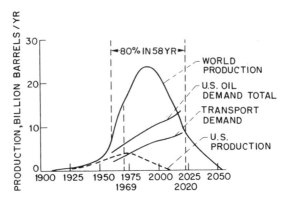

FIG. 3. Oil production 1900 to 2050 [4].

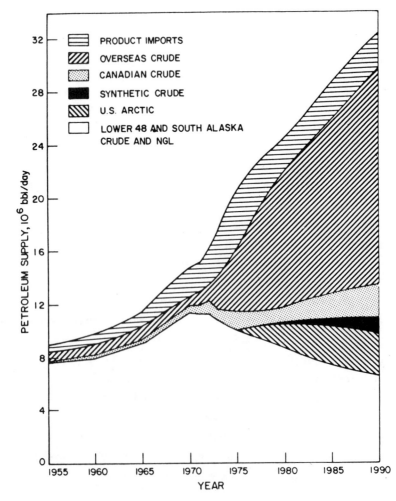

FIG. 4. U.S. petroleum supply 1955 to 1990 [5].

However, no matter how precise the estimates are, the supply of gasoline will be critical. A necessary deduction from the facts that only coal or atomic power are available in satisfactory quantities is the proposition to switch to electric propulsion whenever possible. The energy conversion is thereby switched to central power stations which can operate on those more abundant fuels, with higher efficiency and with better-controlled side effects on the environment.

A study of the future (1980) impact of the conversion from gasoline motor vehicles (GMV) to electrically powered vehicles (EMV) was made with the surprising conclusion that the EMV may actually use more total energy [6].

This is correct. Power plant efficiencies range from 30 to 40%, battery recharging is only 60 to 70% efficient, and with other losses added, the total conversion may be 15 to 20%. Compared with diesel engines (25%) and Otto motors (10 to 15%), these figures do not promise any energy savings [7].

Still, the type of fossil fuel used in power plants is in large supply. Gasoline is not; it must be imported to a large extent. In sulfur oxide production, the EMV is indirectly a heavier pollutor than the GMV, but on all other counts it is environmentally far superior.

The pollution caused by the burning of sulfur-containing fuels is power industry-wide and must be solved sooner or later; therefore this negative argument will disappear.

The scarcity of petroleum products is only a part of the energy supply problems of the world. The reliance on coal and atomic power to supply future energy demands in the United States will be more understandable after studying Figs. 5 and 6 [8].

The situation in 1970 is pictured in Fig. 5. On the left side the energy sources are listed; on the right side appear the users and the percentages of used and wasted energy. The transportation field is one of the least efficient users, over three-fourths of the energy input is rejected as heat. The electric power industry is also inefficient, mainly due to conversion and line losses.

Figure 6 shows the projected situation for 1985, assuming that the energy demand will increase considerably. (There are serious doubts that

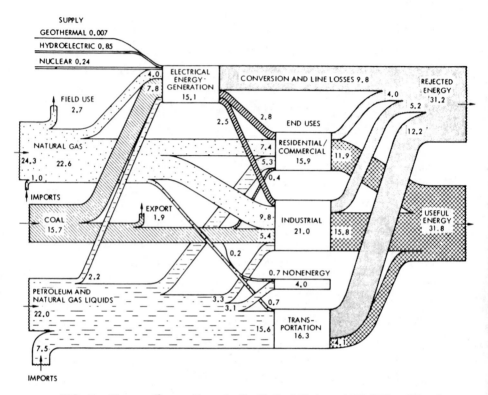

FIG. 5.   Energy flow pattern in the United States, 1970 [8].   All values are in units of 10^{15} Btu.   Total production = 71.6 × 10^{15} Btu.   (Courtesy of Lawrence Livermore Laboratories.)

it can be prevented!)   The petroleum imports are higher (see also Fig. 4), because transportation uses may have doubled.   The power generation via nuclear plants has not increased as hoped for.   Coal production is projected to be 50% higher.   The obvious remedy is to channel some of the coal and atomic power into transportation: the dotted line.

Two possibilities exist:

1.   Produce synthetic fuels which can be used in combustion engines; gasoline, alcohols, ammonia, hydrogen are possible candidates for the operation of improved engines.

2.   Produce electricity and run the nation's transportation system on electric power, as for instance, the European railroads are doing (over 75% completed in Germany).   For personal mobility, use

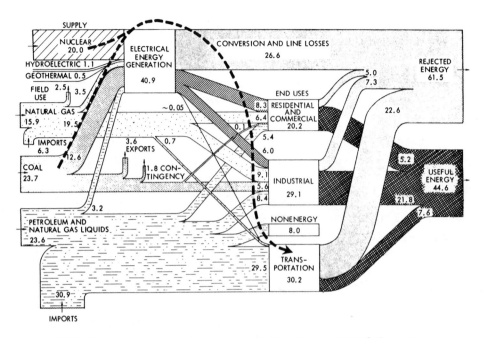

FIG. 6. Energy flow pattern in the United States, 1985 [8]. All values are in units of 10^{15} Btu. Total production = 122×10^{15} Btu. (Courtesy of Lawrence Livermore Laboratories.)

electric cars, not big vehicles but small commuter cars for intracity (or between suburbs) transportation.

The future scenario to be produced by choice 2 puts an extreme load not only on the producers of electric power, on distribution systems and electrical engineering, but also on batteries. Lead batteries as they exist are marginal in capacity and life expectancy, and better systems must be developed. This will take decades; in the meantime we will have to use lead-acid cells.

One of the future possibilities includes the use of chemicals to produce electricity directly in fuel cells, at 50% efficiency, avoiding the loss at the central station and most of the distribution losses in high-tension lines. Hydrogen may be such a favorite fuel, sent to the consumer in pipelines or transported as liquefied hydrogen to larger customers (present railroad tank cars contain 27,000 gal of liquid hydrogen).

Hydrogen is available through coal conversion and has been used as constituent (40%) of the so-called city gas in Europe for heating and cooking for decades, testifying for the safety of its public usage. Hydrogen as fuel for cars with combustion engines is unlikely to succeed as long as it has to be transported in steel cylinders. With 10 to 15% efficiency values for heat engines, the weight increase is prohibitive. As fuel for automotive fuel cells with 50 to 60% efficiency, even compressed tank supplies become feasible (see Sec. 4.2.3).

Storage batteries are required with all these concepts as assist systems (hybrid concept), and the question of providing them in large enough quantities at low enough cost puts the emphasis on battery research and development. In the meantime, industry must rely on lead-acid systems, and there may not be any other secondary battery type qualifying for the task even in the future.

1.3. The Social and Economic Aspects of Electric Vehicles

The technology to build short-distance electrically powered utility and personal vehicles is available now. The delivery van and the second car can tolerate the limited range (50 miles) and the low speed (50 mph) without losing their usefulness. Lead-acid batteries for those cars exist.

The acceptance of small cars with some inconvenience (no air conditioning, no power assist during braking or steering) has been proven by the increase in recent small car sales figures. Economy and low maintenance requirements are far more impressive attributes now than they were 5 years ago. It may become economically advantageous to introduce electric commuter buses, school buses, taxicabs, and service vehicles to our crowded cities. Actually, such vehicles will have to precede the electric personal cars.

Few people can afford a "fashionable" car model at twice the price of an average all-purpose automobile. A low-cost electric vehicle for driving within the local community and shopping centers will have appeal to the homemaker and to the commuter going to the nearest train terminal.

It could change the living conditions in modern housing areas (condominiums) with limited space for parking.

It would change the electric power requirements for recharging overnight. No doubt the environment would be cleaner and quieter. Small vehicles would also conserve energy. The big automobile would be used only
by professionally traveling persons (salesmen, etc.), and a few people going on vacation trips or needing it for special occasions a few times a year.
Maintenance, insurance costs, and depreciation would be minimized by
renting such a car.

Figure 7 shows automobile usage in the United States from the viewpoint of the driver who uses his car only 3,000 miles per year and the average owner who drives about 10,000 miles per year. For 80% of the first
group the average daily mileage is 10 miles, in the second group 60% drive
only 25 miles per day [9]. Those two groups could use electric "city"
vehicles—one type could be a low-cost car for shopping, the other an automobile resembling a small four-seat passenger car for traveling to and
from work.

Conflicting opinions about the market potential of electric vehicles will
be presented in Sec. 6. The discussion is deferred to after the technical
section because the key question is the purchase price of an electric car of
acceptable performance (for its job) and of high engineering standards.
Only a mass-produced vehicle can combine quality, low cost, and adequate
return on investment.

To judge the possibilities it is necessary to examine the state of the
technology more closely. It may also help to learn from the mistakes made
during the history of electric vehicles.

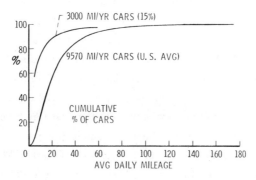

FIG. 7. Automobile usage in the United States [9].

2. THE HISTORY OF ELECTRIC VEHICLES (1837-1960)

The dream of the horseless carriage with electric power was probably first
experienced by the inventors of the electric motor (Faraday, Barlow, 1823;
Davenport, 1837; Anderson, 1839). However, a suitable battery did not
appear until Plante invented his lead battery in the 1860s.

Around 1870 Sir David Salomon propelled a carriage this way, and in
1881 an electric tricycle was built in France with already considerably improved rechargeable lead batteries.

In the United States the year 1888 saw F. M. Kimball's and P. W.
Pratt's attempts to develop electric cars. The first successful vehicle
which received attention in America was constructed by W. Morrison of
Des Moines, Iowa (1891). It was a two-seater "runabout." At that time
the lead battery had an energy content of about 9 Wh/kg (4.1 Wh/lb).

The Morrison car participated in an automobile race over 54 miles between Chicago and Evanston, but lost due to heavy snowfall on Thanksgiving
Day, 1895. However, electrics improved rapidly and soon began to beat
their combustion engine competitors [10]. Speed records were first
achieved in 1899: a cigar-shaped electric car, La Jamais Contente,
reached 66 mph over a 2-km stretch in France. The Riker Torpedo car
set an American record in 1901 for the measured mile of 63 sec. It was a
"battery on wheels." In a road race in 1900 on Long Island, a Wright Electric won the 50-mile event in 2 h, 3-1/2 min ahead of a steam and a gasoline car.

The electric vehicles which were built subsequently (1900 to 1915)
offered many different shapes and designs. In 1912 about 100 manufacturers
of electric cars built 6,000 electric passenger cars and 4,000 commercial
cars annually in the United States. The vehicles were slow, 10 to 15 mph,
and could travel 30 to 50 miles. The handling was simple, attractive to
ladies, who did not like the complicated, constantly breaking-down gasoline
cars. The cars were clean and quiet, but the cost was high: $3,000 for a
Columbia Victoria built by the Electric Vehicle Company in 1901 (this car
had a separate chauffeur's seat in the back) compared to $2,600 for early
W. C. Baker cars (1902 to 1910). Later models were luxury cars with cut

FIG. 8. 1914 Detroit Electric Coupe, personal car of Mrs. Henry Ford. Original cost $3730, speed 5 to 20 mph, range 30 miles. (Henry Ford Museum, Dearborn, Mich.).

glass windows and stuffed plush seats, available at prices which only few could afford ($5,000). Figure 8 shows a picture of a 1914 Detroit Electric, made by Anderson Electric Car Co., Detroit, Michigan, 1907 to 1939.

The development of electric cars [11] made use of many features which were later adopted for combustion engine cars: enclosed bodies, steering wheels, drive shafts, differential gear (which was later replaced by an electric differential motor designed by H. E. Day and C. P. Steinmetz).

Thomas A. Edison tried to develop a better battery for the electric car. He thought that the iron-nickel system would have the power output to compete with gasoline cars. In order to increase the range of electric cars, hybrid vehicles were designed. The 1917 Woods electric car combined a gasoline engine with electric propulsion. This system is considered quite appropriate today, as we will see later!

The Ford Model K, built in 1906, was the first competition to electric vehicles. It provided a challenge for the mechanical capabilities of male drivers! However, the combined effect of 50% lower cost, the wide range and, since 1912, the electric starter for combustion engines ended the "electrical vehicle era" rapidly.

TABLE 2

Production of Passenger Cars
in the United States (1899-1933) [10]

Year	Electric	Gasoline	Steam
1899	1,575	936	1,681
1904	1,425	18,699	1,568
1909	3,826	120,393	2,374
1914	4,669	564,385	
1919	2,498	1,649,127	
1924	391	3,185,490	
1929	757	4,454,421	
1933	0	1,560,599	

Table 2 shows U.S. production of passenger cars per year, 1899 to 1933 (electric, gasoline, and steam powered [10]).

In Germany the interest in electric vehicles began later than in the United States; it was coupled there with the appearance of a lighter lead battery of about 19 Wh/kg capacity around the turn of the century [12]. A rush to build electric buses and cabs ensued. In 1910 some bus and taxi companies used fleets of hundreds of vehicles. In Berlin and several other cities their operation was so successful that pay charging stations, battery loan and exchange stations were made available. As transport vehicles for beer companies, the mail, and city services, electric trucks were used in larger numbers (20,000) in World War I and later, past 1933 (at that time there were no such vehicles in the United States). After World War II, the government aided the development of buses and trucks by setting standards for battery sizes, and exchangeability was possible for vehicles between 3/4 and 6 tons. In 1954-1955, Auto Union converted one hundred 1-ton trucks to electrics. After that time a rapid decline began, accelerated by a tax policy which included the weight of the batteries in the "tax weight" of the vehicle in 1955. The successful operation of modern battery-operated trains is described in [13]).

In Great Britain the developments in electric traction show a continuity from the early days of electric road vehicles, before World War I, until today [14]. The situation in Great Britain was always favorable for quiet electric utility vehicles; they were used in large numbers for the postal service, delivery of milk and bread, and street-cleaning purposes. Without counting electric industrial lift trucks, their number increased even after 1953, to over 40,000 licensed road vehicles. Of course, the high price of gasoline in all European countries made the operation of electric vehicles more economic there than in the United States.

The personal electric car was unsuccessful in Great Britain, and it needed the revival efforts of the 1960s to induce some manufacturers again to produce small electric cars.

In the 1960s, a new wave of electric vehicle promotion started worldwide, as a reaction to the pollution problem caused by the internal combustion engine. In the 1970s the energy crisis became the overriding concern [15].

3. MODERN ELECTRIC VEHICLES

3.1. Power Requirements

3.1.1. Basic Consideration of Power Losses

A vehicle consists of a power source, a chassis, a transmission train, and wheels. The power source for an electric car is usually a battery, but it may also be a hybrid system consisting of an electric power source and an engine, or a flywheel, or it may (at least for some time period) derive its power from an overhead or track supply.

In this first section only the vehicle is considered. The general equation which summarizes the relationship between the propelling forces and the forces resisting the vehicle movement is

$$F_t = Wk + W \sin \alpha + Ma + F_w \tag{1}$$

where F_t denotes the total force to propel the vehicle, W is the weight of the vehicle (mass × gravity constant), k is a constant, M is the mass (w/g) (where $g = 32.2$ ft/sec^2, the gravity constant), α is the inclination, a is the acceleration factor, and F_W is the wind resistance.

It can be seen that the first three terms, Wk = rolling resistance, W sin α = hill climbing, and Ma = accelerating behavior, include the mass of the vehicle in a linear relationship. The fourth term, the air resistance, contains the velocity in a square function, as will be seen later. A car traveling at a steady speed on a level road is concerned only with the rolling resistance and the air resistance (dissipated energies). Energy put into climbing or accelerating is (partly) recoverable.

3.1.1.1. Rolling Resistance

The rolling resistance consists of two parts: the losses in the chassis (wheel bearings and transmission and drive train) and the rolling friction losses, due mainly to the tires. The first part amounts to less than 15% of the total rolling losses and is a constant for a given vehicle; it will therefore be ignored in the following calculation.

The rolling friction F_r is a function of W and the retarding force of the pneumatic tire, which in turn is dependent on the road conditions (to a very high degree) and on the grade. A change from a smooth concrete road to a sandy road may double the factor c, but a 10% incline changes the rolling resistance less than 1%:

$$F_r = Wc \cos \tan^{-1} b \tag{2}$$

whereby b = grade factor in percent. The constant c for a small car of 2,000 lb weight is 0.015 to 0.0175. This represents 35 lb of friction per ton weight. Converting from a force to power by multiplying with the velocity, one obtains for 30 mph: $0.0175 \times 2,000$ (lb) $\times 40$ (ft/sec) $\approx 1,400$ lb-ft/sec. With 1 hp ≈ 550 lb-ft/sec, the losses amount to 2 to 3 hp. At 50 mph, the loss is 6 hp.

There are ways to reduce the tire losses. Table 3 shows tire performances and losses for a small vehicle traveling at 50 mph for seven different types of tires, some of them experimental [16].

Figure 9 illustrates the horsepower losses for a 2,500-lb vehicle as a function of vehicle speed. The top curve is for a production 6.70×13 tire, the lower for an experimental 6.70×13 tire (see Table 3). The losses are cut in half [17].

TABLE 3

Tire Performance and Horsepower Loss at 50 mph [16]

Design rating	Ride	Wear	Stability	Traction	Total horsepower loss, 2,500-lb vehicle
Standard 6.70-13	100	100	100	100	6.0
Belted radial	80	175	95	105	4.8
Special compounding	80	140	95	90	3.8
Gauge reduction	80	130	90	90	3.6
Reduced deflection	75	140	95	90	2.8
Low aspect ratio	65	150	100	95	2.5

3.1.1.2. Wind Resistance

The wind resistance can be expressed as

$$F_w = (0.5\,\rho)\,C_d A V^2 \qquad\qquad (3)$$

with ρ = air density, C_d = drag coefficient (for an average car $C_d \approx 0.6$), and A = the frontal area (about 20 ft^2). With ρ = 0.08071/32.2 in mass units and velocity V in miles per hour, one obtains

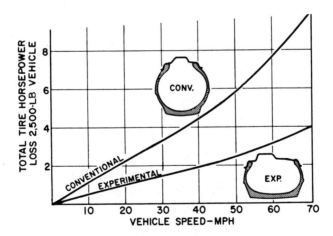

FIG. 9. Total tire horsepower loss versus vehicle speed (2,500-lb vehicle) [17].

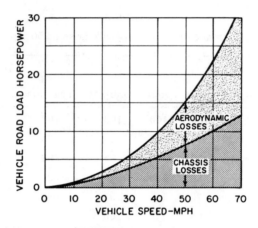

FIG. 10. Vehicle road load horsepower versus vehicle speed [17].

F_w = 0.00256 C_dAV^2 (lb); multiplying by V and converting to units of horsepower,[*]

$$hp = \frac{F_w V}{375} = \frac{C_d AV^3}{146,625} \qquad (4)$$

Figure 10 shows the results as a function of the vehicle speed [17] and the relations to the total chassis losses. (At 30 mph, hp = 0.6 × 20 × 27,000/150,000 = 2.3 hp). For comparison it can be stated that a 1930 Ford Model A had a C_d of 0.83, a modern passenger car has C_d = 0.6, and for a streamlined aerodynamically built sports car, C_d = 0.25.

3.1.1.3. Hill Climbing

The retarding force F_g during climbing a grade is also proportional to the weight:

$$F_g = W \sin \alpha = W \sin \tan^{-1} b \qquad (5)$$

if α is the angle of incline and b is the grade in percent. For a 2,000-lb car, a 4% grade (sin $\tan^{-1}$ b = 0.04), and a speed of 30 mph, the power re-

———————

[*]Metric conversion factors: speed: 1 mph = 1.61 km/h; acceleration: 1 ft/sec^2 = 0.305 m/sec^2; force: 1 lb (weight) = 0.453 kg (weight); g = 980 cm/sec^2 = 32.2 ft/sec^2; energy: 1 hp = 76.04 mkg/sec = 0.745 kW = 550 ft-lb/sec = 42.42 Btu/min; work: force × time; 1 ft-lb = 0.138 m-kg.

quirement is about 3,200 ft-lb/sec or nearly 6.5 hp. The gravitational energy of the car increases steadily at that rate.

3.1.1.4. Acceleration

There are two types of acceleration which are to be considered: the acceleration of the total mass of the vehicle (gain in kinetic energy) and the increase of the energy of the rotating masses (wheels, brakes, armature, axles). The first type takes the large portion (approximately 90%) of the energy reauired to get the vehicle to a certain speed. The energy consumption of the rotating masses is a fixed feature of a particular car; it is neglected in the following considerations. For the electric car in stop-and-go city traffic, the translatory acceleration of the masses of the car therefore takes the energy (E), expressed by

$$E = MVa = MV \frac{dV}{dt} = \frac{d}{dt} \left(\frac{MV^2}{2} \right)$$

where a = acceleration; M = mass; with M = W/g (g = 32.2 ft/sec^2), converting to weight (W) in pounds, V in miles per hour,

$$E = \frac{1}{375} \left(\frac{W}{g} \right) Va \text{ (hp)} \tag{6}$$

Examples: Assuming a car of 2,000 lb weight and an acceleration of 3 ft/sec^2 at a speed of 30 mph, the requirement is (2,000/32)(3)(30)(1/375) = 15 hp. At a speed of 50 mph, it would reauire 25 hp to speed up the car further at the same acceleration. Table 4 lists energy levels needed at different driving speeds for a 2,000-lb car.

TABLE 4

Energy Reauirement for Various Acceleration
Conditions Calculated with Eq. (6) for a 2,000-lb Car

Speed (mph)	a = 3 ft/sec^2	a = 6 ft/sec^2	a = 9 ft/sec^2	Acceleration energy
5	2.5	5	22.5	E = hp =
10	5	10	45	0.166Va
30	15	30	135	(Eq. 6)
50	25	50	225	

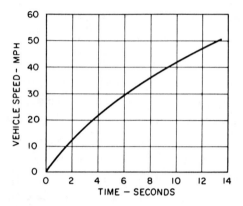

FIG. 11.   Vehicle speed versus time (compact car).

Figure 11 shows how a typical experimental plot of vehicle speed versus time looks for a compact car.   The vehicle reaches 30 mph in 6 sec,
50 mph in 13 sec.   Figure 12 shows the derivate of the experimental curve
from Fig. 11, the change of velocity with time.   Using Fig. 11, Fig. 12,
and Eq. (6), it is possible to tabulate the energy requirement for acceleration versus elapsed time.   In this particular case the cumulative power (in
hp-sec) for the time span 0 to 13 sec, corresponding to the attainment of a
speed of 50 mph by a 2,000-lb car, is calculated.   Table 5 presents these
data.

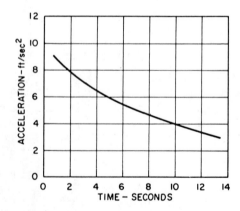

FIG. 12.   Acceleration versus time (compact car).

TABLE 5

Power Requirement for the Acceleration

of a 2,000-lb Car from 0 to 50 mph. Data from

Figs. 11 and 12 Used as Input for Eq. (6). $\Delta E = 0.166Va$

Time (sec)	Speed (mph)	Acceleration (ft/sec^2)	Incremental power (hp-sec)	Cumulative power (hp-sec)
0-2	7	9	21	21
2-4	17	7	40	61
4-6	26	6	53	114
6-8	32	5	55	169
8-10	39	4.2	56	225
10-12	47	3.3	52	277
12-13	50	3	25	302

Note the conversion factor: 1 hp = 0.745 kW.

3.1.2.   Comparison and Reduction of Power Losses

If we take the rolling resistance at 30 mph (2 hp), calculate the requirement for 1 mile driving at 30 mph (4 hp min), and compare it with the
energy consumption for accelerating to 30 mph in 6 sec (2 hp min), then it
becomes understandable why stop-and-go traffic wastes so much energy
even with a lightweight, 2,000-lb car. Acceleration from 0 to 50 mph in
13 sec is equivalent to 3/4 miles of driving at 50 mph. Yield signs instead
of stop signs would be really beneficial!

It is obvious that weight reduction is the primary goal to seek for an
economic car. Tire loss reduction is next. Aerodynamic design follows
closely (for 50 to 60 mph driving). Figure 13 shows a comparison between
a conventional compact car and an optimized (electric) car with respect to
energy consumption at 50 mph steady speed [17]. The weight reduction considered possible is 20%, the tires are improved (see Table 3), the styling is
streamlined so that the air drag coefficient is low ($C_d = 0.25$). All these
changes result in 45% energy savings. With some considerations to better
driving habits (less braking, slower acceleration, economic planning of
routes, avoiding upgrades and stop signs), a 50% cut is definitely feasible.

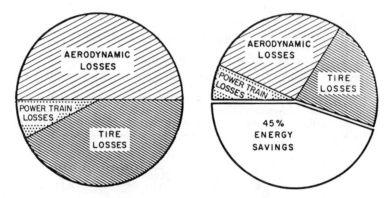

FIG. 13. Expressway driving at 50 mph—distribution of energy consumption [17]. Left: conventionally designed car, right: optimized car.

If we need only a city vehicle, which usually does not go much faster than 30 mph, the savings are truly remarkable. Figure 14 gives a comparison for driving at 30 and 50 mph [1]. The power is expressed in kilowatts. This comparison shows clearly that it is not only desirable to have a high-capacity power source for long sustained driving, but also a high power source for giving the vehicle the acceleration at driving speeds, for passing, emergencies, etc. This requirement is expressed by the acceleration bars in Fig. 14.

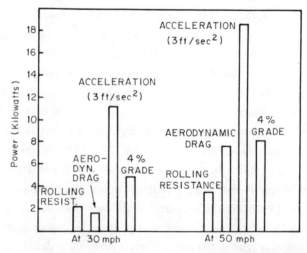

FIG. 14. Power required to wheels for 2,000-lb compact car [1].

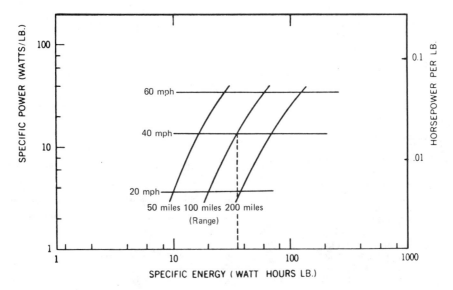

FIG. 15. Specific power and specific energy for power source of
small car [1].

Assuming that we can allow a weight of 500 lb for the source of energy,
it is then possible to construct a diagram which indicates specific power
needs (hp/lb or W/lb) and specific energy requirements (Wh/lb for a battery). The fuel supply—in case of a combustion engine or a fuel cell—is included in the power source weight, since it is a part of the energy-producing
system. Figure 15 shows such a power-density energy-density diagram [1].

It may seem that just doubling the weight of the power source would
solve range and acceleration problems. The pitfalls of constructing "batteries on wheels" have often been experienced and lead to dismal failures
(Sec. 3.1.1.4).

3.2. Vehicle Designs

3.2.1. Types of Cars

If we study the vehicles on the road today and try to convert them to electric
cars (with no concern about the existence of a suitable battery), it is not difficult to arrive at certain vehicle specifications which must be fulfilled to

TABLE 6

Electric Vehicle Specifications [18]

		Family car	Commuter car	Utility car	Delivery van	City taxi	City bus
Assumptions							
1. Acceleration to	(mph)	60	60	30	40	40	30
2. in	(sec)	15	30	10	20	15	15
3. Range	(mi)	200	100	50	60	150	120
4. Seats or payload	(lb)	6	4	2	2,500	6	10,000
5. Loaded weight[*]	(lb)	4,000	2,500	1,700	7,000	4,000	30,000
6. Curb weight	(lb)	3,500	2,100	1,400	4,500	3,500	20,000
7. Weight assignable to propulsion, energy storage, controls	(lb)						
a. Conventional construction		1,250	750	500	1,400	1,250	5,000
b. Lightweight construction		1,750	1,050	700	2,000	1,750	7,000
8. Frontal area	(ft^2)	25	18	18	42	25	80
9. Drag coefficient		0.35	0.25	0.25	0.85	0.35	0.85
10. Elec. transmission efficiency	(%)	82	77	72	79	76	85
Derived parameters							
11. Max. power delivered by motors	(kW)	70	22	12	49	36	135
	(hp)	94	30	16	66	48	180
12. Max. output of power source	(kW)	85	29	17	62	47	159
13. Max. velocity	(mph)	100	80	65	56	77	55
14. Delivered energy	(kWh)	100	20	8	45	75	300
15. Stored energy	(kWh)	122	26	11	57	99	353
16. Weight of motors, transmission, and controls	(lb)	348	118	82	259	210	615
17. Weight assignable to power source	(lb)						
a. Conventional construction		902	632	418	1,141	1,040	4,385
b. Lightweight construction		1,402	932	618	1,741	1,540	6,385
Power source requirements							
18. With conventional construction							
Energy density	(Wh/lb)	135	41	26	50	96	81
Power density	(W/lb)	94	46	40	55	45	36
19. With lightweight construction							
Energy density	(Wh/lb)	87	28	18	33	64	55
Power density	(W/lb)	60	31	28	36	30	25

[*]The loaded weights given for the cars and taxi are not the maximum which they are capable of carrying but rather reflect typical usage.

satisfy the user. Such vehicles would be commercial successes, and they would be immediately accepted by the public. Such a study was already made in 1967 by Arthur D. Little, Inc. [18], and it is still worthwhile to repeat the results and discuss the types of cars established in this survey. Table 6 lists the characteristic features of six vehicles, making some normalizing assumptions first and then deriving parameters for the power train, motors, and power sources.

In a graphic display these six vehicle types can be classified by one parameter which is closely connected with the use of the vehicle and its power consumption, usually the acceleration behavior. Figure 16, also from Ref. 18, shows the standardized acceleration curves of those automobiles.

The figures of this survey have been extensively discussed over the years, and arguments for and against them have been made. They are undoubtedly correct if we wish to replace gasoline vehicles to which we are accustomed and which we like for their speed, acceleration, and reasonably low cost. The survey was made to find low-polluting electric equivalents for the internal combustion car.

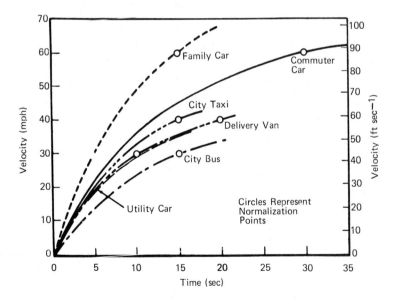

FIG. 16. Standardized acceleration curves for various vehicles [18].

In the years after 1970, the emphasis changed considerably. Low
energy consumption is now more important, and the need for a low-powered
gasoline automobile became an acceptable truth even for the automobile
manufacturers. Increasing gasoline prices make the public conscious of
the energy wasted with a 5,000-lb automobile in city traffic and, considering the automobile usage indicated in Fig. 7, an electric city vehicle becomes a far closer reality with today's lead battery than the data of Table 6
would suggest. The "commuter car" and the "utility car" with a slightly
relaxed acceleration specification can be produced with present technology,
and their availability becomes a capital investment and public information
problem. Manufacturing such cars on a large scale is necessary to permit
the cost to come down to an acceptable level.

Section 5 will give many examples for prototypes of "city vehicles" and
also for delivery vans, buses, and other utility vehicles based on lead battery power systems.

Many private persons have converted suitable small gasoline-powered
cars into electric cars by removing the engine and replacing it with an electric motor. Then, batteries were stashed anywhere in the car where some
room was available and the result was called an electric vehicle. The
product may have excited the hobbyist, and it must be stated that many val-
uable results were obtained from such work. However, the costs of such
vehicles were often stated too low (personal labor does not count in such
cases), legal requirements were not always fulfilled and—naturally—the
performance was usually somewhat exaggerated (by overlooking negative
data). The author has himself built several models and is personally
aware of those dangers of overenthusiasm. This attitude will help in judging reported efforts in the electric car business.

However, the willingness of amateurs to build and test personal electric vehicles should be nourished. As it has been in the amateur radio
world for decades, there is no substitute for the endurance and resource-
fulness of a dedicated hobbyist.

There is no large-scale-produced electric city car (four-seater) on the
market, and only a few vehicles can be bought, at twice the price of a (superior) gasoline car. An electric car-building kit may be the answer to an
existing need. Such a "kit" should contain all the components (frame,

chassis, motor, power train), selected and prefitted to a standard battery set which could be purchased locally. The assembled vehicle should be a completely roadworthy car with a 115-Volt charging unit, nothing essential missing, except perhaps the final paint job.

The mass production of parts and the shifting of the assembly work to the customer may keep the total cost within the price range of a small gasoline car and make many do-it-yourself people happy. A good and safe product would pay the builder for his effort by providing low cost and avant garde local transportation. The use of standard-type batteries would make replacements easier and again, lower the cost, if quantity buying (in groups or clubs) is adopted.

3.2.2. Correct Electric Car Design

The modified conventional vehicle is not the answer to the production of electric cars which perform optimally with batteries. The adding of, for example, 1,000 lb of lead-acid batteries (after removing about 400 lb of internal combustion engine components) stresses the structural limits of the frame and overloads the suspension system. Brakes become unsafe and driving hazardous due to changes in the center of gravity of the vehicle.

A good electric car must be designed "from the frame up." In the following a brochure about electric vehicle system design [19] is quoted extensively.

The "frame design" allows the battery to be at the center of the car in a sealed compartment from which it can be removed as a whole for battery exchange or servicing. This feature assures a low center of gravity and provides an important point for safety in case of an accident. Also, battery compartment ventilation is separated from the passenger compartment.

Figure 17 shows the passenger car (2 + 2) compartment design. The design of Ref. 19 prefers a front wheel drive arrangement, with the control elements (SCR chopper) also in front, in an exchangeable package. The general design of the car body makes it useful as a commercial vehicle (delivery van or taxi) or an "urban car."

Figure 18 shows the basic design of such a vehicle. In Sec. 5.1 a two-seater passenger car is described. The quick exchange of the battery pack

FIG. 17. Electric vehicle with central battery compartment—four-passenger car [19].

would make it easy to replace exhausted batteries on a service station concept. Loaned batteries instead of owned batteries would give the advantage of better battery maintenance. The changes in the designs during three generations of this "from the frame up" electric vehicle are very educational [20]. The first vehicle was a "battery with four wheels"—a rolling test bed with a gross weight of 3,500 lb, the battery weighing 1,500 lb (41%). The second car (two-seater sports car) weighed 2,580 lb, the battery 750 lb (29%). The final model (four-seat production prototype) weighed 2,670 lb, battery 850 lb (32%), and had the interchangeable chassis for a van or a taxi. The specifications for a commuter car as given in Table 6 are rather closely met. The city taxi of Table 6 is a far heavier vehicle for six persons; it better fits the designs of a minibus or a delivery van chassis.

A remarkable design effort which went along the same universal duty concepts as [19] but concentrated on lightweight body construction, use of plastic components wherever possible, sandwich construction with side-exchangeable batteries, and low cost for small production series, was started in Germany (1971) by a large combine of related industries [21]. Figure 19 shows the basic design concept. The vehicle performance will be

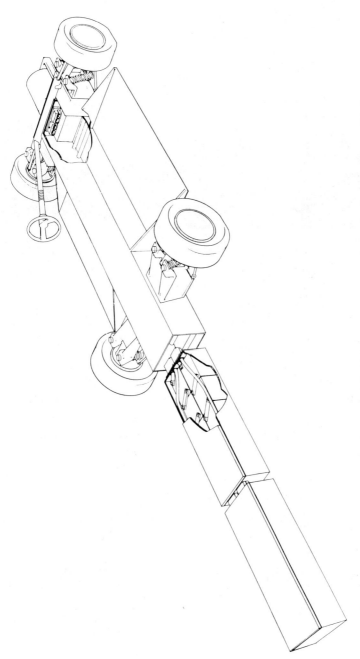

FIG. 18.   Electric vehicle with central battery compartment—battery exchange [19].

FIG. 19.   Electric van, side view; battery compartment door is visible [21].

discussed in Sec. 5.2.   The vehicle curb weight is 2,100 kg, battery weight 870 kg, and useful load capacity 1,000 kg.   This design comes close to the "delivery van" of Table 6.

Buses for city transportation have been designed along the lines of all-electric vehicles with exchangeable battery sections or battery trailers or as hybrid vehicles, for instance a diesel engine coupled with a generator, rectifier, and a large lead-acid battery.   The latter design follows the concepts of diesel-electric locomotives.   A more detailed performance description will be given in Secs. 5.1 and 5.2.

The only car type which has defied all attempts of designers aimed at an electric vehicle is the "family car" of Table 6.   It just cannot be done. There is no battery in sight which can duplicate the performance of the internal combustion engine in the fast acceleration-high speed-long range area.

Some of the "new batteries" discussed in Sec. 4.2 may be able to provide power sources for such vehicles in perhaps 10 to 15 years.   However, it is safe to say that these new power sources will require completely different designs for which we do not wish to attempt a prediction today.

3.2.3. Direct Current Electric Motors

Two types of dc motors are commonly used: the shunt motor and the series
motor. The difference is that the shunt motor has the field circuit and the
armature circuit in parallel connection to the battery and the currents can
be regulated independently if desired. The series motor has the field circuit and the armature in series, and the same current flows through both.

Some motors have both series and shunt windings for special purposes
(a shunt winding on a series motor can, for instance, prevent the runaway
of an unloaded series motor, which tends to increase its speed over the
limit and destroy itself). Some special windings help in reduction of the
commutator-brush firing (interpole winding), and another type (pole face
winding) reduces the armature inductance to prevent magnetic field distor-
tions.

Figure 20a shows a cross-section diagram of a dc motor with all the
windings mentioned, although some may not be used in every application
[22]. Figure 20b shows the corresponding circuit arrangements of the
windings and the armature. Other dc motors are permanent magnet motors
(used in sizes of a few kilowatts), disk motors (AXEM machines) which have
the windings mounted on an insulated disk and can be highly overloaded, and
printed circuit motors (very small).

Series field dc motors have the very desirable characteristics of high
torque at low speeds and, furthermore, are simple, reliable, and easy to

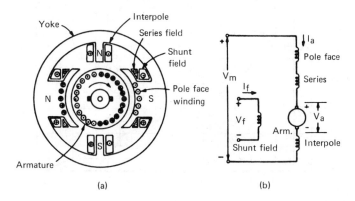

(a)   (b)

FIG. 20.   (a) Cross section, windings, and (b) circuit of a dc motor [22].

control by voltage variations, resistance changes, and field weakening (reduction of magnetic flux).

The speed-torque relationship for a dc motor is given by the equations

$$n = \frac{V - IR}{C_1 \varphi} \tag{7}$$

and

$$M = C_2 \varphi I \tag{8}$$

where n = motor speed (rpm), V = terminal voltage, I = motor current, R = motor resistance, M = torque, φ = magnetic flux, and C_1 and C_2 are proportionality constants. Since armature and field winding are in series, and the magnetic flux is itself a function of the current, Eq. (8) can be written as

$$M = C_2 f(I) I \tag{9}$$

Figure 21 shows the characteristics of a dc series-wound motor with a nominal power of 8 kW (at 2,000 rpm) [23]. It should be noted that the power curve has a maximum (18 kW) while the torque, as a linear function of the current, is highest (45 kpm) at $n = 0$. Also, for low-speed, high-torque demands it is beneficial to use a gearbox (50%, 75% reduction) be-

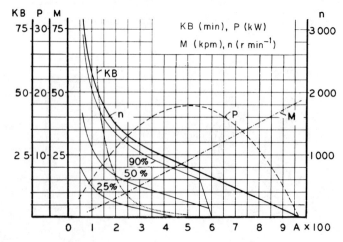

FIG. 21. Characteristics of an 8-kW (2,000-rpm) dc series-wound motor [23].

cause the current must be limited to a safe maximum value. The speed-reduction device shifts the peak power to the left, and a more uniform power can be delivered over the vehicle speed range with less heat losses $(i^2 R)$.

The nominal power is that which the motor can produce for 1 h until the maximum temperature of the windings is reached (usually $65°C$). For higher loads, shorter times are required. In Fig. 21 the temperature buildup time curve is drawn as $KB = f(I)$ and indicates that this motor can be loaded for 1 h with 122 A, 30 min with 156 A, 5 min with 296 A, and 1 min with 500 A.

This (nominal) 8-kW motor can fulfill the requirements of Table 5. At the highest incremental power demand (55 hp-sec for 2 sec = 20 kW-sec), the peak power is utilized. The 2,000-lb car can be accelerated to 50 mph in 13 sec. Four percent hill climbing at 30 mph is possible continuously (see Fig. 14, rolling assistance, aerodynamic drag, and grade climbing add up to 9 kW). For 5 min an incline twice as steep can be driven without reaching the dangerous heating limit.

Direct current vehicle motors are usually not ventilated, to prevent water or salt spray from entering. They are permanently lubricated. Additional air cooling increases the nominal power limit but not the overload capability. Oil cooling of the motor would increase the latter considerably. The motors available have a specific weight of 7 to 10 kg/kW, at speeds of 1,000 to 3,000 rpm. Higher speeds decrease the weight, but brush maintenance becomes more frequent. Changing the efficiency from 85 to 75% can reduce the weight 25%, but this is not advantageous with a battery power source which is limited in capacity.

Low voltages are inefficient because of the brush-resistance losses. The highest possible voltage should be used, limited only by the commutator construction (diameter, isolation). For highway electric vehicles, 72 to 96 V are most frequently used. Golf carts and forklift trucks operate at 24 to 36 V. A few designers of modern electric cars go up to 240 V; even to 480 V for exceptional performance (speed records, etc.), but the practicality of very high voltages is questioned.

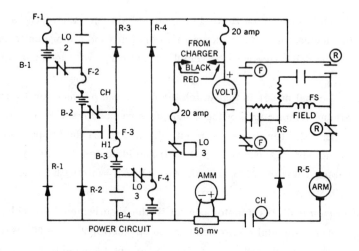

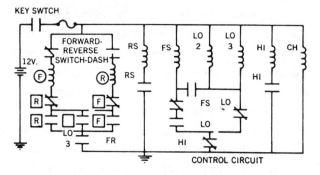

FIG. 22. Diagram of conventional controls: battery switching with contactors [24]. Note: □ Denotes contactor auxiliary switch. Control circuit switches marked RS, FS, HI, FR, LO are cam-actuated.

3.2.4. Speed Controls for dc Motors

3.2.4.1. Battery Switching

Battery switching is a type of speed control which is efficient, but requires heavy contactors. It was done in the Henney Kilowatt car of the 1960s and is again used in the small Vanguard vehicle (see Sec. 5.1). Figure 22 shows the diagram of the Henny Kilowatt switching system, including the charging connections [24]. There are few losses in this system (except for the lowest speed), because all the power goes into the motor.

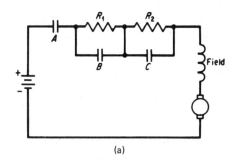

(a)

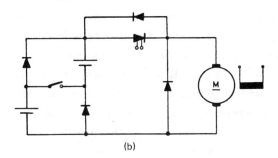

(b)

FIG. 23. Motor controls. (a) Resistance switching. (b) Battery switching [31].

3.2.4.2. Resistor Control

Resistor control is used with most golf carts; it is very simple, but not efficient. It makes use of the fact that the speed of the motor is a linear function of the voltage [see Eq. (7)]. In most of these vehicles, resistances are used only to get the car to its operating speed (usually low, 10 mph) and from then on the battery is connected directly and the vehicle speed is a function of the road load, grade, etc. Figure 23a shows a resistor step control circuit. Only switch A connects series elements and is in danger of burning up through dc arcing. A diode parallel to its terminals can effectively prevent this danger on switch opening. Switches B and C just short out the resistors. For reversing the vehicle motor, a mechanically operated switch arrangement is used, or it is done by gears. Regenerative braking can be used by incorporating a double-pole, double-throw relay switch changing the field-armature polarity relationship. A mechanical relay interlock is required, otherwise faulty switching can destroy the circuit by battery shorting. Figure 23b shows solid state battery (parallel-series) switching [31].

A combination of battery switching and resistor application is used in some older delivery vehicles, mail trucks, etc. The use of two motors (series/parallel) can improve the efficiency greatly, but is more expensive.

3.2.4.3. Speed Converters (Mechanical, Electrical)

The use of a conventional gear box is a very efficient (and low-cost) way to use the electric motor always close to its peak power output (see Fig. 21). The author has used it with excellent results, keeping even the clutch and flywheel of the (converted) car in the drive train. In combination with one small starting (limiting) resistor, it provides a smooth and fast way of moving the vehicles from standstill to any speed within its range (slowly, or drag race-style). The flywheel, when revved up, can act as a hybrid acceleration device, helping to keep the battery peak current down (see Sec. 4.2).

The E.S.B. Sundancer, a prototype electric sports car (see Sec. 5.1) uses a two-speed synchromesh transmission mounted directly on the transaxle. Another mechanical device with high efficiency is an infinitely variable belt drive. The MK-16 commuter car of McKee Engineering Corp. (see Sec. 5.1) uses one in combination with an 8-hp series-wound dc motor [25]. The motor is arranged on the rear wheels; the batteries are inserted in a central compartment, similar to the design shown in Fig. 17.

a. Torque converters. Torque converters are used in some electric vehicles to avoid gear shifting, or at least to assist shifting. The losses of any fluid connection are higher than those of mechanical drives, but some recent models seem to be satisfactory and find acceptance in combination with electronic controls (see EVA Car, Sec. 5.1).

b. Motor-Generator-Motor Electric Converters. Motor-generator-motor converters have been used for many years in locomotives, in the wheel drives for construction equipment, but not in small cars. A special combination is used in ac drives (see later).

3.2.4.4. Field Control

Field control is very suitable for shunt dc motors combined with electronic
controls, but it can also be used with many series-wound motors where the
series field has two windings (four terminals). Reducing the magnetic flux
by switching (or partly shortening) the field windings, also called field weakening, increases the speed.

The effect is clear from Eq. (7), where the flux φ is inversely proportional to the speed n. Field reduction is a low-loss speed change, simple
to effect by a contactor across a part of the field winding. Field weakening
is not possible over a wide range and is most effective at the high-speed
end of the motor characteristic, where it can serve as an "electric over-
drive." The author has had excellent experience with it in the speed range
of 35 to 55 mph (see Sec. 4.2).

3.2.4.5. Solid State (Electronic) Controls

Solid state electronic controls are the modern devices which provide a safe
and smooth motor speed control. They are usually efficient, but care must
be taken not to use them in the wrong mode of operation; they are very inefficient when very short, very high current peaks are drawn from the battery.
The I^2R losses in the power source and in the connections can amount to
serious losses. With shunt motors the separate field control provides additional flexibility of circuit design; with series motors a two speed gear box
can improve the efficiency considerably.

In evaluating electronic controls, testing should be done only with real
(not simulated) batteries (of limited size) and one should not be deceived by
a good current efficiency. The watt-hour efficiency is decisive; it can only
be obtained correctly with instruments that properly show peaks (oscillo-
scopes) or suitably integrate load profiles (thermocouple meters correctly
indicate $I_{rms} = I_{peak}$ (duty cycle)$^{1/2}$). This is important for the size selection of the semiconductor devices (SCRs) themselves.

It is important to understand the functioning of the basic variable-width
pulse current circuit to arrive at a smooth, low-loss, dc motor control.

Figure 24 shows the action of a "chopper" producing a combination of
pulse width and frequency modulation. The ratio of the time "on" to "off"
is changed, with the result that the average voltage across the motor can be

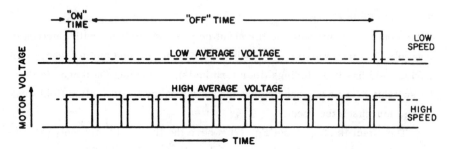

FIG. 24. SCR-chopper waveforms [26].

varied. The change is stepless. The switch is an SCR (silicon controlled rectifier), a solid state device which is controlled by an electronic circuit which turns the SCR on or off in response to the resistance setting of a potentiometer. Figure 25 shows the basic circuit. The diode D ("free-wheeling" diode) prevents the high-voltage buildup across the field in the SCR-off position by carrying the inductive current during that open period and reapplying it to the armature (as stored energy), thereby smoothing out the pulses. The duty cycle of such a chopper is from 20 to 80% of the power range. The basic circuit is called the Jones commutation circuit [26]. For practical performance data see Table 7 [27].

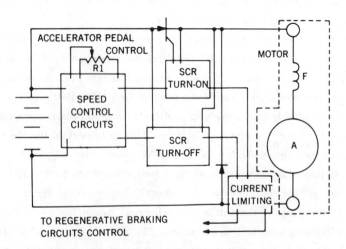

FIG. 25. Block diagram of solid state control circuitry [24].

TABLE 7

Comparison of Power Loss and Efficiency

of Resistor and Chopper Power Controller [27]

Percent of full motor speed	Resistance-controlled battery power consumption (kW)	Efficiency (%)	Power loss (kW)	Chopper battery power consumption (kW)	Efficiency (%)	Chopper power loss (kW)
0	10	0	10	1.2	0	1.2
10	10	10	9	2.2	40	1.2
20	10	20	8	3.3	60	1.3
30	10	30	7	4.3	70	1.3
40	10	40	6	5.3	75	1.3
50	10	50	5	6.3	80	1.3
60	10	60	4	7.3	83	1.3
70	10	70	3	8.3	85	1.3
80	10	80	2	9.3	86	1.3
90	10	90	1	10.3	88	1.3

Vehicle system: 100-V battery, 50-mΩ motor, 50-mΩ battery, 100 A of average motor current.

Note: most of the losses in the chopper system are due to motor and battery resistance.

3.2.4.6. Regenerative Braking

Figure 26 shows how motor braking can be done without using contactors, which are troublesome with respect to maintenance. For the motor to act as a generator, the series field must be reversed (e.g., by switches). Since this is undesirable, an additional field coil, which provides a shunt field (low current), could be used [26]. The amount of the current flowing back into the battery (I) is given by

$$I = \frac{E_g - E_2 - E_B}{R} \tag{10}$$

where E_g is the voltage generated by the motor, E_2 is the voltage across SCR-2 and E_B is the battery voltage, and R is the sum of all resistances

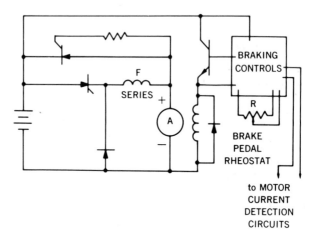

FIG. 26. Circuit diagram for controlled regenerative braking [24].

in the circuit. Also,

$$E_g = K\varphi V \tag{11}$$

indicating that E_g is a function of the flux φ and the velocity V. K is a
constant. As the vehicle slows down, the voltage drops and mechanical
brakes must be engaged. Increasing the shunt field when the speed drops
can counteract this effect somewhat (automatic control).

"Dynamic braking," which allows the current to be fed into resistors,
can brake the vehicle to a lower speed than regenerative braking, but the
resistors must be of large size if, for instance, a longer hill must be driven
down. The heating effect could be used to heat the car in winter.

In general, small vehicles do not save more than 10% in (slow) stop-
and-go city driving by using regenerative braking. The author found that
the additional circuitry was heavy and rather expensive, and one additional
12-V battery was a more reliable way to increase the driving range of his
small car at the same weight penalty. The method the author experimented
with was to use a contactor for reversing the field (no special motor was
available to try the shunt method) and feeding the current into a low-voltage
(12-V) battery (Ni-Cd) from which the energy was transformed (by inverter)
up into the main (84-V) car battery over a longer time period, thereby reducing the size and weight of the auxiliaries as much as possible. Of course,

for driving in a hilly area and for a heavier vehicle, the outlook for regenerative braking is far more advantageous, especially with sophisticated
electronics. A 30,000-lb bus carrying 40 passengers may gain 25 to 50%
by regenerative braking in city traffic [24]. Actual performance studies
with a 23.3-ton Man bus confirmed the calculations [28].

3.2.5. Alternating Current Motors

3.2.5.1. Induction Motors

Induction motors permit a high rotor speed because of the squirrel cage
arrangement of conductors imbedded in steel. Motors of 13,000 to 20,000
rpm are feasible and can provide about 1 to 2 hp/lb of motor weight. Overload capability is only 200%, but there is no brush to wear out. The maintenance requirements are low. For cooling of the motor (and the electronics), oil circulation was provided in experimental General Motors Corp.
vehicles (see Sec. 4.1). The electronic circuitry is complicated and more
expensive than for dc motors [29,30]. For each phase two thyristors are
required. A simplified diagram is shown in Fig. 27 [27]. The use of regenerative braking is possible.

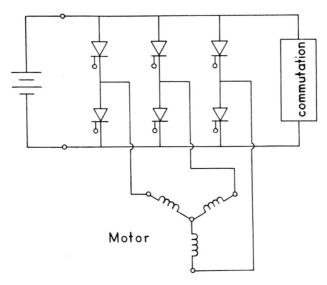

FIG. 27. Specific circuit of a three-phase ac motor [27].

TABLE 8

Electric Control and Transmission System, Electrovair II [24]

Motor Output (peak)		100 hp	74.6 kW
Weight			
Motor		130 lb	
Gear box		25 lb	
Total			155 lb
Electronics			
Inverter assembly		235 lb	
Control circuits		24 lb	
Cooling system		80 lb	
Cables and other hardware		56 lb	
Total			395 lb
Total motor and control system			550 lb
Efficiency of Control and Transmission System	(est.)		> 90%
Cost of Control and Transmission System			
Today	(est.)		$15,000
Future large-volume production	(est.)		200
Cost of Motor			
Today	(est.)		400
Future large-volume production	(est.)		100

Table 8 shows the weight and output figures [24] for the system of the General Motors Corp. Electrovair, a silver-zinc battery powered car.

3.2.5.2. Synchronous Motors

Synchronous motors have slip rings, but no commutator; brush wear is therefore low. Alternating current power is supplied to the "stator" to provide a rotating magnetic field in which the "rotor" moves. The rotor carries the windings which provide the magnetic polarities. Solid rotor motors are manufactured only in small sizes.

3.2.6. Accessories

Electric vehicles need all the standard accessories which modern automobiles provide. The best way to solve this necessity is to have a regular 12-V battery on board. The charging of this battery can be done with an inexpensive alternator from the motor drive train and/or through a separate

115-V ac/15-V dc system, which provides a taper-current charge each
time the main battery is charged from the line supply. A dc-dc inverter
can also be used, but is relatively expensive.

A main disadvantage of an electric car is the lack of a heating system
in winter. The best way to solve this problem is to use a small propane or
gasoline heater, as is commercially available for vans, campers, etc.
Only fuel cell-operated vehicles can have the luxury of a caloric (catalytic)
heater operating directly from the fuel supply (hydrogen or alcohol). These
vehicles could also provide (gas) air conditioning in summer.

3.3. Conclusions About Vehicle Concepts, Optimized Car
      Designs, and Existing Technology

Considering the use patterns of the present-day automobile, a small electric car could fulfill most of the requirements of personal transportation.
In that sense, the electric car should not be a "second car," but the "first
car" in use. The construction of a four-passenger urban car is within the
known technology; but its performance differs greatly from the gasoline vehicle to which everybody has become accustomed. The idea of energy conservation must govern all design parameters. Optimization of electrical
propulsion systems for road vehicles means minimizing the total weight of
the vehicle and matching the limited capacity of the available battery (lead-
acid battery) to it. In order to make fullest use of the battery, electric
drive systems must be further developed to increase efficiency, especially
in the low-speed, high-torque range, and to cut down the presently high
costs [31].

Experience to date confirms the theoretical calculations of battery vehicle performance very closely, and the design of vehicles should be done
with all considerations to produce a modern, useful electric vehicle, know-
ing its limitations [32].

The amateur car builder, converting existing vehicles to electric cars,
may play an important role in the exploration of ways to overcome the cir-
culus vitiosus of high cost because of limited production and vice versa [33].
For instance, the hybrid vehicle, a car with a lead-acid battery and a con-

ventional (small) engine, does not suffer from the range problem of the pure electric car, but is, within city limits, a battery-powered car with all its good features. Hybrid vehicles are, of course, well known in larger sizes (buses), and considerable interest is shown in them, but at present no small vehicle is sold anywhere. A 6- to 8-kW motor generator (200 lb) can supplement an 84-V, 100-Ah lead battery system (350 lb) in a 2,000-lb car, and the result is a 200-mile driving range in city traffic with minimal gasoline consumption (60 mpg or better) and no fear of getting stuck with an exhausted battery or of not being able to drive certain short stretches of high-speed highway (55 mph). The author has had ample experience with a hybrid combination of the stated generating power and battery weight relationship (see Sec. 4.2) and found it satisfactory. Experiments with a 3.5 kW commercial ($500) motor generator were abandoned because of the handling and noise problems which this engine (designed for outdoor use) created [34]. However, there is no doubt that small, quiet engines can be built (or present designs properly modified [35,36]).

Rapidly increasing gasoline prices will soon make the additional costs of a dual power system bearable. For an example, see Sec. 6.3.

4. BATTERY SYSTEMS

4.1. Lead-Acid Batteries

This section gives only a short description of the lead-acid battery system from the viewpoint of the person interested in electric vehicles. The reader is directed to Chapter 1 for the theoretical background and detailed information about the construction and performance of the batteries.

Practical information about sizes, maintenance, charging, and repair (when possible) are published by the battery manufacturers and can be obtained from their service manuals. Many of the following data and descriptions are taken from battery handbooks and are referenced accordingly.

Definitions, standards, and specifications for lead-acid batteries are published by the Battery Council International (BCI), formerly the American Association of Battery Manufacturers (AABM), and the Society of Automotive

Engineers (SAE). They are updated when needed and are available in most libraries.

4.1.1. Electrochemical Information

The theory of galvanic elements may be found in textbooks such as Ref. 37. The electrochemical processes are illustrated in the schematic diagram of Fig. 28. The overall chemical reaction can be written as follows:

$$2PbSO_4 + 2H_2O \underset{\text{discharge}}{\overset{\text{charge}}{\rightleftarrows}} Pb + PbO_2 + 2H_2SO_4 \tag{12}$$

In the manufacturing process a lead oxide paste consisting of PbO and/or Pb_3O_4 mixed with sulfuric acid is pasted onto a lead grid (usually a Pb-Sb alloy) and charged ("formed") in sulfuric acid solution. Porous PbO_2 is produced on the positive plate and spongy lead on the negative plate. PbO_2

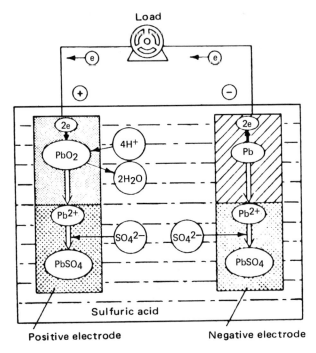

FIG. 28. The discharge process in a lead-acid storage battery [37].

and Pb are good conductors (4×10^{-3} Ω-cm and 2×10^{-5} Ω-cm), and sulfuric acid is also a highly conductive electrolyte (1.3 Ω-cm). $PbSO_4$, the discharge product on both electrodes, is an insulator. The plates must have a large inner surface (10 m^2/g) based on a high porosity (60%), or the coating with $PbSO_4$ would soon block the available electrode areas. Sulfuric acid is used up when the battery is discharged; the state of charge can therefore be determined by measuring the specific gravity of the electrolyte. "Battery acid," 35% H_2SO_4, has a specific gravity of 1.260 at 20°C; a fully charged battery usually shows this value when tested with a hydrometer. A temperature correction must be considered: plus or minus three digits in the third decimal place for each 10°F lower or higher. A discharged battery has its sulfuric acid concentration lowered to about 15% (sp gr 1.110).

Table 9 shows the relationships among specific gravity, state of charge, and cell voltage [34]. While the theoretical equivalent capacity of PbO_2 can be calculated to 4.47 g/Ah (0.475 cm^3/Ah) and that of Pb to 3.86 g/Ah (0.34 cm^3/Ah), it is obvious that in reality the amount of sulfuric acid is capacity-limiting for the system.

Table 10 shows the electrochemical equivalent of sulfuric acid based on Eq. (12). Since we cannot go under a specific gravity of 1.1 (for conductivity reasons) and not much over 1.30 (for self-discharge and stability

TABLE 9

Steady State Open Circuit Voltage (RT)

State of charge	Specific gravity of H_2SO_4 electrolyte	Cell voltage
100%	1.260	2.11
87	1.240	2.09
75	1.220	2.07
62	1.200	2.05
50	1.180	2.03
37	1.160	2.01
25	1.140	1.99
12	1.120	1.97
0	1.100	1.95

TABLE 10

Electrochemical Equivalent of Solutions of Sulfuric Acid at $23°C$

Specific gravity	Ampere-hours per liter	Specific gravity	Ampere-hours per liter
1.040	17	1.220	100
1.060	26	1.240	110
1.080	35	1.260	120
1.100	44	1.280	130
1.120	53	1.300	141
1.140	62	1.320	151
1.160	71	1.340	162
1.180	81	1.360	173
1.200	90		

reasons), the differential value of capacity is theoretically 100 Ah/liter; the practical value depends on cell construction and discharge rates.

4.1.2. The Materials in a Battery

It is very instructive to discuss the make-up of a typical commercial lead-acid battery [34]. If one buys a typical 12-V (6-cell) dry-charged auto (starter) battery [38] and makes a weight breakdown, he may find:

Battery: Total weight: 47 lb, rated capacity: 84 Ah (20 h)

Cell: Inside space of each cell: $19 \times 4 \times 16.8$ cm
          7 negative plates, weights: $142 \times 7 = 994$ g
          6 positive plates, weights: $176 \times 6 = 1,054$ g
          12 separators, weights: $7 \times 12 = 84$ g

Each dry negative plate (15.5×13.5 cm) contains 56.5 g of lead grid, 76.5 g of active (porous) lead.

Each dry positive plate (15.5×13.5 cm) contains 72.4 g of lead grid, 103.6 g of active PbO_2 powder.

The sulfuric acid amounts to 850 ml or 1,080 g of H_2SO_4 (sp gr 1.28; 150 ml are above the plates, 400 ml between, and 300 ml in the pores (measured by draining after filling).

The theoretical ampere-hour capacities calculated from the active materials amount to 139 Ah for the lead and 136 Ah for the PbO_2. From the H_2SO_4

weight it can be calculated that the rated discharge capacity (84 Ah) would change the original 37% or $5\underline{M}$ H_2SO_4 (sp gr 1.28) to a 12% or $1.3\underline{M}$ H_2SO_4 (sp gr 1.08). The sulfuric acid is therefore utilized in conformity with the "rules" stated above. The active materials are only utilized to the extent of 60% at the 20-h rate.

A complete material list is given in Table 11.

The following points should be remembered:

The active materials make up only one-third of the total weight; their utilization is about 60% even at a slow discharge rate.

The sulfuric acid weight is also one-third of the battery weight.

The remaining weight is distributed among the grids, collectors, and cases.

4.1.3. Commercial Lead-Acid Batteries

Commercial batteries are classified with respect to ampere-hour capacity at a given discharge rate. The "rate" is the quotient of capacity over time and has the dimensions of current. C/20 rate means that the battery is used at a current which leads to a complete discharge in 20 h.

An 84-Ah battery should operate at a current of 4.2 A for 20 h. The same battery does not give, for example, 3 h of service at a 28-A load current. Due to less efficient usage of the active material (earlier coating with a blocking sulfate layer, diffusion hindrance, etc.), the discharge time

TABLE 11

Material (%) in the Lead-Acid Battery (47 lb = 21.3 kg) [34]

		Weight (g)	Percent
1.	Active material		
	PbO_2	3,710	17.3
	Pb	3,200	15.0
2.	Grid (current collector)	5,000	23.5
3.	Electrolyte		
	H_2SO_4 solution ($5\underline{M}$, sp gr 1.28)	7,025	33.0
4.	Connecting bars and terminals	1,015	4.93
5.	Plastic case	1,300	6.10

will be shorter. In such a case the capacity of the battery can only be determined experimentally. Assuming that the actual value would be around 2 h, the statement would read: this battery is a 56-Ah battery at the 2 h rate: C/2 is then 28 A. The same battery may have a capacity of only 40 Ah at the half-hour rate (80 A).

Not only the lower utilization of active materials, but also the increasing electrolyte resistance at the end of discharge contributes to the shorter operating time: the battery is useful only above a certain voltage, for example, 1.8 V, and is severely damaged if discharged below. Low temperatures aggravate the situation.

Figure 29 shows typical ampere-hour yield curves for lead-acid cells at various temperatures. This situation is encountered in the operation of electric vehicles, and the need for establishing a different rating for vehicle batteries is obvious because the 20-h rate found in most catalogues does not mean much. However, the "cranking power" in minutes (listed for the auto-

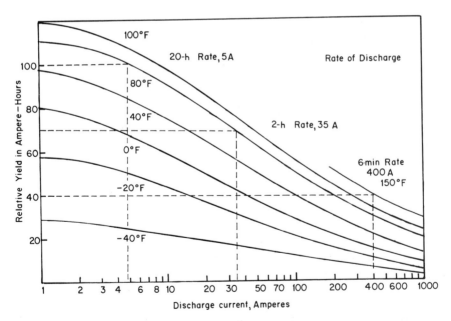

FIG. 29. Rate of discharge versus ampere-hour capacity (yield) at different temperatures. (Courtesy of E. Y. Weissman.)

mobile starter battery) is indicative of the maximum load capability, especially at low temperatures.

4.1.3.1.   The Automotive SLI (Starting-Lighting-Ignition)
               Lead-Acid Battery

The automotive SLI (starting-lighting-ignition) battery features a thin-plate
construction to optimize the high-rate drain characteristics, thereby increasing specific power (watts per unit weight and volume).   These batteries,
particularly those fabricated with polypropylene cases, have excellent
energy densities, generally higher than other lead-acid types: from 12 to
22 Wh/lb (28 to 48 Wh/kg) and from 0.8 to 1.6 Wh/in.^{3} (50 to 100 Wh/liter).
These batteries, because of high-speed production techniques and the use
of automated equipment, have the lowest initial cost per watt-hour of any of
the lead-acid systems.

In all cases the batteries are available either as fully charged,
electrolyte-filled units, or in the dry-charged state (the electrodes are
fully charged and the battery is ready for use after the electrolyte is filled
in).

A comparatively recent innovation has been the introduction of a polypropylene container as a replacement for the traditional hard rubber case
in automotive batteries.   This permits thinner walls, resulting in a greater
number of plates, more sulfuric acid, and better impact strength.   The
weight savings are 8 to 10%.

Table 12 shows the general features of SLI automotive batteries [39].
For most applications the density of the acid is 1.265.   In tropical areas
the acid used is generally of a lower specific gravity, about 1.225, to minimize grid corrosion and separator deterioration.   Separators deteriorate
more rapidly at higher operating temperatures and acid gravities.

The 12-V version of a first-line (high-quality) SLI battery produced in
the United States is characterized by the data shown in Table 13 [39].

The low-temperature (0°F = -17.78°C) cranking performance provides
an indication of a battery's high-rate capability.   By definition, the cranking performance at 0°F is the number of amperes a battery can sustain for
30 sec at 0°F, at a voltage not lower than 1.2 V per cell.   This rating has
been established by the SAE (Society of Automotive Engineers) and the BCI

TABLE 12

SLI Automotive Batteries: General Features [39]

Open circuit voltage (V)	2.05-2.10
Open circuit battery voltage (V)	6, 8, 12
Nom. operating voltage at 20-h rate (V)	1.98
Nom. end-of-chg. voltage at 20-h rate (V)	2.53
Recom. chg. rate	C/20 or higher—do not exceed electrolyte temp. of 125°F or overcharge recommendation.
Recom. overchg. at 20-h rate (%)	5 - 40
Recom. dischg. temp. (°F)	70 - 90 (21 - 32°C)
Functional dischg. temp. (°F)	-40 - 140 (-40 - 60°C)
Optimum chg. temp. (°F)	50 - 115 (10 - 46°C)
Permis. chg. temp. (°F)	-40 - 125 (-40 - 52°C)
Recom. trickle chg./float/finishing rate	Trickle chg. C/100—continuous overcharging not recommended
Recom. storage temp., wet (°F)	-40 - 115 (-40 - 46°C)
Permis. storage temp., wet (°F)	-40 - 120 (-40 - 49°C)
Recom. storage temp., dry (°F)	-40 - 115 (-40 - 46°C)
Self-dischg. rate/mo. at R.T., wet (%)	5 - 11
Self-dischg. rate/mo. at R.T., dry (%)	1 - 5
Wh/lb	12.7 - 21.8 (28 - 48 Wh/kg)
Wh/in.^{3}	0.79 - 1.6 (48 - 98 Wh/l)
Impedance (Ω)	0.0010 - 0.0024
Cycle life (cycles)	150 - 250
Electrolyte	Sulfuric acid solution—sp gr 1.265 - 1.300

TABLE 13

Capacities and Dimensions of 12-V SLI Batteries [39]

BCI group size	Engine CID up to:	Cranking perform. at 0°F (A)	Reserve cap. (min. at 80°F)	Plates per batt.	Shipping wt. (lb) Wet chgd.	Dry chgd.	Elect. (qt)	Overall dim. (in.) L	W	H	Ref. 20-h rate (Ah)
22F	355	355	90	66	36	25	4.6	9-1/2	6-7/8	8-1/4	61
24	450	450	138	78	45	31	5.5	10-1/4	6-7/8	8-3/4	81
24F	450	450	138	78	45	31	5.4	10-3/4	6-7/8	8-3/4	81
27	515	515	165	90	52	35	6.5	12	6-7/8	8-3/4	93
27F	515	515	165	90	52	35	6.5	12-1/2	6-7/8	8-3/4	93

(Battery Council International). As a general rule, the cranking performance in amperes is on a one-to-one relationship with the CID (cubic inch
displacement) of an automobile engine; that is, if the CID of an internal combustion engine is 312 (5.1 liter), the battery selected for starting that engine
should have a cranking performance of no less than 312 A.

The reserve capacity rating has been established by the reserve capacity
test. This test is conducted as follows: A fully charged battery is discharged at 25 A to a cut-off voltage of 1.75 V per cell. The number of minutes the battery operates under those conditions constitutes the reserve
capacity. The reserve capacity provides a measure of the length of time an
automobile can operate in the event of a failure of the charging system, with
its head and tail lights and its electric windshield wipers turned on.

a. High Temperature. Although the power output of a lead-acid battery is
greater at higher temperatures, persistent operation at elevated temperatures reduces cycle life, primarily through the accelerated corrosion rate
of the grid members.

b. Cycle Life. The cycle life of an automotive, thin-plate battery is between 150 and 250 cycles, when the depth of discharge is 50% or greater,
and the temperature 80°F. The cycle life would be greater if the depth of
discharge were less. Under the type of service for which automotive batteries are specifically designed, the life ranges from 2 to 5 years. The
ordinary duty cycles encountered in automobile ignition service are high-
rate, shallow discharges, and the charging is voltage-regulated and
temperature-compensated, the setting of the regulator for a 12-V battery
being generally 14.4 V.

A graph displaying the relationship between cycle life and depth of discharge is shown in Fig. 30. The depth of discharge in automobile SLI service is usually less than 20%, unless an excessive use of accessories
causes an occasional deep discharge cycle.

c. Self-discharge. The self-discharge rate of a lead-acid battery is
temperature-dependent. At room temperature the self-discharge rate for
a new battery ranges from 5 to 11% loss per month. For the dry-charged
battery, the rate of self-discharge is about 1 to 15% loss per month, depending on sealing and the addition of oxidation inhibitors. Self-discharge of a

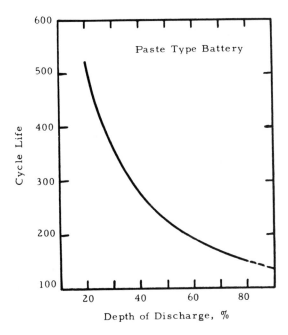

FIG. 30. Cycle life versus depth of discharge (SLI automotive battery) [40].

wet lead-acid battery is caused primarily by the dissolution of antimony from the grids and its subsequent deposition on the negative plates, where it causes electrochemical reactions that discharge the lead plates. An additional cause of self-discharge is the presence of impurities in the electrolyte, e.g., heavy metals such as silver, copper, nickel, or platinum.

Dry-charged batteries must be sealed, otherwise they self-discharge through the reaction of oxygen in the air with the negative plates. This reaction is accelerated by temperature. Dry-charged batteries should be stored at a temperature between 60 and 90°F. Figure 31 shows the rate of decrease of both the specific gravity and voltage within the 90-day test period.

Comparing the curves of Fig. 31 with the state of charge data of Table 8, it can be seen that a cell may have lost 30% of its initial capacity over a standing time of 3 months at room temperature. Elevated temperatures cause far higher losses.

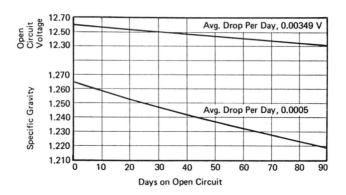

FIG. 31.   SLI automotive battery, open circuit voltage and specific gravity versus time [39].

4.1.3.2.   Traction (Motive Power) Batteries

Table 14 shows the general features of a motive power battery used for traction purposes such as materials, mine locomotives, tractors, shuttle cars, floor-cleaning equipment, personnel carriers, and transport vehicles (in airports), lawn mowers, and golf carts. The capacity is based on the 6-h discharge rate, with a cut-off voltage of 1.70 V per cell, and at a specific gravity of 1.280 to 1.290.

The energy density of traction batteries at the 6-h rate is between 8.6 and 11.0 Wh/lb (19-24 Wh/kg) and between 1.0 and 1.4 Wh/in.^{3} (60-85 Wh/ liter). In addition to the industrial-type motive power cells, specially designed batteries are available for powering golf carts, personnel carriers, and other applications requiring both excellent mechanical endurance and high energy outputs. Specific dimensions and performance data on these are presented in Table 15.

Processing methods for the active material used in motive power cells are essentially the same as those used in manufacturing other lead-acid batteries; however, the cell construction is different to assure a longer cycle life. The positive plates are vertically and horizontally wrapped with a glass-sliver tape and fiberglass mats which reduce shedding of the positive material under abusive operating conditions. Additional protection is provided by encasing the positive plates with perforated plastic envelopes. The edges of these envelopes are solid, reducing the probability of the oc-

TABLE 14

General Features of Traction Batteries [39]

Open circuit voltage at 1.280 sp gr at 77°F (V)	2.12
Nom. operating voltage (V at 6-h rate)	1.94
Typical end-of-chg. voltage (V)	2.55 - 2.65
Recom. initial chg. rate to a dischgd. batt.	22.5 A per 100 Ah of capacity at 6-h rate
Recom. finishing rate (A at 6-h rate)	5A
Recom. overchg. (%)	5 - 15
Recom. dischg. temp. (°F)	70 - 110 (21 - 43°C)
Functional dischg. temp. (°F)	0 - 110 (-17.8 - 43°C)
Optimum chg. temp. (°F)	75 - 110 (24 - 43°C)
Permis. chg. temp. (°F)	50 - 110 (10 - 43°C)
Recom. storage temp., wet chg. (°F)	30 - 77 (-1 - 25°C)
Permis. storage temp., wet chg. (°F)	-40 - 110 (-40 - 43°C)
Recom. storage temp., dry chg., (°F)	32 - 100 (0 - 37.8°C)
Self-dischg. rate/mo. at R.T., wet chg. (%)	7 - 10
Self-dischg. rate/mo. at R.T., dry chg. (%)	2 - 4
Wh/lb at 6-h rate	8.56 - 11.0 (19 - 24 Wh/kg)
Wh/in.³ at 6-h rate	1.08 - 1.37 (66 - 84 Wh/l)
Cycle life (cycles)	1000 - 2000
Electrolyte	Sulfuric acid solution—sp gr 1.280 - 1.290 at 77°F

TABLE 15

Capacities and Dimensions of High-Energy Traction Batteries [39]

BCI group size	Voltage (V)	Cap. (min.) at		Dischg. rate (A)	Reference 20-h rate (Ah)	Calc. 1-h rate (Ah)	Plates per batt.	Ship. wt. (lb)	Elect. (qt)	Overall dim. (in.)		
										L	W	H
Polypropylene												
GC2	6	100	at	75	220	110	63	63	6.5	10-3/8	7-1/8	10-7/8
GC2	6	82	at	75	180	90	51	57	7.0	10-3/8	7-1/8	10-7/8
27	12	150	at	25	90	55	78	53	6.1	12	6-13/16	8-3/4

currence of "moss" shorts. As a further precaution against shorting, the
bottom edge of each positive electrode is inserted into a plastic shield.

Synthetic microporous separators are used between plates of cells.
Their ribbed surfaces are positioned against the positive electrodes, providing channels for uniform electrolyte flow.

A comb-shaped element protector is located above the plates to prevent
top "moss" shorts, and to prevent damage to the elements when hydrometers, thermometers, and watering devices are used.

The sediment bridge is undercut so that sediment may settle uniformly
below the element. This increases the sediment reservoir and permits
better use of the available space and less chances of shorting plates.

Cell jars are molded of high-impact plastic material, with molded
rubber and heat-bonded seals. The lightweight construction leads to a
higher energy density.

a. Discharge Characteristics [39]. Figure 32 shows a typical discharge
diagram of a traction cell at 77°F. Figure 33 shows the behavior at temperatures from -50 to +110°F. Figure 33 indicates that the battery performance at freezing temperatures is considerably lower than at room temperatures. For higher rates than the 6-h discharge shown, the losses are very
much increased. This feature is in contrast to the relatively good behavior
of the automotive SLI battery built for high-current output even at lowest
temperatures. The reason for this difference is not an electrochemical

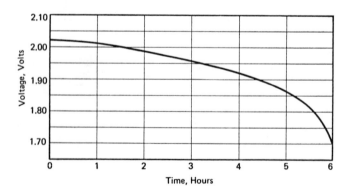

FIG. 32. Voltage versus time diagram of a traction cell (6-h rate, 77°F) [39].

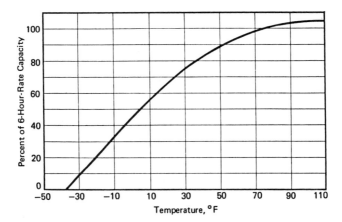

FIG. 33. Performance of traction cell, capacity versus temperature [39].

one; it is mainly a result of the lower plate number (higher resistance and increased polarization).

b. Self-discharge. The self-discharge of a wet, fully charged battery is about 7 to 10% per month at room temperature. For dry-charged batteries, the self-discharge rate is between 2 and 4% per month. Because self-discharge is accelerated at higher temperatures, batteries should be stored at room temperature or below. Dry-charged batteries should also be stored at low humidity.

A fully charged battery unused long enough to have a specific gravity drop of 0.025 points should be charged at the finishing rate until the specific gravity remains constant for 4 h. A low charging rate is necessary because of the formation of a hard, gritty sulfate on the negative plate, a condition which closes the pores of the plate.

c. Cycle Life. The cycle life of a traction battery is usually given as 1,000 to 2,000 cycles, even for deep discharge conditions. For train batteries, which use a special type of electrodes in which the active mass is supported and even confined in tubes (armored, clad-type), the cycle life can be as high as 4,000 cycles for 30% depth of discharge. Figure 34 shows a typical diagram of cycle life versus depth of discharge [40]. The difference between SLI batteries and traction batteries is approximately a factor of 10 in cycling capability (compare Fig. 30 with Fig. 34).

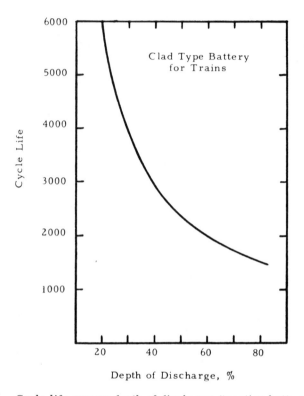

FIG. 34. Cycle life versus depth of discharge (traction battery) [40].

4.1.3.3. Electric Vehicle Batteries

The "propulsion battery" especially constructed for the use in on-the-road electric vehicles is probably not yet produced by any manufacturer. The technology to manufacture a battery especially for electric vehicles exists today, a statement which is enhanced by the SLI type and the traction type on the opposing ends of present batteries (see Sec. 6.4). There is obviously not enough demand at the present time to invest the capital for developing a really optimized battery for electric vehicles—better cycle life than the SLI type, but better heavy load characteristics than the traction-type battery. An argument can be made for the suggestion of using both in the same vehicle in parallel sets (hybrid operation). The SLI battery (which costs about one-third to one-half as much as the traction battery) will prevent the damaging action of cell reversals in the traction battery on peak loads. It must be realized that the 1,000 cycles (or more) claimed for a train or golf cart battery are not obtainable in the electric car application. Especially with chopper circuits for speed control, demanding high peak currents (of short duration, but causing considerable i^2R losses), the picture shifts in the direction of the high-quality starter battery as providing a more economical service.

"Electric vehicle batteries" offered by some manufacturers must therefore be critically judged with respect to their "origin." The batteries listed in Table 14 are labeled "electric vehicle batteries," which may give some indication but no assurance as to their suitability in on-the-road electric automobiles. The diagram showing Ah capacity versus various discharge rates of such an "electric vehicle battery" is shown in Fig. 35. It demonstrates the attempt to obtain a performance superior to general traction batteries.

The curve is essentially the same as the 80°F curve in Fig. 29; the extrapolation to the 20-h rate would give approximately 120 Ah capacity (100%). However, there is no commitment to any cycle life number for the 1/2-, 1-, or 2-h rate, which is encountered in stop-and-go city driving situations.

The electric vehicle batteries as shown in Table 14 are primarily designed for golf carts, for relatively low loads (75 A for the 6-V, 25 A for the 12-V version). Figure 35 (which shows the Ah capacity yield versus rate) does not show the watt-hour losses caused by the relatively high internal resistance of the golf cart batteries. They can be estimated from the voltage losses shown in Fig. 36 for a 100-Ah battery subjected to high discharge rates (in amperes).

It should be noted that the state of discharge of the battery of Fig. 36 was achieved by discharge at the 6-h rate, prior to the testing. Of course, lower temperatures aggravate the situation. However, "electrical vehicle batteries" are better than the curve in Fig. 33 indicates for "general motive power" batteries. At -30°F about 30% of the room-temperature capacity is still available. This feature again points to the efforts of the battery manufacturers to produce a compromise with available (technological) means at acceptable cost levels; using present large-scale production methods is essential from the economic point of view.

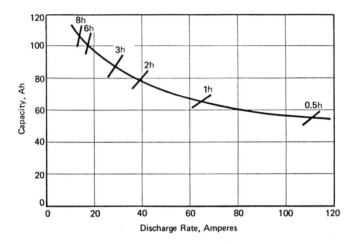

FIG. 35. Capacity versus discharge rate (electric vehicle battery):
100-Ah battery at 77°F [39].

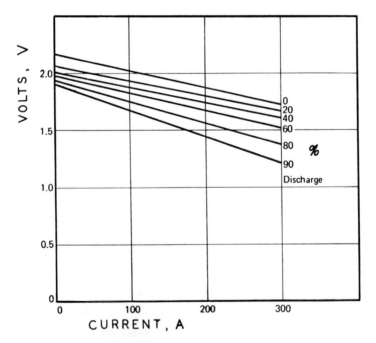

FIG. 36. Cell voltage and current versus percentage of discharge:
100-Ah electric vehicle battery at 77°F [39].

4.1.4. The Charging of Lead Batteries

4.1.4.1. General Rules

In order to recharge a storage battery it is necessary to pass direct current through the cells (in opposite direction) for a time needed to replace the ampere-hours discharged plus an excess of 5 to 20%, depending on the previous rate of discharge, age of the battery, temperature, etc. Excessive gassing and overheating (over 125°F) must be avoided. A high voltage or excessive charging rate can damage the voltage-control equipment.

Figure 37 shows the typical charging curve of a six-cell starter battery if charged at the 20-h rate. The change in specific gravity of the electrolyte is noted.

Figure 38 shows the change in end-of-charge voltage with temperature, again at the 20-h charge rate.

In practice, however, the constant-current charging method (20-h rate) is not used; it is actually a waste of time, since the batteries are capable of accepting very high currents at the earlier, lower state of charge, but as the end of charge approaches, the proper "finishing rate" becomes all-important for the life of the batteries.

Repeated excessive overcharge is harmful in several ways. The positive grid severely corrodes, causing a loss in electrical conductivity and loss in capacity. Increased temperature aggravates the situation. Hydrogen and oxygen gas evolved during overcharge, with its resultant turbulence,

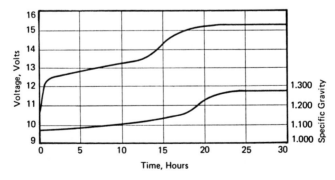

FIG. 37. Six-cell automotive battery, charge voltage at 20-h rate versus specific gravity [39].

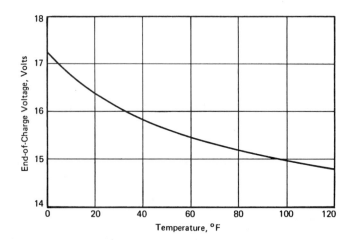

FIG. 38. Six-cell automotive battery, end-of-charge voltage at 20-h rate versus temperature [39].

tends to dislodge material from the plates; the escaping hydrogen and oxygen are an explosion hazard. Decomposition of water, and the high acid concentration thereby produced, is harmful to the plates and separators, besides requiring frequent water additions. The amount of overcharge a battery can safely accommodate is dependent on its history. A new battery can sustain a longer overcharge than one that has had long use.

A battery that has been stored for long periods develops a coarse-grained sulfate structure on both the positive and negative plates. Such batteries should be charged only at rates no higher than the 20-h rate.

Charging becomes even more difficult when a battery is stored in a discharged state for a long time. To avoid such problems, batteries should be charged as soon as possible after a discharge, and not be left in a discharged condition for more than 24 h, or below 32°F (0°C). The electrolyte of a discharged battery freezes easily.

4.1.4.2. Methods of Charging

a. Constant-current, Timer-controlled Charging. The constant-current, timer-controlled method is usable only for long charge periods (20-h rate and lower) and is rarely used in practice, except for equalization of charge

in a series of batteries, or for the reactivation of batteries which have not
been used for a long time.

b. Constant-potential Charging. The charge acceptance of a battery in a
low state of charge is very much different from a battery which is half
charged or nearly fully charged. Figure 39 shows the charge rates possible for a 100-Ah battery as a function of battery voltage. Higher rates
would damage the battery because of heat development—but the electrochemical system could accept them if cooling would be provided. The
ampere-hour rule assumes that there is a linear relationship between state
of charge (in percent) and amperes accepted (A/100-Ah capacity). A 50%
discharged battery of 100 Ah full capacity can then be charged with 50 A, a
25% discharged battery with 25 A, and so forth. However, considerable
deviations are observed with the newer lead-calcium systems.

c. Modified Constant-potential Charging. The modified constant-potential
method is used for multiple batteries with either a motor generator or a

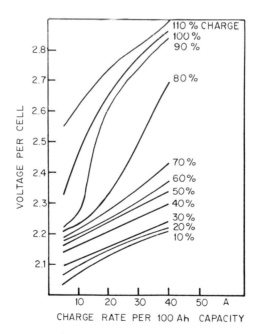

FIG. 39. Charge-rate characteristics of lead cells [41].

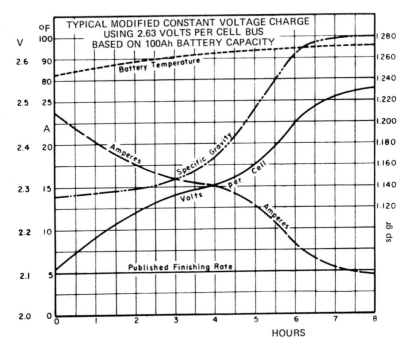

FIG. 40. Modified constant-potential charging characteristics [41].

rectifier power source. It is wasteful to a certain extent because ballast resistors are used, but it is flexible and adaptable to the time available for charging. The danger of overcharge is reduced since the charge current inherently tapers off. Figure 40 shows the voltage, current, specific gravity, and temperature characteristics during a typical modified (restricted) constant-voltage charge [41].

d. The Two-rate Method. The two-rate method uses a two-resistor arrangement to switch the rate of charge, usually when 2.7 V per cell is reached.

e. The Taper-charge Method. The taper-charge method is similar to the modified constant-voltage charge, but no resistors are used. The power source (generator) provides the correct current/voltage characteristic by a ferro resonance, or reactance limitation. The latter adapts better to the batteries at different temperatures, of various age and maintenance condi-

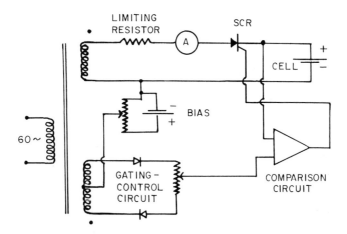

FIG. 41. Battery charger with resistance-free voltage control [43].

tions; the finishing current is kept within closer limits, in spite of the possible differences in battery characteristics [42].

f. The Resistance-free Voltage-sensing Charge Method. The resistance of free voltage-sensing method is a recently studied method which uses a pulse current charge with voltage sensing between the pulses, thereby eliminating the variable (age, construction, state of charge dependent) internal resistance of the battery [43]. It can be used in combination with the previously mentioned resistive regulated methods (b, c, d). Figure 41 shows the principal circuit diagram of the charger.

4.1.4.3. State-of-charge Monitoring

For an electric vehicle it is very important to monitor the discharge of the batteries. The simplest method is the use of an ampere-hour indicator, as produced by several manufacturers, based on various integration techniques: motor-driven counter (Sangamo Electric Co.), mercury or cadmium salt coulometer (Sprague Electric Co.), or electronic integration circuits based on capacitor charges or other memory devices (Curtis Instrument Co., Bisset-Berman Corp., Gould, Inc.) [44].

Unfortunately, the use of counting discharge devices depending on the elapsed time/current summation has a big disadvantage: it does not consider the changes in the battery depending on age. Temperature compensation can be built in, rate consideration (lowering the capacity available) can also be programmed; a 10 to 20% overcharge is added each time of recharging. Still, these devices fail to sense the condition of the battery and must be reset frequently if any reliability is essential.

A foolproof method of sensing has not been found in spite of serious work and efforts (especially for space and military applications where the state of charge information is essential to fulfill the mission).

Pulse current tests have been suggested and developed [45,46], but the the resulting equipment was too large for mobile use and not reliable under all conditions encountered in the field.

Specific gravity sensing seems to be the most logical state-of-charge testing procedure. However, when measured electrically, the resistance goes through a minimum value during discharge (sp gr 1.200 to 1.300, depending on temperature), and the reading must be confirmed by a second information (e.g., load pulse-voltage drop test, preferably resistance-free). An instrument of that type, small enough to be incorporated into the instrument panel, was developed for testing the German BMW electric car [47].

The loaded voltmeter test is still the simplest way of judging the condition of the vehicle battery, and its reliability is generally fair.

A simple state-of-charge indicator is obtained by using a dc meter with two coils, one acting on the moving in response to the voltage, the other in response to the voltage drop caused by the current across a shunt, but in the opposite direction, thereby compensating the voltage drop under load when the voltage is fully charged, but showing the increasingly overriding voltage drop when the battery gets discharged [34]. The meter indication is, of course, valid only when the car is moving and a reasonable load is being drawn from the battery. As experience shows, this simple device shows clearly when one of a series of several batteries is nearly discharged or some single cells reverse. This is the most important indication for the electric car driver: nondetection ruins the battery irreversibly. None of the integrating coulometric or elapsed time measuring devices show it, specific gravity indicators only if individual cells are monitored. Figure 42 shows the instrument circuit [48].

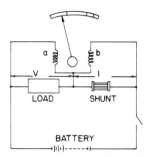

FIG. 42.   Circuit of loaded voltmeter state-of-charge indicator [40].

4.1.4.4.   Safety Considerations During Charging

Stationary chargers are usually well grounded and equipped with a separation transformer at the power-line input to prevent any accidental "hot" connection to the device charged.   Small electric vehicles with on-board chargers usually follow the same safety precautions, plus an additional ground cable.   When the vehicles are larger, the power transformer needed for the charging requirements (15 to 20 A) gets large and heavy, and it is therefore easy to understand why amateur designers of electric automobiles are looking for ways to avoid the transformer and charge directly from the 115-V ac line.   The on-board circuitry then becomes extremely lightweight and inexpensive; all that is needed (if the vehicle has a 72-, 84-, or 96-V battery) is a two-phase rectifying circuit (four diodes) and a limiting resistor to avoid blowing fuses in the home circuits which may often be laid out only for 15 A.

The danger with such circuits is that the vehicle battery is then connected to the 115-V supply line.   Any accidental short (wetness!) with the vehicle body, which is only poorly grounded (rubber tires!), makes it dangerous to touch the vehicle.   Center grounding of the battery is possible with separation transformer circuits.   The only way to assure that the vehicle is on a good ground connection is to use a three-prong connector plug whereby the third wire is connected to the car body.   Unfortunately, some household circuits have only two-prong receptacles, and even three-prong connections may not be wired properly.   A switch circuit is therefore needed, preventing any power connection to the on-board charger if the car is not on a good ground.

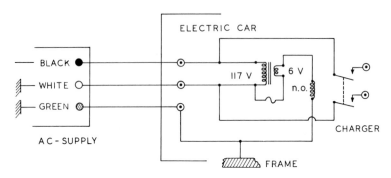

FIG. 43. Switch circuit for on-board charger.

The author has experimented with the protective circuit shown in Fig. 43 and found it to work well. Deliberately wiring the receptacle incorrectly (exchange of black and white) immediately blows the circuit fuse, and the power connection is limited to the very short time period until it blows.

4.1.5. Battery Circuits

Assuming that one would use the 6-V, size GC 2 (180-Ah), golf cart battery from Table 14 as a unit for building an electric car for city use (commuter car), the batteries could be assembled in series as the upper part of Table 16 shows.

The same system could be built from two strings of 12-V batteries of the size 27 (90-Ah) also listed in Table 14. The question arises: what is

TABLE 16

Lead Battery System for Electric Car

No. of batteries	Voltage V	Weight (lb)	Weight (kg)	Ah (1-h rate)	kWh (1-h rate)
6 V ⎧ 10	60	570	260	90	5.4
6 V ⎨ 12	72	684	310	90	6.5
6 V ⎩ 14	84	798	358	90	7.6
12 V, 7 + 7	84	742	328	110	9.2

more beneficial for the vehicle user with respect to performance, life expectancy or cost of the batteries. Offhand, the 6-V batteries are less advantageous, because fourteen 12-V batteries weigh less and have a considerably higher 1-h rating. The reason is the higher number of plates for the 20-h capacity: 17 for 180 Ah and 13 for 90 Ah (in parallel).

On second thought, the added connectors and cable weight needed for two strings must be considered and the lower amount of electrolyte in the 12-V cells. The nominal discharge rate (75 A versus $2 \times 25 = 50$ A) is higher for the 6-V batteries; however, the ampere-minute product ($82 \times 75 = 615$ versus $150 \times 50 = 750$) is better for the 12-V batteries.

Unfortunately, there are not sufficient cycle life data available to compare the performance on that basis, but assuming that the batteries are similar in their construction and that only the plate number is different, the 12-V batteries would be preferred since they would stand up better in a peak-current stop-and-go traffic pattern.

Mathematically the problem of reliability can be stated as follows: r is the reliability of a single cell with respect to an operational time event. s cells in series have the total statistical reliability R_s.

$$R_s = r^s \tag{13}$$

p cells in parallel have a total statistical reliability R_p.

$$R_p = 1 - (1 - r)^p \tag{14}$$

Examples: If 95 cells out of 100 reach the operational time event (e.g., fulfilling the requirement of 1,000 cycles), their reliability $r = 0.95$. Three cells in series have only the reliability $R_s = 0.95^3 = 0.86$, but two cells in parallel have the increased reliability $R_p = 1 - 0.05^2 = 0.997$.

There is the additional combination: series/parallel,

$$R_{sp} = 1 - (1 - R_s)^p = 1 - (1 - r^s)^p \tag{15}$$

and parallel/series:

$$R_{ps} = R_p^s = [1 - (1 - r)^p]^s \tag{16}$$

As an example, the reliability of three cells in series, connected parallel to another series of three cells, $R_{3,2} = 0.98$. The capacity of the series-chain cells is chosen as half the desired total capacity in order to end up with the same ampere-hour figure after paralleling the chains.

The same considerations are valid for the number of plates in a cell. The operational life expectancy (cycles to a given end point) of a cell with 33% more plates is increased about 10% (with the same materials used). On the other hand, a battery with six cells (12-V battery) has a lower reliability than a three-cell (6-V) battery.

Further reliability considerations can be applied to the possibility of interconnecting the series and parallel strings to form a network. The reliability increases again, due to the bypasses available when batteries fail in certain network positions. For space power supplies these theories have been applied, but for electric car batteries the network may get too complicated—and too heavy with connecting cables.

4.1.6. Improvements in the Lead-Acid Battery

4.1.6.1. Present and Near Future

A better compromise between the SLI battery and the golf cart battery may be achieved if the on-the-road vehicle is optimized; this is a question of commercial expediency.

Near-future improvements are just a continuation of already existing trends in battery technology. The goals of development efforts to improve the vehicle battery are as follows:

1. Lower specific current densities: thinner plates, larger surface plates (grooved, slit, etc.).
2. Lower cell resistance: better grid design, intercell connectors, bipolar plates, closer electrode spacing, separators with open structures.
3. Construction changes: lightweight grids (Al?), plastic cases, redesign of terminals.
4. Better electrolyte utilization: optimal plate porosity, circulation of electrolyte (e.g., by thermal effects).

5. Capability for faster recharging: heat-removal improvement, charger adaptation (accurate detection of state of charge).

6. Improved cycle life and insensitivity to deep discharge: improved electrode support.

7. Better charge efficiency and charge retention: modification in plate composition, new alloys, special additives.

Figure 44 shows the weight distribution (percent) of components contained in present and near-future Japanese lead-acid batteries. Both starter- and traction-type batteries are included in this comparison. These batteries were previously compared in cycle life (Figs. 30 and 34). It should be noted that the main improvements consist of an increase in the electrolyte volume (clad types 1, 2, 3) and in an improvement of the utilization of the active materials (fractions a and b in the section of PbO_2 and Pb in all types).

The claims for energy densities (Wh/kg) listed in the headings of Fig. 44 need some careful examination: they are not based on the same power rating (W/kg). To get the true picture, Fig. 45 must also be consulted. The same batteries (nos. 1 to 6) are shown, but now with their energy/power output relationship.

4.1.6.2. Distant-future Improvements (Possibilities)

The energy per weight probably cannot be changed much; after reduction of the heavy grid structures, the limit of the capacity is reached. Improvement in power output is a function of the plate number which can be built economically (in competition with the required collector weight).

Functional improvements, however, seem to be possible. It is well known that a reduction of the specific gravity of the electrolyte improves cycle life and corrosion resistance. Twenty to 80% improvements have been observed with low (1.240) density electrolytes [51]. On the other hand, the capacity depends on the sulfuric acid concentration. The solution to this dilemma is to provide an outside tank with extra electrolyte and circulate it through the vehicle battery. Technically, such electrolyte circulation systems are well known (See Sec. 4.2), and the resulting parasitic losses can be kept very low during operation, reduced to nearly zero by a proper design of the pumping system during idle periods (air traps). It is

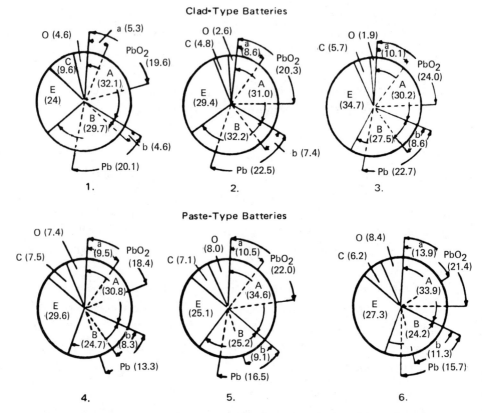

FIG. 44. Present and near-future lead-acid batteries [49]. 1. Present battery for train (23.8 Wh/kg). 2. Improved battery A for electric car (38.4 Wh/kg). 3. Improved battery B for electric car (45.2 Wh/kg). 4. Present automotive battery (37.2 Wh/kg). 5. Improved battery A' for electric car (47.2 Wh/kg). 6. Improved battery B' for electric car (60.0 Wh/kg). Weight distribution in percentage. Lead-acid batteries recently developed in Japan for electric cars. A, cathode plate; B, anode plate; C, case; E, electrolyte; O, other material; Pb, active material in the anode; PbO_2, active material in the cathode; a, amount that the PbO_2 utilized for discharge; b, amount that the Pb utilized for discharge.

also well known that the best power performance can be obtained at elevated temperatures ($45°C$); it is also claimed that some silver and cobalt additives are beneficial during overcharging.

The circulating electrolyte may provide battery cooling in summer, at high loads, and heating in winter time, if a heat exchange system is em-

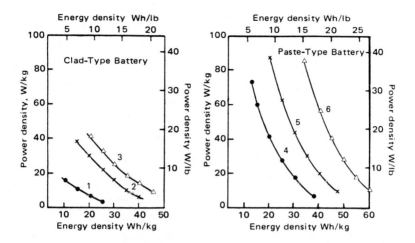

FIG. 45. Energy density versus power density of the batteries from Fig. 43 [50]. 1. Storage battery used now for trains. 2. Improved battery A for electric cars. 3. Further improved battery B for electric cars. 4. Ordinary automotive battery. 5. Improved battery A' for electric cars. 6. Further improved battery B' for electric cars.

ployed. A propane heater is used in luxury-electric vehicles today (it can also operate an air conditioner).

It is clear that such improved lead battery systems require a completely new and more expensive design. This may come about if the other future battery systems (which will be discussed in the next sections) cannot actually deliver what they promise today. At least it is satisfying to know that the end of the road has not necessarily been reached for the lead-acid battery.

4.2. Other Battery Systems

4.2.1. Introduction

If we consider electric automobiles as energy-saving vehicles, we may be right or wrong, depending on the type of battery used. Electric cars with rechargeable batteries are not more efficient than present gasoline engine cars: about 12 to 15% fuel utilization (sometimes less).

The power plant may be 40% efficient, transmission lines (in town) and transformation of power at the most 80%; the charge and discharge efficiency of batteries may be 60% (77 × 77), and finally the motor and speed controls 80%; result: a total of 16% efficiency, with optimized situations presumed!

Only electric vehicles with primary batteries (fuel cells) as power sources can have a significantly better fuel utilization. The reason is, as will be explained later, that a galvanic element is not a heat engine and therefore not subject to Carnot's law, which limits the efficiency of power plants and combustion engines to a theoretical value of approximately 60%, with practical values much lower (depending also on size).

A battery can convert the energy of its chemicals with over 95% efficiency theoretically, and over 60% practically. An electric car with a fuel cell power source (e.g., a hydrogen-air battery) can have an overall efficiency of 40 to 50%, three times higher than a car with rechargeable batteries, or a gasoline vehicle.

Another factor of importance is the refueling principle. Gasoline engines can fill their tanks in minutes. Batteries need hours for recharging, and while it is possible to exchange batteries in a short time, the task is a considerable one, requiring special equipment and a spare battery available from a "loan system" which must be maintained on a service station basis. Fuel cell systems have refillable tanks which can be restored to full capacity as fast as a gas tank can be pressurized or a reservoir can be filled with a liquid fuel.

With the capability to refuel a generator (or a fuel cell battery) goes the high capacity (energy content) of a system in which power generation and fuel supply are separate and the range of a vehicle with such a system depends on the size of the fuel supply, not the power producer.

Storage batteries have all their chemical energy stored in their plates. Doubling the capacity means doubling the weight and size. A fuel battery needs only to add an incremental portion of its weight to double the capacity. As an example, a 1-kWh lead battery weighs approximately 50 lb. A 20-kWh capacity requires 1,000 lb.

A fuel cell battery (H_2-air) capable of producing 1 kW continuously may weigh 100 lb (including accessories for operation), but the fuel needed for each kilowatt-hour produced weighs only 6 lb/kW (calculated for hydrogen, stored as compressed gas at 2,000 psi, including cylinder weight, assum-

TABLE 17

On-Road Vehicle Requirements (Summary)[*]

	Range (km)	(miles)	Speed (kmph)	(mph)	Energy density (Wh/kg)	(Wh/lb)	Power density (W/kg)	(W/lb)
Family car	322	200	97	60	190	87	132	60
Commuter car	161	100	97	60	62	28	68	31
Delivery van	97	60	64	40	73	33	79	36
City taxi	241	150	64	40	141	64	66	30
City bus	193	120	48	30	121	55	55	25

[*]From Table 6.

ing 50% cell efficiency). The required 20 kWh therefore are produced by a
power system weighing 220 lb. The requirements of gasoline-equivalent on-
the-road electric cars are summarized in Table 17. The data are from the
previously presented Table 6; the small "utility car" (shopping car), considered to be satisfactory with present-day lead-acid batteries, is now
omitted.

Using these energy density and power density requirements, it is now
possible to judge batteries as "possible candidates" for vehicle batteries,
with the automobile performance characteristics as the basis.

There are three well-established (alkaline) storage battery systems
commercially available. Table 18 compares their characteristics with the
lead-acid system.

TABLE 18

Energy Densities, Power Densities,

and Cycle Life of Rechargeable Batteries

	Energy density (Wh/kg)	(Wh/lb)	Power density (W/kg)	(W/lb)	Cycle life (cycles)
Lead-acid	22–35	10–16	22–176	10–80	300–2,000
Nickel-iron	22–26	10–12	66–88	30–40	2,000–5,000
Nickel-cadmium	26–44	12–20	220–660	100–300	1,000–3,000
Silver-zinc	44–176	40–80	220–440	100–200	100–200

The present nickel-iron system [52] used for traction purposes (mine locomotives) is too heavy for on-the-road cars, however the cycle life and durability of this sytem are excellent and much work is going on to improve the capacity [53]. An experimental Ni-Fe battery operating a passenger car was demonstrated recently [54]. Nickel-cadmium [52] and silver-zinc [55] systems are surpassing the lead-acid system in capacity, but their cost is too high. Several experimental and record-breaking vehicles have been constructed with silver-zinc batteries, proving the high energy (and power) density of this system. A powerful low-cost system would be the manganese dioxide-zinc (alkaline) element (35 Wh/lb, 77 Wh/kg); its principal rechargeability is assured, but limited to specific conditions and is presently not sufficiently explored [56].

4.2.2. Future Rechargeable Vehicle Batteries

Rechargeable batteries with twice and up to five times the energy density of the lead-acid battery are under development. They are discussed in the following sections.

4.2.2.1. Nickel-Zinc Batteries

The nickel-zinc system has an open circuit potential of 1.7 V, a theoretical energy density of 170 Wh/lb, and presently an achieved energy density of 30 to 40 Wh/lb. Over 200 cycles have been obtained [57]. Nonsintered, bonded nickel oxide plates of low cost have been developed [58]. The cycle life is limited by the tendency of the zinc electrode in alkaline electrolytes to grow dendrites during charging and short circuit the electrodes. These troubles are shared with all alkaline cells with zinc anodes and are subject of applied research, reflected in an extensive literature. The overall reaction in Ni-Zn cells with KOH electrolyte is

$$2NiOOH + Zn + 2H_2O \rightleftarrows 2Ni(OH)_2 + Zn(OH)_2 \tag{17}$$

The theoretical capacity ($1e^-$) of $Ni(OH)_2$ is 0.289 Ah/g.

Figure 46 shows performance characteristics of the nickel-zinc batteries [59] involved in a performance-demonstration program of the U.S. government. Nickel-zinc batteries developed for the National Aeronautics and Space Administration (NASA) showed a very good cycle life, mainly

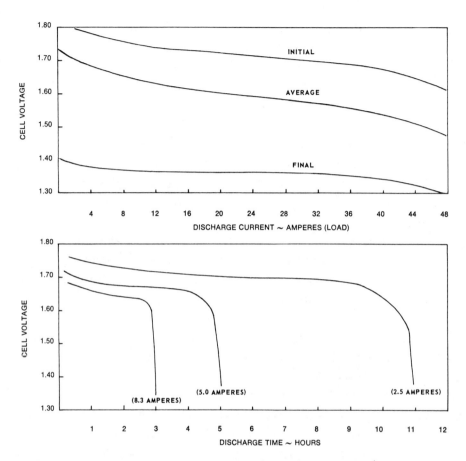

FIG. 46. Performance of nickel-zinc batteries [59].

because of the use of special separators [15, 60]. Such batteries are now tested in delivery vehicles of the U. S. Mail Service as part of the program set up by the Energy Research and Development Administration (ERDA).

4.2.2.2. Metal-Air Systems

a. Zinc-Air Batteries. Among high-energy batteries the zinc-air batteries have been the subject of particular attention. The advantages of zinc as anode material compared to others are of electrochemical nature: reactivity and stability, low cost, and some rechargeability. The theoretical specific energy of the zinc-air couple is 480 Wh/lb; one-fourth of this value is achieved in practical batteries.

Primary zinc-air cells are "old technology"; they have found use as low-current signal batteries for railroads, navigation lights, etc. High-rate batteries have only been developed over the last 10 years, as a result of efforts in the fuel cell field. The cathodes (air electrodes) needed improvement, especially in the form of better structures (thin electrodes) and better catalysts.

For a vehicle application the air-zinc battery must be rechargeable. There are, in principle, two ways of doing this: electrically and mechanically. The electrically rechargeable system is preferable. Figure 47 shows the main processes going on in an air-zinc cell during discharge and charge.

Unfortunately, there are difficulties with both electrodes. The zinc electrode is soluble as zincate and redeposition from the solution is not easy, mainly because zinc trees are formed, which often short the cell. A separator helps, but is not an ideal solution for the problem since it increases the cell resistance. A carbon air cathode is oxidized when brought up to a high charging potential; even porous nickel electrodes are damaged. The oxygen evolved during charging must be removed from the cell. Avoiding the cathode stability problem by charging with the help of a third electrode makes the use of a switching mechanism necessary.

At General Dynamics a system was devised which avoided some of the difficulties by circulating the electrolyte [61]; the zinc deposition was made coherent instead of spongy, and the air electrode was an oxidation-resistant

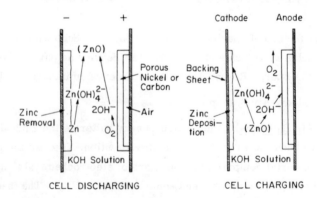

FIG. 47. Air-zinc cell discharge and charge process.

porous nickel plaque which was prevented from drowning by applying a positive air pressure on the gas side. Oversaturation of the electrolyte was
avoided by precipitating zinc oxide out of the electrolyte.

At the Laboratoires de Marcoussis, France, The Compagnie Générale
d'Electricité is conducting development work on a recirculating-slurry
zinc-air battery with either built in or external recharging facility.
110 Wh/kg energy density and a lifetime of 2,000 operating hours are
claimed. Special features are a very high electrolyte flow, the addition of
colloid stabilizers to increase the concentration at which the zincate solubility limits the capacity. Zinc powder is used, a mechanical recharge is
possible. The air electrodes are tubular-networks.

The mechanically rechargeable system has been developed primarily
for portable military power sources. General Motors Corp. built a vehicle
with mechanically exchangeable zinc electrodes [63]. The vehicle performed well with its 20-kW battery, but the exchange of the plates took
several hours.

A zinc-air system developed by the Sony Corp. of Japan utilizes powder
zinc as fuel and regenerates the electrolyte continuously. The vehicle was
operated over a test route extending over 500 km with quick "refills." The
electrolyte, which is saturated with zincate, can be regenerated (outside of
the battery); the reclaimed zinc is also reusable. Sony Corp. built an electric car with two units; each produces 3 kW at 100 V [64]. The power density is 23 W/kg and the energy density is 64 Wh/kg including fuel for 5 h.
For better acceleration the vehicle used a nickel-cadmium battery in parallel with the zinc-air batteries (hybrid principle).

Zinc-air battery efforts to date are summarized in Table 19 [65].

The problem of CO_2 removal to preserve the alkalinity of the electrolyte is another restriction of the use of zinc-air cells.

Zinc-oxygen batteries do not suffer from the problems of electrolyte
carbonation. A closed zinc-oxygen system which stores the oxygen formed
during charge in high-pressure cylinders is listed in Sec. 4.2.2.3. The
operation of this system seems to be very simple as it is described in the
literature; however, the zinc recharging from alkaline solution is not easier
than in the previously discussed systems.

TABLE 19

Performances of Metal-Air Automotive Battery Prototypes [65, 69, 70]

Company	Type of battery	Characteristics of prototype	Energy density (Wh/lb)	Wh/kg	Power density (W/lb)	W/kg
Gulf General Atomic	Circulating electrolyte, zinc, externally rechargeable	20 kW 19 kWh 36 V 700 lb	27	60	28	62
Sony	Pulverized zinc as fuel and externally regenerated	3 kW 100 V 510 lb (5 h) 725 lb (10 h)	30 (5 h) 41 (10 h)	66 (5 h) 90 (10 h)	10	22
General Motors	Zn-air, mechanically rechargeable	20.5 kW 27 kWh 648 lb	42	92	31	68
		1 kW 21 kWh 72 lb	30	66	20	44
Swedish Natl. Def. Co.	Iron-air rechargeable	ML-1B 1 kW	12	26	20	44
Zaromb	Al-air, externally regenerated	2 kW	50	110	20	44

b. Cadmium-Oxygen Systems. Cadmium-oxygen systems have been studied for space applications, and it was found that several thousand cycles are possible. Unfortunately, the voltage of that system is too low to be of interest for vehicle batteries.

c. Iron-Air or Iron-Oxygen Systems. Iron-air or iron-oxygen systems have gained publicity when high current output iron electrodes were reported to be successful in nickel-iron batteries [53, 66]. Some work in this direction is continuing in Germany [67, 68] and in Sweden [69].

d. Aluminum-Air Cells. Aluminum-air cells with alkaline electrolytes have been planned up to kilowatt sizes. The reaction products are collected and removed; they can be regenerated outside of the battery [70].

e. Sodium-Air Systems. A sodium-air system suggested by Atomics International would use the well-known Na-Hg electrode as intermediate anode in a molten salt cell which operated slightly above the melting point of sodium ($97.5°C$); it can therefore be considered a (relatively) low-temperature cell. On discharge, sodium is deposited in a Na-Hg electrode and is available for the alkaline-air cell. NaOH is a reaction product and must be removed into a storage tank. On charge, the NaOH is electrolyzed; oxygen evolves from the air cathode. Sodium travels back to the Na storage anode [71].

4.2.2.3. Sealed Metal (Oxide)-Gas Systems

Table 20 lists the theoretical voltages and energy densities of various metal (oxide)-gas systems in comparison with the H_2-O_2 regenerative system. The last two are relatively new and especially Ni-H_2 found preferred use in space systems (COMSAT) for the following reasons: H_2-O_2 is a two-gas system with separation problems. Zinc-oxygen suffers from zinc dendrites and cadmium is a scarce material. Results of design studies indicated the possibility of reaching 88 Wh/kg energy density with the Ni-H_2 system (compare with 40 Wh/kg for Ni-Cd). Figure 48 shows the principal design of a metal gas rechargeable system [72,73].

4.2.2.4. Zinc-Halogen Systems

a. The Zinc-Chlorine Hydrate System. Zinc can be plated out of zinc chloride solutions very well, and a rechargeable system avoiding alkaline

TABLE 20
Metal (Oxide)-Gas Systems

System	Theoretical Voltage E	Energy density Wh/kg	Wh/lb	Actual Voltage V	Energy density Wh/kg	Wh/lb	No. of cycles achieved
Hydrogen-oxygen	1.229	3,652	1,660	0.88	44-110	20-50	200
Zinc-oxygen	1.645	1,078	490	1.30	110-220	50-100	200
Cadmium-oxygen	1.210	506	230	0.85	55-95	25-43	500
Nickel oxide-H_2	1.358	389	177	1.25	55-93	25-42	2,000
Silver oxide-H_2	1.398	523	238	1.10	60-132	35-60	600

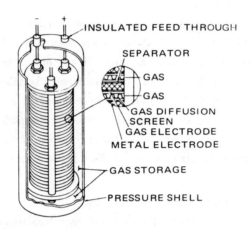

FIG. 48. Metal oxide-gas element [72].

electrolytes should be a big step forward. Air electrodes perform poorly
in neutral electrolytes, therefore chlorine is substituted. The problem of
storage of Cl_2 gas is circumvented by handling chlorine hydrate instead
[74].

The system has a high theoretical energy density—377 Wh/lb—low material cost, and a high cell voltage (2.12 V). Chlorine hydrate ($Cl_2 \cdot 6H_2O$)
can be formed by bubbling chlorine gas under cooling ($-\Delta H = 80$ cal/g).
Chlorine hydrate is stable as long as it is kept in a well-insulated cooler.

A fully charged battery has zinc on the anode plates and Cl_2 hydrate in
the tank. To discharge the battery the circulation pump is started and
chlorine is released in the (warmer) battery. Figure 49 shows the $Zn-Cl_2$
hydride system and its components in the process of being charged.

A prototype system [75] was tried out in an experimental car (Vega
1971, modified). Five 40-V modules were connected in series for a 200-V
100-A power plant to operate the car. Each 40-V module weighed 250 lb
(zinc weight: 113 lb). The chlorine hydrate storage must be added to these
figures. The finished car had a curb weight of 4,289 lb, 47% of it (2,010 lb)
attributed to the battery and its chemicals. The vehicle could still transport
its two passengers at speeds up to 66 mph at the end of a 150-mile run at
the International Speedway in Irish Hills, Michigan.

Occidental Petroleum-Gulf-Western Industries (with Gould, Inc.) continued the development of this system. Considerable progress was made in

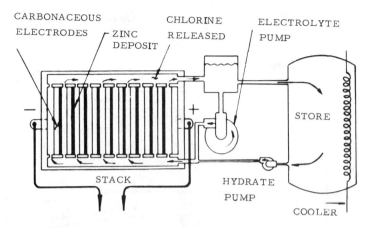

FIG. 49.   Zinc-chlorine hydrate system [74].

improving the chlorine electrodes and at the present time (1976) funds for the building of large batteries are provided by the Energy Research and Development Administration (ERDA) and the Electric Power Research Institute (EPRI).

The safety of the cooling system for the hydrate must be ascertained. Loss of cooling means free Cl_2 gas, which may build up to high pressure when confined in the system or cause poisoning if released.

b. The Zinc-Bromine System. Its theoretical energy density is 196 Wh/lb. The basic principle is the electrolysis of a zinc bromide solution using carbon electrodes [76] or titanium electrodes [77]. Zinc is plated out on the anode; bromine is absorbed on the surface of the large surface cathode, forming a (not defined) graphite-bromine complex. General Electric Co. is developing a similar system.

The same inventor [76] proposed an iron-iron chloride redox system as a rechargeable battery for electric vehicles and for load-leveling purposes in connection with power plants.

4.2.2.5. High-temperature Systems

a. The Sodium-Sulfur System. Ford Motor Co. has been carrying out research on a sodium-sulfur cell since 1966 with the objective of developing an electric vehicle battery. The basic principle is to separate the molten reactants, sodium and sulfur, by a sodium ion-conducting ceramic mem-

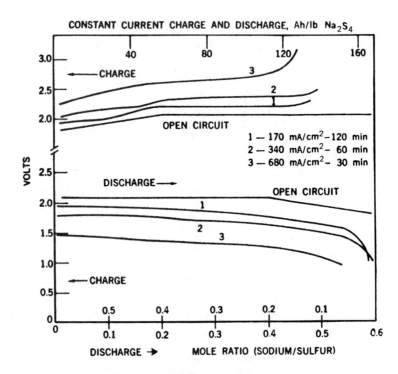

FIG. 50.   Charge-discharge characteristics of the Na-S cell operating at 350°C [78].

brane of β-alumina.   Graphite felt makes the sulfur conductive; sodium polysulfide, once formed, is an ionic conductor.   The difficulties encountered were connected mainly with the membrane.   Means to minimize cracking and increase life were found after several years of development [78].

Figure 50 shows the charge-discharge characteristics of the Na-S cell. Ford proposed a 200-W cell as the building block for a vehicle battery (100 cells: 20 kW, 150 to 200 V).   The battery operates at about 350°C; 2,000 cycles and a 7-month "hot life," 43 Wh/lb and 93 W/lb, were reported.

The Dow Chemical Co. has developed a sodium-sulfur battery which uses a sodium ion-conducting glass in the form of hollow fibers as the separator between sulfur and sodium.   The large interface area is achieved by mounting thousands of such glass capillaries into bundles of tubes.   The actual current density then needs only to be in the 2 to 3 mA/cm^2 range. This construction also makes a safer cell, since the glass fibers are flexible and resistant to impact damage [79]. Energy densities ranging from 176 Wh/kg for small cells to 297 Wh/kg for large cells are predicted.

The advantage of sodium-sulfur cells is seen in the unique combination of high energy density (theoretically 346 Wh/lb) with very low-cost materials and good rechargeability (80% voltage efficiency).

The largest foreign program is in the United Kingdom. British Rails has tested both tubular and flat-plate cells and has a 1-kW battery operational. The Electricity Council has developed a 960-cell, 50-kWh battery, the modules for which are shown in Fig. 51a. The battery was road-tested in a Bedford van (Fig. 51b) in November 1972. The present battery is rated at 15.5 kW average output, 29 kW peak power, and weighs 800 kg. The energy density is therefore 63 Wh/kg and the power density is 36 W/kg. The energy density is expected ultimately to reach 200 Wh/kg, with a life in excess of 1,000 cycles.

The Japanese program is part of a government-sponsored electric vehicle program. Yuasa is testing single cells and seven-cell units. Life times are in the order of 1,000 h (166 cycles) at 350°C. The energy density delivered is about 110 Wh/kg [15].

Sodium-sulfur cell development in France [77] and Swiss efforts are also concentrating on long life operation of rechargeable cells. All sodium-sulfur projects face the same key problem today, namely deterioration of the β-alumina electrolyte after 1,000 to 2,000 h at operating temperature. This deterioration is ascribed to a variety of causes and has led to a number of proprietary "fixes." Economic success will ultimately depend on finding inexpensive ways to produce the desired ceramic and to fabricate large batteries [15]. A sodium-sulfur cell built by TRW is shown in Fig. 52 [80].

b. High-temperature Lithium Systems. General Motors Corp. studied a wide variety of promising high-energy couples—Li-S, Li-air, Na-Cl_2—and developed a high-performance system on the basis of Li-Cl_2 [81]. The highest would be Li-air, but no suitable air electrode seems to exist today. There exists only an experimental Li-Cl_2 cell. Its open circuit voltage is 3.5 V, with an operating voltage of 3.2 V at very high current densities (4,000 A/SF). The drawbacks are 650°C operating temperature with chlorine supplied to a porous carbon electrode. A design which made the Li-Cl_2

FIG. 51. (a) Sodium–sulfur battery modules. (b) Na–S battery-powered van. Experimental battery and test vehicle, Electricity Council Research Center, Great Britain, 1972.

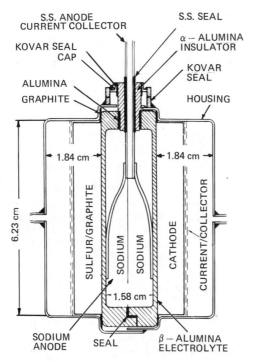

FIG. 52. Sodium-sulfur cell built by TRW for the U.S. Army [80].

cell a stackable unit (for higher voltages) was developed at the GM Defense
Research Laboratories.

Table 21 gives a comparison of theoretical energy densities and also
indicates the goals for development of Li-molten salt batteries [82]. If
these values can be realized, the internal combustion engine may become
obsolete.

At the Argonne National Laboratories, lithium anodes were coupled
with sulfur, selenium, and thallium cathodes. Pasted and liquid (molten)
electrolytes consisting of LiF-LiCl-KCl have been studied. The latter
gave over 2,000 charge-discharge cycles. The anode current collectors
were made of porous nickel which were filled with lithium. The cathode
materials were immobilized with porous graphite. Operating temperature
was around $400°C$.

TABLE 21

Comparison of Theoretically Available

Energy from a Variety of Reactant Pairs

Reactant pairs	Li/Cl	Li/S	Na/Cl	Na/S	Zn/Air	Li/Air	H_2O_2
Wh/lb	1,050	1,130	600	300-650	500	2,000	1,600
Wh/kg			1,320	660-1,430	1,100	4,400	3,520

Performance Goals for Lithium Batteries

for Ultimate Use in Military Hybrid Vehicles [82]

Battery,* 1-kW development	(7.5-cm-diam. cells)
Peak specific power	310 W/kg (140 W/lb)
Specific energy (1-h rate)	110 Wh/kg (50 Wh/lb)
Battery, 50 kW, full size	
Peak specific power	550 W/kg (250 W/lb)
Specific energy (1-h rate)	165 Wh/kg (75 Wh/lb)
Cycle life	13,000
Life time	26,000 h
Total weight	91 kg (200 lb)
Peak power	59 kW
Total energy stored (1-h rate)	15 kWh
Total volume	0.057 m^3 (2 ft^3)

*Goals for initial stages of development.

Improved systems use Al-Li anodes and metal (Fe) sulfides [83]. The program at Argonne National Laboratories is sponsored by the Energy Research and Development Administration (ERDA). A similar program at Atomics International, is sponsored by the Electric Power Research Institute (EPRI), Palo Alto, California [84].

Figures 53 and 54 show constructions of these cells.

Summarizing the data for rechargeable batteries of the future, one may produce a more realistic list of performance data and problem areas as shown in Table 22.

4.2.3. Fuel Cell Systems

4.2.3.1. Introduction

Figure 55 compares different galvanic systems in a diagram which considers the behavior of the cells on load [1]. This diagram indicates, e.g.,

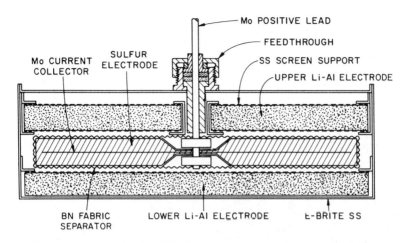

FIG. 53. Argonne National Laboratory Li/metal sulfide cell [83].

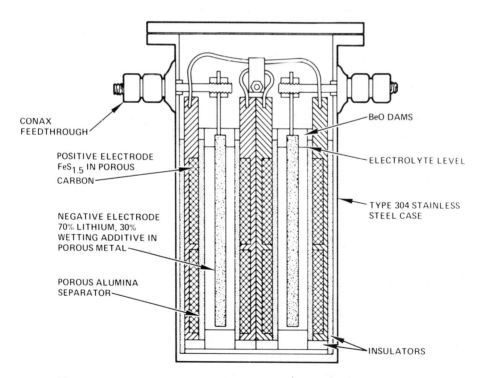

FIG. 54. Atomics International 100-Wh Li/metal sulfide cell [84].

TABLE 22

Future Batteries for Electric Vehicles

System	Projected performance				Problem areas
	(Wh/kg)	(Wh/lb)	(W/kg)	(W/lb)	
Conventional type					
Nickel–iron	44	20	110	50	Weight
Nickel–zinc	66	30	220	100	Cost, life
Metal–gas					
Iron–air	88	40	66	30	Life
Zinc–air	88	40	66	30	Life, complexity
Nickel–hydrogen	66	30	110	50	Volume
Zinc–oxygen	110	50	110	50	Life, cost
Cadmium–oxygen	66	30	66	30	Life, cost
Alkali metal–high temp.					
Sodium–sulfur (β–alumina)	176	80	220	100	Life of separator
Sodium–sulfur (glass)	176	80	220	100	Glass stability
Lithium alloy–metal sulfide	165	75	220	100	Corrosion, materials
Lithium–chlorine	110	50	220	100	Cycle life
Other systems					
Zinc bromine	66	30	110	50	Life
Zinc–chlorine hydrate	165	75	110	50	Accessories, life
Fuel cells (Air–H_2)	Tank limited		110	50	Catalyst cost, H_2–supply
(O_2–H_2)	Tank limited		220	100	Cost, H_2 and O_2–supply

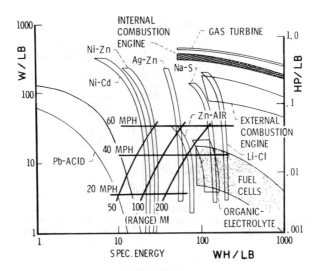

FIG. 55. Specific power and specific energy of batteries [1, 15].

that the lead-acid battery has a capacity of 20 Wh/lb if the power output is
1 W/lb (20-h discharge rate), but the watt-hour per pound figure drops to
half (10 Wh/lb) if the output level is increased to 5 W/lb (2-h discharge
rate). A Ni-Zn or Ni-Cd battery behaves far less sensitively to the load-
level changes, and very high power densities can be reached—approaching
those of combustion engines (right upper corner of diagram). Fuel cells
(in the lower right corner) show very high capacities (energy content), actually limited only by the size of the fuel tank; the 500-lb limitation to the
total weight of the power source restricts the container size. The specific
power output of fuel cells is shown low, in accordance with the state of the
art in 1967. Today it has at least doubled. The power and energy demand
of a 2,000 lb vehicle (as previously shown in Fig. 15) is represented by the
superimposed curves.

    Fuel cells can probably be improved threefold, but an increase in
power output by a factor of 10 is unlikely. To remedy this situation a secondary power source (e.g., a Ni-Cd battery) and a fuel cell system can be
combined to form a hybrid system which satisfies the power- and energy-
density (capacity) requirements. The big advantage of fuel cells is the

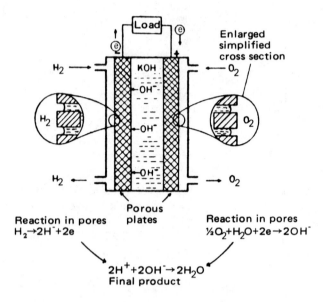

FIG. 56.   Principle of H_2-O_2 fuel cell.

rapid "refilling" instead of electric recharging. With filling stations along
the roads the capacity becomes unlimited.

The other new battery systems pictured in Fig. 55 were described in
Sec. 4.2.2. These power sources may some day replace combustion engines in medium size cars. The hp/lb scale on the right side, multiplied
with the specified power source weight (500 lb), puts, e.g., the rechargeable Na-S battery in the 100-hp class. The Li-organic electrolyte batteries
indicated in the diagram are presently primary cells; they may someday be
made rechargeable or capable of refueling.

The construction and the principle of operation of a fuel cell is shown
in Fig. 56. A discussion of fuel cell electrode properties and their basic
theories can be found in Refs. 85 and 86. Also, the many different types of
fuel cells are explained in these books.

In order to judge fuel cells as power sources for electric vehicles, it
is most important to compare the conversion efficiencies of heat engines
versus that of galvanic cells. The heat produced while burning hydrogen or
coal or oil is equal to $-\Delta H$, the enthalpy change for the combustion reaction. It is the basis for determining the efficiency (E) of a heat engine,

which is also governed by Carnot's law, stating that the fraction of ΔH which can be utilized as mechanical energy is

$$E_{Carnot} = 1 - \frac{T_0}{T} \text{ (in K)}^*$$

T being the operating temperature and T_0 the exhaust temperature on the absolute scale. For $T = 500°C$ and $T_0 = 30°C$, the efficiency is only 0.59 for an ideal heat engine. Practical values are 0.4 for steam turbines, 0.3 for diesel, and 0.2 for Otto motor (usual) automobile engines.

A galvanic cell is <u>not</u> a heat engine, therefore the voltage of a reversible electrochemical cell is proportional to the free energy change ΔG for the cell reaction. These values are related as follows:

$$\Delta G = \Delta H - T \Delta S$$

where ΔS is the entropy change and T is the temperature.

The ideal (maximum) efficiency is

$$E_{max} = \frac{\Delta G}{\Delta H}$$

Since $T \Delta S$ is small compared to ΔH, the quotient is close to 1. For the hydrogen-oxygen reaction it is about 0.95. Sixty percent of the energy can be converted into electricity in technical cells.

Figure 57 shows a comparison of fuel cell and heat engine efficiencies. It also indicates the extent of losses in practical cells and engines.

Cell voltage losses are caused by the cell resistance and the polarization (overvoltage) on each electrode. The maximum power output of a cell is obtained at half of the open circuit voltage (internal and external voltage drop matched), but only in a few cases is it desirable to load a cell to such high power levels. Voltages in the 0.7- to 0.9-V range are preferred for greater efficiency and longer life.

Hydrogen, hydrazine, and methanol fuel cell systems operating at low pressure, ambient temperature, and with alkaline electrolytes have been used in vehicle applications to date. There is no reason to argue that other

---

*K = degree Kelvin. According to the Systeme International d'Unités (the official "SI" system of units), $T°C \equiv TK - 273.15$. The unit in the "K" = $°C$.

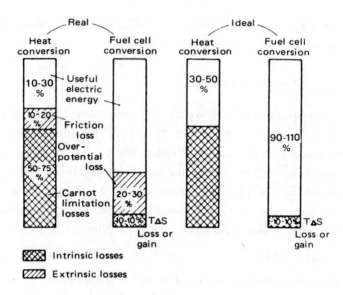

FIG. 57. Comparison of the efficiencies of heat engines and fuel cells
[85].

types may not become practical in the future: medium-pressure H_2 batteries operating at elevated temperatures, like the Apollo space battery;
or ion-exchange membrane cells, as were developed for the earlier Gemini
program. Acidic cells with immobilized matrices operating at 130°C with
phosphoric acid electrolyte or even high-temperature solid electrolyte cells
may become candidates.

4.2.3.2. Hydrogen as Fuel

A Union Carbide Corp. fuel cell system was used in two experimental cars,
a hydrogen-oxygen fuel cell-powered vehicle built by General Motors Corp.
and a hydrogen-air fuel cell-lead-acid battery hybrid vehicle built by the
author.

a. The General Motors Corp. Electrovan. The Electrovan was built to determine the state of the art of fuel cells (1965-1967) as applied to vehicle
propulsion. Hydrogen-oxygen fuel cells were chosen since they were the
only ones available that could provide the needed performance. By using
cryogenic containers it was possible to carry enough fuel and oxidants to
give the vehicle a range comparable to the gasoline-powered automobile [87].

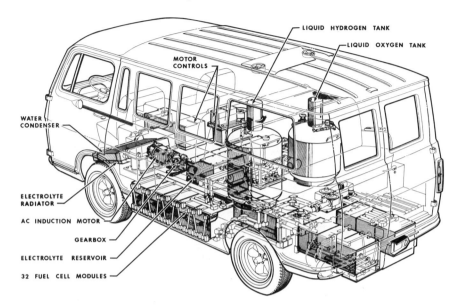

FIG. 58. Cutaway view of General Motors Electrovan [87].

The main bulk of the fuel cell power plant was installed under the floor of the wagon; the oxygen and hydrogen tanks were situated between the two benches in the passenger compartment. Thereby a good utilization of the loading space was achieved. Figure 58 shows a cutaway view of the Electrovan.

The drive motor, built by Delco Products Division, is a 125-hp, four-pole, three-phase induction motor, designed to operate at 13,700 rpm with oil cooling. The motor control system was constructed by the GM Defense Research Laboratories. The silicon-controlled rectifiers in the unit were also oil cooled.

The vehicle fuel cell system was built by Union Carbide Corp. [88] and utilized the "thin electrode modules" which at that time (1965) represented a 4:1 reduction in size compared to previous technology [89]. The basic principle and construction of fuel cell electrodes for alkaline cells with circulating electrolyte are described in detail in the literature [86, 90] and will not be repeated here.

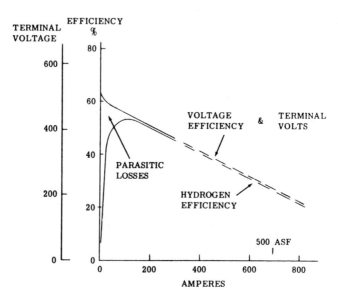

FIG. 59. Fuel cell power plant terminal voltage and efficiency at 150°F [88].

Figure 59 indicates the efficiency of the power plant under different load conditions. Table 23 gives a comparison of performance and weights of the Electrovan and the gasoline-powered wagon [87].

Summary of the Electrovan Experiment. The program demonstrated that the fuel cell technology of 1967 had reached the point where a high-

TABLE 23

Comparison of Performance and Weights;

Electrovan Versus GM Production Van [87]

	Electrovan	GM Van
Total vehicle weight	7,100 lb	3,250 lb
Fuel cell power plant	3,380 lb	
Electric drive	550 lb	
Power train total	3,930 lb	870 lb
Performance, 0-60 mph	30 sec	23 sec
Top speed	70 mph	71 mph
Range	100-150 miles	200-250 miles

output vehicle power plant was technically feasible. The evaluation tests also defined many points which were unsatisfactory:

1. Heavy weight and volume (100% overweight)
2. Short life time (several hundred hours)
3. Complicated, lengthy start-up and shut-down procedures (4 h)
4. Complexity of the system (1 liquid, 2 gas loops)
5. Safety problems: high voltage, liquid H_2, liquid oxygen
6. Costly components and materials ($250 worth of Pt metals/kW)

On the basis of these findings, efforts were started in 1969 to build another vehicle which would incorporate improvements and at least partial solutions to the problems stated.

b. The Hydrogen-Air/Lead Battery Hybrid System [91-93]. The Electrovan operated on pure oxygen. This was considered necessary because electrodes operating on air do not provide the peak current densities needed to give a car the desired acceleration properties.

In response to the results obtained with the GM Electrovan, the following improvements were considered necessary:

1. Reduce the weight by decreasing the number of modules and the need for a 32/160-kW fuel cell power plant. Use a hybrid system: fuel cell-secondary battery.
2. Operate the fuel cell only when the vehicle is moving: only operating time should count.
3. Cut the activating and shut-down procedure to minutes.
4. Use air instead of oxygen.
5. Use compressed hydrogen only.
6. Reduce the number of modules and lower the cost by using Pt metals on anodes only.

In 1966 the author had converted an Austin A-40 to electric propulsion by replacing the gasoline engine with an 8-kW lift truck motor. The power source was an 84-V lead-acid system consisting of two strings of seven 12-V batteries (one set in front, one in the trunk). The resulting vehicle, which retained the four-step gear box and the flywheel (plus clutch), behaved very closely to the expectation of a 2,000-lb electric car, as discussed previously in Secs. 3.1 and 3.2.

In 1968-1969 this car was partially converted to fuel cell power. A 90-V hydrogen-air battery replaced the seven lead-acid batteries in the rear compartment while the seven front batteries remained as before.

The experimental car represented a test vehicle, for discovering any inherent maintenance problems, battery life expectancy, and establishing feasibility of further design improvements as well as durability of materials and accessories. The vehicle was provided with safety features which prevented any trouble due to accidental improper handling. The transportation rules governing compressed flammable gases were carefully observed.

A photograph of the vehicle is shown in Fig. 60. The fuel cell system is mounted in the rear compartment together with all of the necessary accessories for automatic start-up and shut-down, including nitrogen purging.

FIG. 60. Electric car powered by a hydrogen-air/lead battery hybrid system [91].

The lead battery is situated in the front next to the motor. The gas cylinders are mounted on the roof.

No special effort had been made to optimize components. Commercially available accessories (pumps, blowers, etc.) were used throughout the system.

Improvements in electrode technology during the years 1967-1970 could not be included because the batteries built from earlier electrode stock could not be mixed with later ones for the sake of uniformity of the complete power plant, but it should be noted that the improved electrodes would have allowed a considerable reduction in lead battery weight. The improvement consisted in the use of a dual porosity nickel carrier for the active carbon layer, gas and electrolyte sides were reversed [93].

Many problems disappear if a fuel cell battery is operated in parallel with a secondary battery. The start-up time can be stretched to several minutes while the equipment (or vehicle) is operated during that time from the secondary battery. Overloading is effectively prevented because the secondary battery takes over at the minimum voltage determined by the fuel cell characteristics. Figure 61 shows the principal wiring diagram of the hybrid automotive power plant. Seven 12-V lead-acid batteries are connected in series (42 cells), and this bank is connected in parallel with a hydrogen-air fuel cell battery consisting of 15 eight-cell modules in series (120 cells). At low load levels, the fuel cell battery reaches a higher voltage than the lead-acid battery and is therefore able to charge the secondary battery. However, when the fuel cell battery is not in operation, it contains no hydrogen and has a far lower voltage than the lead-acid battery. The latter would now electrolyze water at the fuel cell electrodes and thereby cause wasteful and harmful operation. To prevent this, a diode is inserted into the connecting leads to permit current to flow only in one direction. This is one of the innovations permitting shut-down of the fuel cell battery and emptying of gases (and electrolyte) during nonoperating periods. This feature prolongs the life time of the fuel cells.

During start-up, however, hydrogen reaches the cells at different times and causes the activated cells to reverse the still inactive cells. This results in permanent damage to the "driven" or reversed cells. The reversed cells can be "righted" only by individually applying a pulse cur-

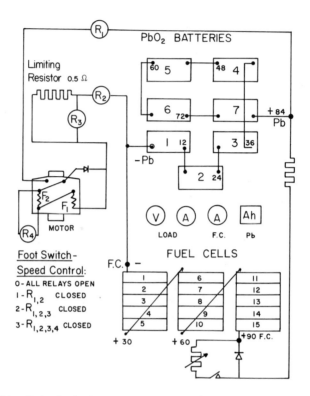

FIG. 61. Principal circuit diagram of an automotive fuel cell hybrid power plant [91].

rent (in the charging direction), a very time-consuming procedure. The difficulty can be avoided simply by applying the lead-acid battery voltage to the fuel cell battery while the anode compartments are filled with nitrogen, by means of a resistor to bridge the diode during that period.

Figure 62 shows a block diagram of the hydrogen-air fuel cell power plant as used in the automobile. Figure 63 is a photograph of the fuel cell system mounted in the rear of the car.

Hydrogen is stored as compressed gas in six cylinders on top of the car. These lightweight tanks are made for aircraft use, scuba diving, and similar purposes. The six tanks weigh 180 lb and contain 660 ft^3 of hydrogen.

Air is moved through the battery by a vacuum cleaner-type blower. The pressure side of the blower is connected to the CO_2 scrubber mounted

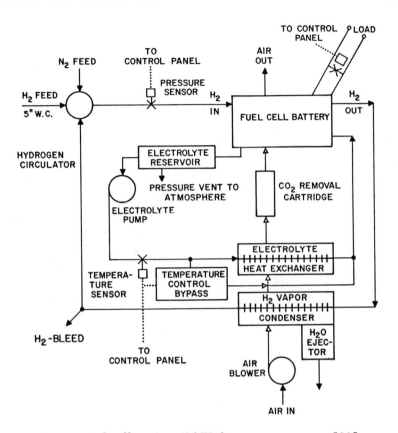

FIG. 62.   Fuel cell system (6 kW) for a passenger car [91].

on top of the battery.   From the CO_2 scrubber the air is directed into three
air-intake manifolds, one for each battery block.   The moisture-laden air
is exhausted to the outside.

The complete start-up and shut-down procedure is sequenced by a six-
position switch on the dashboard.

The Lead-Acid Battery.   The seven 12-V batteries are all located under
under the hood, two in front of the motor and five above it.   The batteries
used were manufactured by Globe-Union, Inc., sold by Sears, Roebuck and
Co. under the trade name "Die Hard."

Dimensions: 10-1/4 in. × 6-3/4 in. × 8-1/4 in. (height)
Weight: 47 lb each (329 lb total)

FIG. 63. Fuel cell system mounted in the rear of the car [91].

TABLE 24

Load Sharing of the Hydrogen-Air/Lead
System Under Different Operating Conditions [91]

Power data			PbO$_2$ battery State of charge	H$_2$-air battery		Operating conditions
kW	V	A		A	A	
11.0	85	125	Charged	80	45	55 mph — Full speed driving with
9.5	80	115	3/4	60	55	52 mph motor field reduced.
8.0	75	105	1/2	40	65	48 mph Battery temp.>60°C
6.5	70	95	1/4	25	70	42 mph
20.0	75	260	Charged	200	60	Maximum power during acceler-
14.0	65	210	3/4	140	70	ation. Fuel cells warming up
12.0	60	195	1/2	120	75	(~45°C)
8.0	90	90	Charged	60	30	45 mph — Driving with full motor
7.0	85	80	3/4	40	40	42 mph field. Battery temp.
6.0	80	75	1/2	25	50	38 mph ~60°C.
5.0	73	70	1/4	10	60	35 mph
0.5	110	5	Charged	0	5	Fuel cell charging into PbO$_2$
2.0	100	20	3/4	0	20	battery. Motor not running.
2.8	95	30	1/2	0	30	Battery temp. >60°C.
4.2	85	50	1/4	0	50	

Note: 60 A from fuel cell = 100 ASF current density.

No. of plates: 78 plates total (13 in each cell) in each battery

Capacity: 84 Ah (20-h rate)

Performance of the Automobile. Table 24 was compiled from instrument readings while the vehicle was moving. Road performance and driving range can be readily estimated from checking its columns.

Fuel Consumption. As closely as can be estimated, the hydrogen consumption rate closely follows that predicted for driving at mixed speeds (suburban mode). For 100 miles the pressure drops to about half of full pressure (2200 psi). The total available hydrogen volume is 660 ft^3, therefore the figure for hydrogen consumption is 330 ft^3/100 miles, or the driving range is approximately 200 miles.

The hydrogen consumption per mile is expected to be lowest at a steady (40 mph) speed for a long uninterrupted distance. Under such conditions about 75 A current (at 80 V) are required, to which the fuel cell contributes about one-half to two-thirds until the lead batteries drop below 25% remaining capacity (or 25% state of charge). Such a situation will be encountered if more than 2 h of continuous driving are planned. If there is then no recharge pause available, the fuel cells will have to supply nearly all of the power (6 kW).

Operating Costs. The figure of 330 ft^3 of H_2/100 miles is only a tentative estimate for city driving. On this basis, the cost per mile of driving is 1/2 cent/mile if we assume the cost of hydrogen in large quantities (1975) as 30 cents/lb (220 ft^3). Liquid hydrogen storage may be considered for supply stations.

The hydrogen refilled from steel cylinders, of course, is more expensive (up to 10 times) than hydrogen at the plant or at large filling stations. The automobile could also be operated on liquid ammonia plus a catalytic converter. The reduction in weight would also be favorable by comparison. A 15-min time limit for getting the converter on "full stream" seems technically feasible. Only one cylinder instead of six would be needed.

Conversion Efficiency. The fuel cell battery operates at about 0.8 V/ cell. This means that the electrochemical efficiency (0.8/1.23) is 60%, based on $H_2 \rightarrow H_2O$ (liquid) and the overall thermal efficiency (83% of this) is about 50%.

Conclusions. The purpose of this project was to demonstrate (1) that an electric automobile could be powered by a fuel cell/lead-acid battery hybrid system and retain the advantages of both battery types, and (2) that such a system can be easily started and shut down and is no more complicated than any other assembly of batteries with such common accessories as pumps, motors, and simple electrical circuits.

These facts were demonstrated and the electric automobile showed in driving tests that it behaved essentially like the gasoline engine-powered automobile, except that its top speed was somewhat limited.

To date (1975) the vehicle has been driven about 12,000 miles in local traffic, and of the fifteen 6-V stacks originally installed, 10 are still performing quite well. Three stacks show mechanical leakage. The concept of long life through intermittent operation seems therefore confirmed. A 10- to 12-kW fuel cell power plant would be desired to cut down the lead battery weight and to assure continuous high-speed driving.

Summarizing the pertinent energy and power density figures, the following data are listed:

Power and Energy-per-Weight Figures.

Fuel cell system: 6 kW (at 100 ASF) output: 370 lb (~60 lb/kW); 33-kWh (actual) capacity; 90 Wh/lb; Efficiency of conversion: ~50%.

Lead-acid batteries: 16 kW peak output; weight: 330 lb (~20 lb/kW); 3.4 kWh (actual) capacity (at 50 A); 11 Wh/lb.

c. Hydrogen/Oxygen-Lead Battery Hybrid (A Study). A detailed study of the feasibility of an electrically powered minibus was made by researchers at the Siemens A. G. Laboratory in Germany [94]. The standard driving cycle was used to show that at the present state of development of the H_2O_2 fuel cell system as built by Siemens A. G. [95], it is technically possible to use such a power plant in parallel with a conventional secondary battery and achieve satisfactory vehicle performance. Figure 64 shows a 33-cell-block, representing a 2-kW H_2-O_2 battery module for building any larger power plant. Figure 65 shows the characteristic curves of a 6-kW fuel cell battery at 80°C. Figure 66 shows the payload of a 3.6-ton minibus with a maximal motive power of 20 kW as a function of the driving distance (1 mile = 1.6 km). Very interesting are the projections of weight, volume,

FIG. 64. Two-kilowatt hydrogen–oxygen fuel cell battery with accessories (Siemens A. G.) [95].

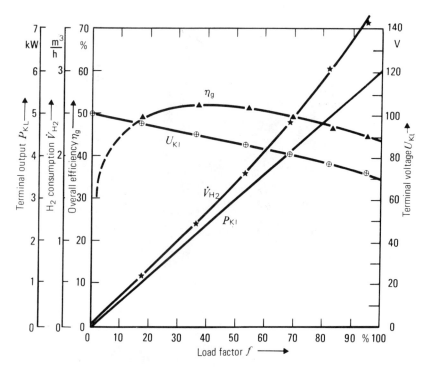

FIG. 65. Characteristics of a 6-kW H_2/O_2 fuel cell at 80°C [94].

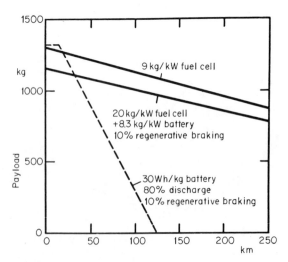

FIG. 66. Payload of a 3.6-ton minibus with a maximum motive power of 20 kW, as a function of driving distance [94].

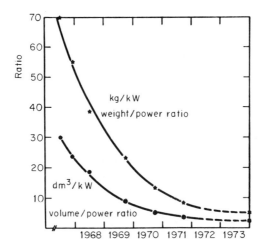

FIG. 67.   Volume/power ratio and weight/power ratio of fuel cell
blocks as a function of development time [94].

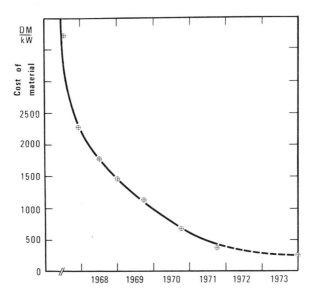

FIG. 68.   Cost of production materials for fuel cell battery blocks
(2 kW) as a function of development time (2.5 DM ≈ $1.00) [94].

and material cost of such 2-kW modules. Safe figures are available for the
technologically well-established batteries of 1971/1972. The dashed lines
represent research estimates. No noble metals are used in these batteries.
Figures 67 and 68 indicate the progress made over the last 15 years.

d. The Hydrogen-Air Battery System Evaluated by the Institut Francais du
Petrole (A Study). A fuel cell system proposed by the Institut Francais du
Petrole would make use of a hydrogen-air battery as the only power sup-
plier. As an alkaline system it would have to remove carbon dioxide from
the air. First, regenerable amine decarbonization was suggested; later,
solid absorbers were also considered (soda lime). The hydrogen supply
could be either a methanol reformer system or a compressed-gas storage
system [96].

The conclusions were that it seems feasible to operate a city car on a
power source of about 10 to 12 kW output at the cost of gasoline, if hydrogen produced in a large-scale industrial plant is used. The life of the batteries was not considered a limiting factor.

e. Aerospace Fuel Cell Systems (H_2O_2). Hydrogen-oxygen fuel cell systems have performed excellently in the space program [97]. The General
Electric Gemini system (from 1964 on) was used in the early years of the
space flights. Thirty-two cells were built into a single module of 1 kW output. A special characteristic was the use of ion-exchange membranes [98].

The more spectacular Apollo program, which produced the manned vehicles for the moon flights, used a H_2/O_2 system built by Pratt & Whitney
Aircraft Co. based on the earlier Bacon electrodes [99].

The Apollo system consisted of three units, each producing 560 to
1,400 W at 31 to 27 V. The maximum output was 2.3 kW at 20.5 V. It operated at elevated temperatures ($250°C$) and medium pressures (30 psi$_a$).
The water produced could be used for drinking purposes by the astronauts
[100]. Fuel cell technology may be determined by achievements in the
specialized aerospace industry.

NASA needs a fuel cell power supply for the space shuttle, a reusable
vehicle that will transport crews and supplies between Earth and orbiting
vehicles. The design calls for a 6- to 10-kW unit with a specific energy
output of 75 to 220 W/kg and a life time of 10,000 h. General Electric Co.

TABLE 25

Aerospace Fuel Cell Performance Gains [103]

System	Gemini	Apollo	Space shuttle
Power level (kW)	1.0	1.4	5-7
Specific weight (kg/kW)	30.8	77.0	13.6-5.4
(lb/kW)	68	170	30-12
Specific volume (m^3/kW)	0.051	0.167	0.034-0.006
(ft^3/kW)	1.8	5.9	1.2-0.21

and Pratt & Whitney Aircraft Co. are engaged in a technology evaluation program.

General Electric Co. uses an ion-exchange membrane cell similar to the previous Gemini fuel cell, but much improved [101].

Pratt & Whitney Aircraft Co. developed an advanced 7-kW fuel cell system which improved the specific weight and volume figures nearly a magnitude over the Apollo system [102].

Table 25 shows the aerospace fuel cell performance gains achieved during the progress of the various space programs [103].

f. The Transport and Safety of Hydrogen. Hydrogen cylinders of the industrial, welded type are heavy (150 lb for 220 ft^3 of hydrogen) and not suitable for gas transport on a vehicle. Deep-drawn steel tanks of the scuba type weigh only one-half to one-third of the industrial cylinders with the same pressure ratings, so that a fuel capacity of about 200 Wh/lb can be achieved for such a commercially available system (see Sec. 4.2.3.2).

The safety of compressed-gas storage is in the same category as gasoline: the same rules apply for its transportation on board a vehicle. Safety-melting plugs (in case of a fire) are required to release the pressure at an elevated temperature. Federal regulations require that the steel tanks must be pressure (test) rated at three times the operating pressure and rechecked every 5 years.

Fiber-reinforced tanks are far lighter than steel tanks. Some spherical models are available but not approved for general use.

The limits of flammability of hydrogen in air are 4 to 75% (compare with CH_4: 5 to 15%). The upper range is not likely to be reached in any environment that is not completely closed: hydrogen rises extremely quickly. A hydrogen flame is colorless and does not ignite flammable material by heat radiation, as other carbonaceous fuels may do. The explosive limits of 18 to 60% are also unlikely to occur in unconfined spaces because of its rapid diffusion rate: a spill of 500 gal of liquid hydrogen in open air is dissipated below the explosion limit in 1 min.

Ignition (spark) sources must be eliminated (the car must have grounding strips). The fact that city gas or manufactured gas (50 to 60% H_2) is routinely used for cooking and heating in many countries should remind us that hydrogen is not as dangerous as is commonly thought. Gasoline vapor has a lower flammability limit of 1.5%! However, due to its lower molecular weight, hydrogen leaks more readily than other fuels.

The use of hydrogen as fuel for combustion engines has been tried successfully. Special mixing precautions are required to prevent backfirings [104]. However, it is not conceivable that future automobiles should use hydrogen in such a wasteful way (12 to 15% efficiency) when fuel cells could do it at 50% efficiency.

The transportation of liquid hydrogen on vehicles is probably risky for public vehicles. However, means of storing and distributing liquid hydrogen are well developed and used in industry. Tank cars transporting liquid hydrogen are operated reliably and safely; they could supply filling stations throughout the country [105].

Metal hydrides as hydrogen reservoirs could play the role of storage and distribution for vehicle applications. Several alloys with a wide range of properties have been synthesized and studied at the Brookhaven National Laboratory [104]. They are exothermic, i.e., heat is evolved when hydrogen is absorbed and the hydrogen can be recovered by lowering the pressure below or raising the temperature above the adsorption process. Magnesium-nickel alloys and the hydrides of V, Nb, and their alloys were investigated. A special group of hydrides was discovered by the Philips Laboratories at Eindhoven: the AB_5 alloys where A is a rare earth metal and B is Fe, Co, Ni, or Cu [106]. Seven H atoms are adsorbed per AB_5 unit. Lighter alloys of possible use are FeTi and the hydrides of aluminum [107].

g. Hydrogen from Ammonia. Ammonia, NH_3, is a gas that is easy to liquefy and to transport in low-pressure cylinders. It is used in huge

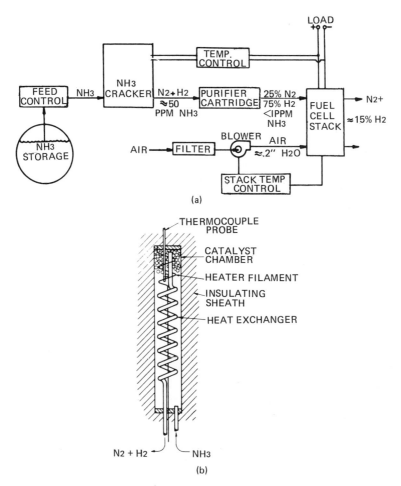

FIG. 69. Ammonia-H_2/air fuel cell system. (a) System flow schematic. (b) Ammonia cracker. [108]

quantities as a fertilizing chemical; the cost is low (even if it is produced from hydrogen) because of the convenience of an existing distribution system. Hydrogen can be produced from ammonia in a catalytic cracking tower; the resulting mixture of 75% hydrogen and 25% nitrogen can be used directly in any hydrogen-air fuel cell, down to about 6% H_2. The effluent gas can be used for heating the cracker unit [108]. Figure 69 shows the schematic of an NH_3-H_2/air system.

4.2.3.3. Hydrazine as Fuel

The principle of a hydrazine fuel cell involves the decomposition of hydrazine, N_2H_4, into nitrogen and hydrogen at the negative electrode by means of a catalyst (Ni, Pd)—similar to the decomposition of NH_3 discussed before, except that it occurs more easily and at room temperature [109]. The hydrogen is utilized; the nitrogen is exhausted.

a. Hydrazine-Air/Ni-Cd Battery-Operated Motorcycle. In 1966, to demonstrate that hydrazine-air batteries are capable of operating a vehicle over long distances, the author fitted a motorcycle with a hydrazine-air/nickel-cadmium hybrid system. Figure 70 shows the motorbike. The two 800-W, 16-V hydrazine-air batteries are mounted in front, the rechargeable batteries in the saddle bags [93].

This vehicle also uses the principle of battery usage only during vehicle operating time. When the battery is not active, the electrolyte is kept in a stainless steel tank under the battery. The hydrazine is supplied on demand

FIG. 70. Motorbike powered by a hydrazine-air hybrid system [93].

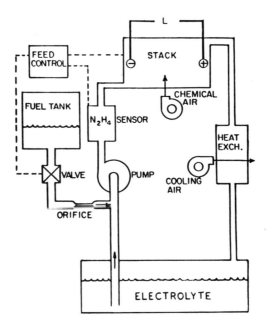

FIG. 71. Diagram of the hydrazine-air battery system used on the motorcycle [109].

by a sensing device activating a fuel injector. The system is schematically shown in Fig. 71 [109].

The 16-V Ni-Cd batteries consist of 15-Ah cells which can supply a peak power of 1 kW for short periods (acceleration). Together with the fuel cell battery, a maximum output of over 2 kW can be obtained, giving the motorbike the performance of a small gasoline-powered vehicle. At 20 mph (on level roads), the bike uses only the average power supplied by the fuel cell battery. The range is 100 to 120 mpg of 64% hydrazine solution (0.8 kWh/lb).

The weight of the complete vehicle is 208 lb; the motor is a 0.75/2-hp compound motor with speed regulation obtained by field switching. The nickel-cadmium batteries have been exchanged with lead-acid and silver-zinc batteries on an experimental basis. The matching of different systems must be done by considering the voltage-current characteristics. Under average load conditions the fuel cells should supply all the power. During heavy load conditions the secondary batteries take over; during idling or

during low-drain periods, the fuel cells must charge the secondary batteries.

b. Hydrazine-Air Battery-Operated Truck. Monsanto Research Corp. in
cooperation with the U.S. Army Engineer Research and Development Laboratories, developed a hydrazine-air fuel cell module, which when multiplied provides power for an M-37 Army truck. The module consists of
2 × 70 cells (0.8 V each). The rated current density is ~100 A/SF. The
twin decks produce 28 V, 5 kW. Size is 9 in. × 9 in. × 22 in., weight is
95 lb with electrolyte. Specific weight is ~20 lb/kW. Thermal efficiency
is about 50% [110].

The M-37 truck was supposed to take eight modules (40 kW) of the
hydrazine-air battery. Later it was modified to take only four modules
(20 kW) and operate parallel with a lead-acid battery in hybrid fashion.

Figure 72 shows a schematic diagram of the power plant for the M-37
Army truck [110].

The use of hydrazine in batteries for military vehicles was based on
the hope that the cost of this chemical could be reduced from about $8/gal

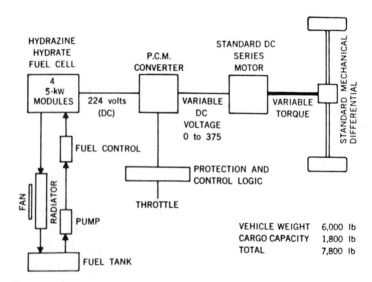

FIG. 72. Schematic diagram of the power plant of the M-37 Army
truck [110].

to the level of gasoline in large-scale production. This expectation turned out to be wrong.

c. Hydrazine Cell/Lead-Acid Battery Car. At the Shell-Thornton Research Laboratories, a DAF 44 saloon car has been extensively modified to study some of the problems associated with building, controlling, and driving a car powered by a fuel cell battery. A prototype hybrid car has been made, using as a power source two banks of 120-cell hydrazine/air fuel batteries in conjunction with six conventional 12-V lead-acid accumulators. The car has been demonstrated successfully on several occasions, and its performance has been measured [111].

The performance of the car, in its present form, falls between that of today's internal combustion engine vehicles and that of secondary battery-powered ones. However, unlike more conventional electric cars, its ranges under town driving conditions and at its steady cruising speed are not limited by the quantity of stored electricity.

Table 26 compares the performance of the hydrazine fuel cell with the original gasoline-powered car.

The flow diagram of the 120-cell fuel battery system is shown in Fig. 73. The load-sharing action of the hybrid system can be seen from Fig. 74,

TABLE 26

Performance of a Hydrazine-Air/Lead Battery Hybrid Car [111]

	Original DAF car	Thornton electric DAF car
Weight, kg	940	1,380
Top speed, mph	~69	~50
Acceleration, mph/sec		
At 20 mph	~3.8	2.6
At 30 mph	~2.9	1.3
At 40 mph	~1.6	0.5
Time, sec		
Needed to go 10 to 30 mph	5.3	8.3
Power consumption, kW		
At 20 mph		7.5
At 30 mph		10
At 40 mph		16.5

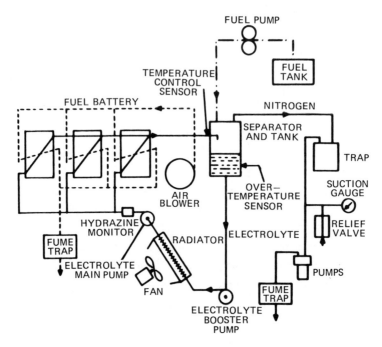

FIG. 73. Schematic diagram of the power plant of the Shell DAF-44 automobile [111].

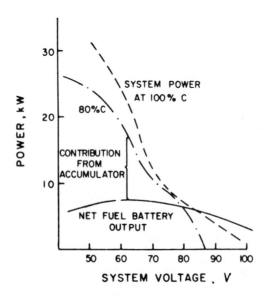

FIG. 74. Load-sharing diagram: fuel cell/lead battery in the Shell car [111].

which shows separately the power output of the fuel cell and from the lead battery under maximum (100%) power condition.

The present cost of hydrazine is of course also prohibitively high for a civilian vehicle; furthermore, the fact that hydrazine is a poisonous substance even in very low quantities (liver, kidney damage) makes it unlikely that this fuel will be used in commercial cells.

d. The Alsthom-Exxon Fuel Cell System (N_2H_4/H_2O_2). Exxon, U.S., and Alsthom, a French company, produce very compact battery stacks by using nonporous electrodes separated by semipermeable membranes. Hydrazine and hydrogen peroxide, dissolved in the electrolyte, are circulated by powerful pumps through the narrow cells. Catalysts are "glued" to the electrode surfaces by adhesive resins. Power densities of 1.8 ft^3/kW and 55 lb/kW are claimed [112].

Figure 75 shows a hydrazine-hydrogen peroxide fuel cell system. The use of two liquids simplifies the cell design and allows a very thin construction. Methanol can also be used in similar cells, and hydrazine could be avoided. The development of a fuel cell system for automotive propulsion could follow. The question of finding cheap catalysts is a main part of the program.

4.2.3.4. Methanol as Fuel

a. Direct Cells. A direct methanol-air fuel cell battery demonstrator unit was developed by Esso Research and Engineering Co. It operated with sulfuric acid as the electrolyte, to which 3% methanol is added, controlled by an automatic methanol analyzer [113]. Corrosion was the most severe problem with the H_2SO_4 system, and it was finally decided to go the reformer route with alkaline or acidic hydrogen cells.

b. Indirect Cells. The Shell Research Laboratory in Thornton, England, developed a 5-kW hydrogen-air battery with alkaline electrolyte which was mounted on a small truck. The hydrogen was produced in a reformer from methanol. The separation of the reaction products was accomplished by a Pd-Ag membrane (diffusion) system. The overall conversion efficiency was only 20 to 25% [114].

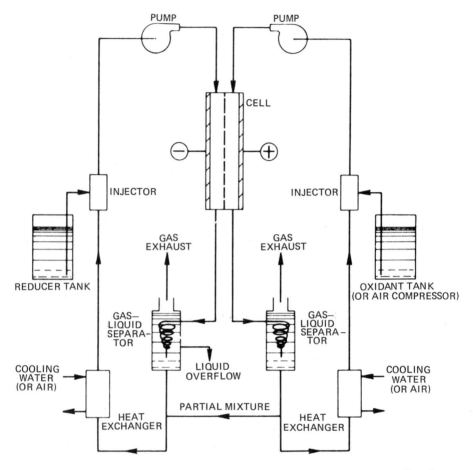

FIG. 75. Hydrazine-hydrogen peroxide fuel cell system (Alsthom) [112].

4.2.3.5. Hydrocarbons as Fuel

a. Low-temperature Direct Cells. The idea of using hydrocarbon (alcohols excepted) in fuel cells, without a preceding reformer step, has stimu-
lated much research and development work, but to date no system of practical utility has evolved. Either the current densities were too low (10 mA/
cm^2 at 0.5 V) or the catalyst costs (50 mg Pt/cm^2) were too high. The
(liquid) decane-air system was studied by Esso Research and Engineering
Co. under U.S. Army sponsorship [115].

b. High-temperature Direct Cells. Extensive work has been done with fuel
cells operating in molten salt electrolytes at high temperatures (from 600
to 1,000°C). These systems can oxidize hydrocarbons directly, without
any need for noble metal catalysts. Short operating times and material
problems were the main obstacles with these systems. They probably will
not be used in vehicles. For power plant applications there may be some
advantages [116].

c. Hydrocarbon Fuel Converter Systems. The most active development of
hydrocarbon-burning systems is being done at Pratt & Whitney for a con-
sortium of gas and utility companies. The name TARGET-PROGRAM (from
Team to Advance Research for Gas Energy Transformation) implies an objective which has a power plant of a size and capability which will serve
apartment houses, commercial establishments, and small industrial facilities as the main goal [117]. The methane is reformed externally with
steam and the product (mainly hydrogen) is reacted in cells employing thin
(paper) carbon electrodes with phosphoric acid (in a matrix) as electrolyte
(130°C).

While these units are built for stationary use, it is interesting to discuss them from the viewpoint of hydrogen-air batteries in connection with
the future hydrogen economy [3] in which all energy ultimately comes from
coal or nuclear power plants. Hydrogen would then be produced by electrolysis and piped to the consumers. Electric cars would then operate on
hydrogen-air fuel cells. The fuel cells developed for the local "house"
energy supply could be modified for vehicular use also, since they are
essentially hydrogen-air batteries. It cannot be predicted today if the ultimate versions will be acidic or alkaline cells.

Figure 76 pictures the hydrogen economy in which all energy comes
from a thermal source. The conversion of coal to hydrogen is another
link in this future scenario [118]. The idea includes the transmission of
energy over pipelines instead of high-tension electric lines, which have
losses three to six times larger [119]. Hydrogen pipes are already a common commodity in the chemical industry.

The first "commercial" fuel cell power plant was designated PC-16
(for Powercel, Pratt & Whitney's trademark) and built to compete with

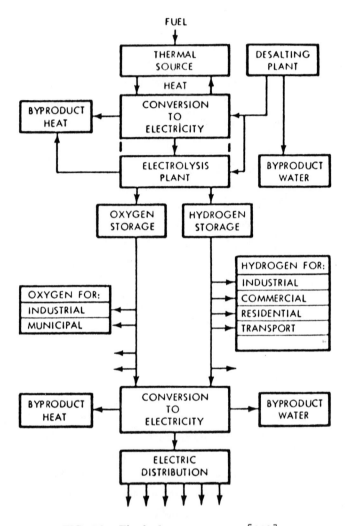

FIG. 76. The hydrogen economy [118].

diesel generators of the 12- to 15-kW size. From 1962 to 1972, $70 million
have been spent on the program and about $9 million are added per year.
Recently another $40 million have been added for the establishment of a
prototype power plant capable of producing 26 MW of electricity. A separate division of United Aircraft Co. was formed.[*]

---

[*]In May 1975 the name of United Aircraft Co. was changed to United
Technology Co.

The forerunner of the PC-16, the experimental PC-11, has been operating at 20 sites throughout the United States in extreme weather situations,
e.g., deserts, high altitudes, etc. The PC-16 is claimed to be one-fourth
the size and weight of the PC-11. Pratt & Whitney Aircraft Co. estimates
a minimum cost of $200 per kW and 40,000 h operating life as required for
economic competition (present-1975-claims: $350 per kW and 16,000 h).
In comparison, large turbines for utility stations cost about $150/kW and
operate for 100,000 h between overhauls. If the acidic low-temperature
system should not fulfill the expectations, a backup program using molten-
salt high-temperature cells exists.

Progress in catalyst cost reduction has been significant in recent
years. Electrodes consisting of thin porous layers of Teflon in close contact with platinum black have been shown to operate at very high current
densities with small amounts of platinum (1 to 2 mg/cm^2) in phosphoric
acid at 120 to 160°C [120]. Thin active carbon layers on carbon paper sub-
strates, catalyzed with platinum salt solutions, were also designed with the
goal of cutting catalyst cost [121].

4.2.3.6. Alkali Metal/Water Fuel Cell

The alkali metal/water fuel cell is a unique system which was recently revealed by Lockheed Aircraft [122]. Lithium or sodium metal is pressed
against a rotating iron disk soaked with water. Normally one would expect
a very violent reaction. However, it seems that the speed of reaction is
not too fast due to a passive film formation on the alkali metal. The Li or
Na establishes a negative potential against the porous steel (positive); the
voltage difference is about 0.5 V at a very high current flow and under co-
pious hydrogen evolution.

While the efficiency of this cell is probably not over 50%, power is
produced while hydrogen is formed. This controllable source of hydrogen
can be the input for a hydrogen-air fuel cell, thereby solving some of the
hydrogen supply problems [123]. Alkali metals are readily available, and
sodium is very cheap (~20 cents/lb). Improved cells use hydrogen peroxide as oxidizing agent.

At the present time this system is studied for marine propulsion applications; it has been suggested as a vehicle power source (Livermore Laboratories).

5. STATUS OF ELECTRIC VEHICLE DEVELOPMENTS

5.1. In the United States

The automobile companies of the United States are often blamed for indifference toward electric vehicles. However, over the years many efforts have been made to ascertain the situation and to judge the "readiness" of electric vehicles to enter the automobile market. The efforts had been futile in the period 1960-1970 because the customer was considered to be inclined to buy only a big, powerful car and it was impossible to provide such a vehicle with practical batteries. Only with the advent of the energy crises did the philosophy change.

In the following many examples of industrial efforts in the electric vehicle field will be listed. The sequence is arbitrary and should not give any impression of predominance or judgment of success. It may show historical trends.

5.1.1. General Motors

General Motors Corp. was leading with "nonconventional" batteries.

a. Electrovair I and II [29] (powered by Yardney Electric Co.; silver oxide-zinc batteries). In 1963 General Motors Engineering staff began an evaluation of the state of the art with the objective of producing an electric car that would include the best available technology in the motor and control field and the highest power output and capacity battery that could be built, namely silver oxide-zinc batteries. The specifications of the gasoline car were the guidelines.

Electrovair I was a converted 1964 Corvair. The problems encountered suggested many basic improvements in the drive system because a smooth torque control was sought, competing with automatic transmissions. Electrovair II was the next model, built on a 1966 body and chassis. Figure 77 shows the vehicle and the location of the major components.

The battery was a 286-cell silver oxide-zinc battery. The open circuit voltage was 530 V, and the current rating was 60 A. The total energy capacity was 25.4 kWh. The weight of the battery was 680 lb. The battery

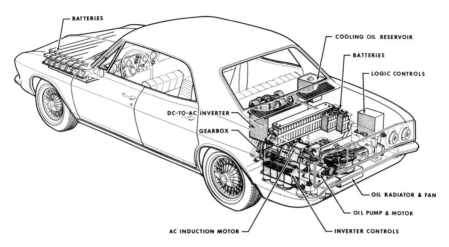

FIG. 77.   Cutaway view of Electrovair II [29].

was split between front and rear compartment.   Cooling was provided by ram-air flow in front, by a blower in the rear.   Peak power was 120 kW.

The motor occupied the place of the engine (in the rear).   It was a 115-hp, four-pole, three-phase induction motor built by Delco Products Division.   Its weight was 130 lb, it was oil-cooled, and it ran at 13,000 rpm.

Control was via silicon-controlled rectifiers to produce three-phase alternating current.   Also, this system was oil-cooled and used a radiator-fan combination (80 lb).

Vehicle data: curb weight 3,400 lb, compared to 2,600 lb of a production Corvair with a 680-lb gasoline engine power train.

Performance: top speed: 80 mph, acceleration 0 to 60 mph: 16.7 sec, 0 to 20 mph: 6 sec.   Range: 40 to 70 miles (60 to 12 mph).

Battery life: short, long recharge time at dangerously high voltage. Costly ($15,000 to $20,000).   Poor low-temperature performance.

Yardney Electric Co. also built their own car, a Henney Kilowatt with silver oxide-zinc batteries (240 lb, aircraft type) [124].   Yardney Electric Co. produces prototype nickel oxide-zinc batteries for vehicles (Fiat 1976).

b.   The Electrovan.   The Electrovan was a light truck operated by a hydrogen-oxygen fuel cell system.   Its performance and construction has been discussed in Sec. 4.2.3.2. A picture was shown in Fig. 58 (compare with hydrogen-air vehicle, Fig. 60).

c. The XEP. The XEP was a 1970 Opel Kadett with six zinc-air batteries and eight lead-acid batteries that supplied power to two motors. The car achieved a range of 90 miles at 55 mph and 150 miles at 30 mph. Top speed: 60 mph. When the battery is discharged, 300 plates must be removed and replaced. The weight of the power train was 1,600 lb. The system has been discussed in Sec. 4.2.2.2.

d. The Two-Passenger Car. A two-passenger car (type 512) was operated in hybrid fashion with a gasoline engine (200 cm^3) producing about 4 kW, feeding into a 72-V, 32-Ah lead battery. The 4/7·5-kW series motor operated the car for 100 miles at 34 mph.

5.1.2. Ford Motor Company

a. The Comuta City Car [125]. The Comuta City Car was a car built by Ford Motor Co. in England and tested in Dearborn, Michigan. It is a two-seater, weight 1,200 lb, had two 24-V (5-hp) motors and a 48-V, 85-Ah lead battery. Maximum speed was over 30 mph, range 40 miles at 25 mph.

b. The Cortina Estate Car [126]. The Cortina Estate Car was a two-passenger car, weight 3,086 lb, top speed 60 mph. The dc motor was rated at 40 hp (100 V, 8,000 rpm, General Electric Co.). Battery: Ni-Cd aircraft type (General Electric Co.). Nine 10-cell batteries of 100 lb each were mounted in steel trays; their 1-h rating was 110 Ah. The driving range obtained was very poor. At 25 mph, 40 miles were obtained. In city traffic, on a prescribed stop-and-go route, only 18 miles were obtained. Acceleration to 30 mph needed only 7 sec. The chopper circuit crested considerable losses at low speeds; a change in drive torque (gear change) increased the efficiency noticeably.

5.1.3. American Motors

a. AMC and Gulton Industries Joint Program. AMC and Gulton Industries had a joint program (1967-1968) to develop a small electric car using a hybrid combination of Gulton's lithium-nickel fluoride (organic electrolyte) batteries and high power density nickel-cadmium batteries of bipolar construction [127]. The expectations for this combination were exaggerated, and the high current densities actually achieved from Li-NiF$_2$ cells were only of short duration.

In the following years American Motors Corp. supplied bodies and chassis combinations to other companies who tried electric vehicles with lead-acid batteries (see, for instance, Electric Fuel Propulsion, Inc.).

b. American Motors General Corp. and Gould, Inc., U.S. Postal Service Contract. A subsidy of AMC was successful in obtaining a $2 million contract to build 350 quarter-ton electric vehicles from the U.S. Postal Service in 1974 [128]. Gould, Inc., designed the total electric system, including the batteries and the chargers. Figure 78 shows the vehicle, which is a modification of the gasoline engine-powered vehicle type of which AM General had built 80,000 for the U.S. Postal Service since 1969. The "Electruck" will be able to go 32 miles, with 300 stops; batteries are guaranteed for 4 years. The vehicles will have gasoline heaters, on-board chargers, etc. Speed: 39 mph, 10% grade capability. Weight: 3,600 lb, 500 lb cargo (50 ft^3). The Gould propulsion systems provide for armature and (>15 mph) field control. Regenerative braking over 15 mph. A dc-dc converter for on-board accessory charging. Battery type: EV 27-66E, 27 cells, 330 Ah. Motor: 263 lb, controller, 100 lb. Off-board, all-weather charger: 250 lb.

Based on a test program in Cupertino, California, started in 1969, the Postal Service reported about the satisfactory operation of a fleet of electric trucks. With 7,200 days of service, down time is only 8 days. Yearly operating cost per vehicle: $384, including 4,533 kWh of electricity, charger cost, etc. The cost of the vehicles is high: $10,000 each. Hopefully, mass production will reduce it [129].

5.1.4. The Battronic Truck Company

The Battronic Truck Co., partly owned by E.S.B., Inc., produced electric delivery trucks for many years.

A "shuttlebus" for 15 persons was tested by the Electric Illuminating Co. in Cleveland for downtown transportation.

FIG. 78. American Motors General "Electruck" designed for the U. S. Postal Service [128].

A "minivan" for city deliveries was built in several variations. The vehicles have 84- to 96-V lead-acid systems with 425 Ah capacity. Gross weight is 6,000 lb with 1,000 lb payload. Operating range: 42 miles at 30 mph. Drive train: 25-hp GE traction motor. Electric utility companies all over the United States were encouraged to buy and test these vehicles. The Electric Vehicle Council started a 100-vehicle buying program in 1972 [130]. The Battronic vehicles are shown in Fig. 79.

FIG. 79. Battronics Shuttlebus and Minivan.

3.1.5. Electric Storage Battery, Inc.

E.S.B., Inc., experimented extensively with passenger vehicles, starting in 1967 with a converted Renault Dauphine of the "Henney Kilowatt" type (see Sec. 3.2.4.1) but with a 72-V system (instead of a 36-V battery) made up from twelve 6-V golf cart batteries (140 Ah). The acceleration was much improved, but the range was still only 35 miles at 40 mph.

The Sundancer was an electric vehicle with a streamlined fiberglass body, designed from the beginning as an electric vehicle. The frame was a part of the battery tunnel, and the set of batteries could be exchanged quickly (see Fig. 18, Sec. 3.2.2). The car is well streamlined and has a sports car appeal. Its appearance is based on a McKee Engineering Corp. design, which looks quite similar to the Electromotion Sportscar, which is built on the same basic concept.

5.1.6. Electromotion, Inc. (formerly a division of Anderson Power Products, Inc.)

An electric highway van was designed "from the ground up" in Electromotion's Transportation System Laboratory (TSL). Its design features were shown in Figs. 17 and 18: tubular steel chassis, aluminum body, front-wheel drive. The battery is an 84-V lead-acid system; the motor (4,000 rpm) delivers 20 hp continuously with solid state control. As a postal delivery vehicle [131], it competes with the Battronic Van, with the Otis Electric vehicle and the (English) Harbuilt Electric truck [132].

A sports car for two passengers, with battery exchange capability. Streamlined, lightweight body and basically the power train of the van. Weight: 2,350 lb, twelve 6-V batteries.

Electromotion, Inc., was a major factor in developing the quick battery exchange concept. It was tried in the Boston area in cooperation with a car dealership. The vehicles have also a charger on board.

5.1.7. Otis Elevator Co.

Otis Elevator Co. has produced electric motors for decades and moved into the electric vehicle field with a mail delivery truck of its own. The Otis Van is conservatively built, like the Battronic truck. It has a 40-mile range; 43 mph is top speed. There are twelve 6-V batteries, and a 30-hp motor. Weight: 9,500 lb. Excels in grade climbing: 20% at 16 mph.

Acquiring the Electrobus Division from Tork-Link Corp., Otis extended its vehicle use for people transportation. In the city of Long Beach,

California, three of the Otis Electrobus vehicles, with 30% increased batteries (2 tons) went into regular service in August 1974. It has a 96-V Exide battery (E. S. B., Inc.) and a 50-hp dc motor.

Otis Elevator Co. pioneered the idea of electric vehicle transportation in large housing developments in Rochester, New York (1974) with an "electric car in each garage," to be used for visits and short shopping trips. The community also has repair and recharge facilities. When Otis Elevator Co. became a part of United Technology Co. in 1975, the building of of the Electrobus was licensed to Electric Vehicle Associates, Inc. (see 5.1.14).

5.1.8. Westinghouse

In 1967 Westinghouse announced the Markette, an electric vehicle for road use, carrying two persons. It was derived from golf carts and personnel carriers which the company manufactured. With a total weight of 1,730 lb, the Markette carried 792 lb of lead batteries (72 V) and operated on two 4.5-kW motors. Top speed: 25 mph; range: 50 miles. The vehicle was later withdrawn from the market.

An electric bus was introduced in 1970. It was very slow and much underpowered for its carrying capacity of 18 persons. It was designed for shopping center use (8 mph).

5.1.9. Copper Development Association, Inc.

The Copper electric van (model III) had a curb weight of 4,620 lb and a cargo capability of 1,000 lb. All drive components are concentrated in the forward section, which is effectively streamlined. Maximum speed is 50 mph. Range at 40 mph is 90 miles; with two stops per mile it is reduced to 57 miles. Grade climbing: 10%—19 mph, 20%—10 mph [133]. Batteries: 36 Exide 6-V batteries of the golf cart type, arranged in three strings of 72 V each; they can be switched in series for charging (an on-board charger is available). Controller: SCR, 500 A, pulse frequency circuit. Later a lower current field control arrangement was adopted.

The Copper Development Association furnished battery-operated six- to nine-passenger vans through a rental agency and promoted their use in inner city stop-and-go traffic. The light van type was successfully used for the Birmingham, Michigan, Water Department for meter readers. An advanced two seater sports car (designed by McKee Engineering Co.) was produced in 1975.

5.1.10. Linear Alpha, Inc.

Linear Alpha equipped production cars with electric drive trains. The Seneca 1975 was sold and serviced by selected Ford dealers. The 1975 Linearvan was sold and serviced through selected Dodge dealers in Illinois.

5.1.11. Sebring Vanguard, Inc.

Sebring Vanguard offers a second car for Florida living conditions (it is also sold in other parts of the United States). It carries two passengers and 200 lb of cargo at 28 mph, and goes 50 miles between charges.

The City Car has a plastic body and is powered by only six 6-V batteries. It is the only electric vehicle which is reasonably mass produced at the present time (2,500/year in 1974). Figure 80 shows the vehicle. Everything, including the control system, is (on purpose) very simple; ex-

FIG. 80. City Car made by Sebring Vanguard [134].

ample: battery switching with contactors, charging from 115-V home outlet, cost = 1/2 cent/mile to operate [134]. It is not a family car to go on
highways with, but it fulfills most local travel needs far better than a golf
cart. Price in 1973: $1,986. Improvements (48 V system) raised the cost
for 1974/1975 considerably [135].

5.1.12. General Electric

In 1968 General Electric built an experimental car with a battery/battery
hybrid. The vehicle had a magnesium frame and a fiberglass polyester
body; it weighed 2,300 lb and could carry two persons plus cargo. Lead
batteries and nickel oxide-cadmium batteries were used in parallel. The
total battery weight was about 1,000 lb. Top speed: 55 mph, range 40 to
50 miles. No further announcements have been made after GE Report No.
68-C-128.

5.1.13. Electric Fuel Propulsion, Inc.

Electric Fuel Propulsion has been very active over the years in converting
many production car types (American Motors line) into electric vehicles.
They range from a two-passenger car (the X-144, based on the 1973 Grem-
lin, $3,950) to a larger car (six-seater, 6,000 lb, 144-V wagon type,
$9,000) [136]. A luxury car, "Transformer I" (Chevrolet body) with a
trailer containing a motor generator charging unit was introduced in 1976.
Cost: $29,000 without trailer. The car is pictured in Fig. 81.

Electric Fuel Propulsion pioneered the establishment of charging stations on intercity roads (e.g., in Holiday Inns). An 80% charge can be obtained in 45 min, a 40% charge in 20 min. On this basis a car can travel
100 miles/day, with a booster charge at noon. During the Clean Air Car
Race in 1970, from Massachusetts Institute of Technology in Cambridge,
Massachusetts, to Pasadena, California, the company was instrumental in
setting up charging stations.

5.1.14. Electric Vehicle Associates, Inc.

Electric Vehicle Associates is a company located near Cleveland, Ohio,
which produces a two-passenger electric sedan in the $8,000-$10,000 range

FIG. 81. Transformer I, a real luxury electric powered car by Electric Fuel Propulsion Corp., Detroit (1976-company photo).

(Fig. 82). The body and chassis are imported from France (Renault). The car features a 10-kW traction motor, SCR control, an automatic three-speed transmission, radial tires, and a heating system (propane). The battery is a 96-V, heavy-duty, 200-Ah lead battery. A charger is on board (110/220 V). The car is considered to be a convenient city commuter vehicle to and from work or to the rail (transit) station. The range is 30 to 40 miles per charge, depending on speed and traffic conditions. It would be less in winter unless some of the propane heat is used to keep the battery

FIG. 82. The Electric Vehicle Association car (1975-company photo).

FIG. 83.   Corbin-Gentry motorcycle [137].   Speed is 32 mph for 40 miles.

warm.   Several of these cars were sold to the U. S. Government and to
Canada.   The company obtained the Electrobus license from Otis.   Recently
it joined Chloride (England) to produce vans and buses.

5.1.15.   Electric Motorcycles

Corbin-Gentry manufactures a lead battery-powered motorcycle for commuter use ($1,900).   Speed is 30 mph for 50 miles (Fig. 83).

   In 1974 a speed record of 165 mph was set by Corbin with a specially
built vehicle featuring Yardney Electric Corp. silver oxide-zinc cells
(120 V, 1,000 A).   The motorcycle's weight was 700 lb, and it used two jet
aircraft starter motors developing 100 hp each [137].

   ECO-Bike (Auranthetic Corp.) is an electric cycle powered by two
120 V lead batteries.   It can achieve a speed up to 30 mph, a range up to 50
miles (at lower speeds), but it is especially suited for in-plant usage.

   For a fuel cell operated motorbike, see Sec. 4.2.3.3. and Fig. 70.

5.1.16.   Electric Boats

Electric boats for fishing purposes, "amateur" converted by installing one
or two 12-V batteries and an electric motor, are found occasionally on
some lakes and enjoyed by fishermen who like the quiet (slow) operation.

Branson Boat Co., a Florida-based company, uses a permanent magnet
(1.5 hp) dc motor and a 24-V battery. At 5 mph the cruising range is 30
miles. The Electra boat is 18 ft long, has a 5-ft, 6-in. beam and a draft
of only 1 ft.

The two-man submersible Snooper, designed for underwater observa-
tion and photographic missions down to 1,000 ft, is built by Undersea
Graphics, Inc., of California. The vehicle has a 48-V (1,200-lb) battery.

Electric passenger boats are not used commercially in the United
States as yet, but they may find some use in the future, following the European example (Germany, Switzerland) in locations where motorboats are
outlawed or just not desired. A 14-passenger boat is reported to be in operation in California (Duffield Marine, Costa Mesa); it uses two 36-V banks
(250 Ah) of batteries and a 3.5-hp motor with a range of 30 to 40 miles.

5.1.17. Government Agencies

a. U.S. Army. The military had no electric vehicle program for personnel transportation. However, considerable efforts have been expended for
electric drives; generator-hybrid systems based on combustion engine primary power are in use in large trucks. In this connection the use of fuel
cell power generation was investigated. The 5-kW hydrogen-air fuel cell
system built by Allis Chalmers for the U.S. Army [138] was not intended
for use in a vehicle. A field power station supplied with hydrocarbon fuels
(steam-converted) was the purpose, but it could have been part of a hybrid
system for a car. The hydrazine-air system was chosen instead (Sec.
4.2.3.3), but abandoned later.

b. National Aeronautics and Space Administration (NASA). The space
agency had not considered electric vehicles until the drop in space support
and the emergence of environmental objectives changed the orientation of
some programs. The energy crisis puts even more emphasis on electric
propulsion and since 1976 NASA serves as testing agency for the Energy
Research and Development Administration (ERDA).

At the NASA Lewis Research Center, the battery group of the power
sources section is now engaged in vehicle battery testing and evaluation of
new systems for electric cars. For the initial studies, two Otis commer-

cial light trucks were used as test beds. Nickel oxide-zinc batteries are
considered closest to realization [15], ahead of the more powerful high-
temperature batteries. Their objectives follow conclusions derived from
the facts shown in Fig. 7.

c. Support. The activity of government agencies changed rapidly from the
(1974) "no funding" aspect for electric vehicles to an extensive support
within the framework of the Energy Research and Development Administration (ERDA). The report of the Federal Power Commission, Bureau of
Power, February 1967, prepared at the request of the Committee on Commerce of the U.S. Senate on the subject of the development of electrically
powered vehicles was largely negative and often referred to by legislators
and industry people. House Bill No. 5470 (McCormack, 1975) brought
finally the long-awaited funding of electric vehicle research and development. *

5.2. Electric Vehicle Developments Abroad

5.2.1. United Kingdom

The manufacture of electric vehicles for specialized purposes such as milk
delivery, street cleaning, ambulances, personnel carriers, etc., is a well-
established business in England. There are more electric road vehicles in
operation in Great Britain than in any other country. In 1967 the estimated
number was 50,000. It may not have increased significantly until very recently, when the gasoline shortage pointed to the need for an alternate city
transportation. However, the production will not increase very rapidly
since no large-scale manufacturing company exists.

A typical truck has a carrying capability of 3,000 lb; it makes 200 stops
on an intercity route over a distance of 20 to 30 miles and a maximum speed
of 15 to 20 mph. Their quietness, low maintenance, and low operating cost
make them ideal vehicles for door-to-door stop-and-go business.

_____

*The "Electric Vehicle Bill" became Public Law 94-413 in 1976, pro-
viding $160 million for vehicle demonstrations (7500 cars) and development
of technology, including batteries and hybrids, over five years.

FIG. 84.   The Enfield 8000 [139].

To cope with the suburban traffic flow, a speed increase to at least 40 mph is required. The usual 72 V, 300-Ah battery system can be upgraded to 96 V, and a modern control system, replacing the "contactors," may be installed.

a. Small Electric Car Developments. Small electric car developments aimed at the lightweight car (two-seater) are evident since 1967. The English Ford Comuta (see Sec. 5.1.2) or the Scottish Aviation Scamp are examples of this class (curb weight 1,000 to 1,200 lb, 48-V, 100-Ah batteries, maximum speed 40 mph).

The Enfield, a small car in the 1,800 to 2,000-lb class able to carry two adults and two children (or some cargo) has been produced in larger numbers, e.g., 61 for the Electricity Council in Fall 1974. This car is shown in Fig. 84. The control system uses series/parallel voltage and field switching. The battery: 48-V lead-acid; the motor: 5 hp at 2,200 rpm. Charger on board. Speed: 40 mph maximum; range: 80 to 90 miles at 30 mph, nonstop; city run: 50 miles. Acceleration: 0 to 20 mph in 6.5 sec, 0 to 30 mph in 17 sec. Hill climbing up to 20%. The price was about

$2,000 in 1973. The power train of the car (and its speed) has been im-
proved in the meantime (96-V battery, 8-hp motor).

b. Development of Larger Trucks and Buses. Development of larger
trucks (2 and 5 tons) and of a 40-mph bus (Silent Rider), was taken up by
the Chloride Motive Power Project Group [139].

c. New Battery Systems. New battery systems have been investigated by
Crompton, Joseph Lucas, Ltd., Chloride and British Motors, Ltd., first
in the direction of metal (zinc)-air batteries (see Sec. 4.2.2.2), later the
development of sodium-sulfur batteries was favored.

Shell Research Laboratories in Thornton demonstrated two fuel cell-
operated vehicles, one with a methanol and one with a hydrazine-fuel supply (see Sec. 4.2.3.4). The Chloride Battery Co. is supporting research
into new batteries at a fund level of over 1 million. Together with the
Electricity Council Research Center it sponsored the sodium-sulfur electric vehicle project discussed in Sec. 4.2.2.5 (see Fig. 51; also [139]).

d. Hybrid Electric Trucks. Hybrid electric trucks have been offered by
Austin-Crompton-Parkinson. Such vehicles achieve an extended range
through the use of a propane-fed engine generator on board. Crompton-
Parkinson also built experimental battery powered buses. The desirability
of hybrid systems is obvious, but it is a question of economics more than
one of technology.

5.2.2. Germany

Electric buses and trucks have the main attention of German developments.
Passenger cars are entirely in the experimental stage, demonstrating
some capability, but there is no urgent need to produce a small car since
city transportation is usually very good in Europe. However, the replacement of the large, slow, and polluting buses and delivery trucks by electric
vehicles is both desirable and seemingly economical. The list of world-
renowned companies which agreed to cooperate in the development of practical and economical vehicles is impressive:

Electricity producer: Rheinisch Westfülische Elektrizitaets Ges.
Chemical company: Bayer (plastic body material)

Construction and design: Messerschmitt-Boelkow-Blohm

Motor and controls: Siemens A. G., Bosch GmbH

Battery manufacturer: Varta A. G.

Automobile manufacturers: Deutsche Automobilges mbH, Daimler-Benz A. G., Volkswagenwerk A. G.

a. Buses and Trucks. Electric bus line in Moenchengladbach: "Seven battery-operated electric buses replace former diesel buses on a 24-mile line connecting the town center with the suburbs. Top speed is 35 to 40 mph. The 360-V, 455-Ah battery pack allows a range of 50 miles" [140]. This recent announcement (February 1975) is the result of a study begun several years ago, aiming at the development of a battery-powered 16-ton (g. v. wt.) Electrobus. Maschinen fabrik Augsburg-Nürnberg (MAN) produced initially two buses with trailers containing the battery. The battery in the trailer can be easily exchanged through a special lift mechanism. The idea of exchanging the whole trailer (4 tons) was considered first, but later abandoned in favor of the battery exchange, which could be done with less labor (1 man) because the bus could not move without the trailer and vice versa, a complicated situation [141]. The power train employed a 90-kW (122-hp) dc shunt (Bosch) motor (4,800 rpm) and a gear box. The electronic control used variable pulse frequency and duration.

Later, another battery-changing system (no trailer) was provided for buses with the batteries mounted in a special chassis frame which allowed a quick automatic exchange. Also the batteries were increased from 4 tons to 6 tons (still 360 V) and a (Siemens) series-wound motor was tested. Altogether 20 buses are planned for the program, including Düsseldorf a test city.

Daimler-Benz A. G. operated an experimental 16-ton hybrid bus (OE-302) with a diesel engine coupled to a 380-V (4,000- to 6,000-lb) lead battery. The range was about 200 miles [142].

A 1-ton city truck, with battery only (144 V, 1,700 lb) was also built (Transporter LE 306, 8,000 lb), but with a range of only 30 to 50 miles [143].

Volkswagenwerke A. G. built a 1/2-ton transporter, weight 5,500 lb, with the same 144-V, 1,700-lb battery and increased the range (at 35 mph) to 65 miles [144].

TABLE 27

Data for a MBB (Messerschmitt-Bolkow-Blohm) Transport Vehicle [145]

Chassis:	self-supporting plastic sandwich structure
Body:	polyurethane (foam)
Weight:	gross weight: 3.3 tons; load: 1 ton
Drive:	dc Bosch motor, 44 kW (60 hp)
Control:	Bosch thyristor, regenerative braking
Battery:	Varta 6 PZF 180, 72 cells (144 V), 1,900 lb
Top speed:	50 mph
Range:	40 to 70 miles
Acceleration:	0-35 mph: 13 sec
Grade climbing:	21%

Progress in body and chassis construction was considerably advanced in the design of the Elektrotransporter Bölkow (Table 27). Figure 20 showed its special design.

A triple hybrid, a Volkswagen bus with a Wankel motor (300 cm^3) generator (15 kW), a 17,500-rpm flywheel (100 lb), and a lead battery (144 V, 400 lb) has been built at the Technical University of Aachen [146].

b. Passenger Vehicles. Surprisingly little has been done in Germany on personal vehicles. Demonstration cars have been shown by various companies (BMW) at exhibitions and even at the Olympics in Munich. Speed records have been set (Opel), but there seems to be no conviction in industry circles that a personal car is economically feasible today. Indeed, for "leisure" the longest average driving range was calculated from statistics, with the obvious difficulties of returning to the starting point without recharge (72 miles R.T.).

Varta Batterie A.G. has studied the average daily driving distances for "business": 11.6 miles; "shopping": 15.6 miles; and "pleasure": 36 miles.

Realizing that power requirements differ according to applications, Varta has developed a lead-acid battery program divided into three classes: heavy duty batteries with high-power characteristics for buses and commercial vehicles [33 Wh/kg (15 Wh/lb), 1,500 cycles], medium-duty batteries

with high power and medium life for small commercial and city vehicles
[40 Wh/kg (18 Wh/lb), 700 cycles], and light-duty batteries with high power
and low life for leisure vehicles of all types [40 Wh/kg (18 Wh/lb), 300
cycles] [147].

c. Fuel Cell Power Sources. A fuel cell power source for electric vehicles has been considered by Siemens A.G. (see Sec. 4.2.3.2.c and Figs.
64 through 68). The Deutsche Automobil Gesellschaft mbH considered for
some years the development of metal-air batteries for electric vehicles
(see Sec. 4.2.2.2.); now it studies Ni-Zn cells [148].

d. Battery-Propelled Rail Vehicles. Battery-propelled rail vehicles have
been in use in Germany for many years [13]. They are used in peak traffic
or on short stretches as commuter trains and have found good passenger
acceptance. About 300 vehicles were operating in 1968.

e. Motorbikes. The Solo-Kleinmotoren GmbH produces an electric
"moped" for local traffic. About 10,000 were sold in 1973 in Europe.
Originally with pedal assist, the vehicle was improved and is now a full-
grown electric bike. A Bosch 750-W permanent magnet motor is used.
Speed: 16 mph; range: 25 miles. It is convenient for factories, hotels,
messenger use, etc. Figure 85 shows the vehicle. Price: $600 (1974).

f. Electric Boats. Ferry and pleasure boats with battery power are used
on some resort lakes where gasoline or diesel engines are outlawed. This
application seems to be nearly ideal from the environmental point of view.

A hydrogen-oxygen fuel cell-powered boat was demonstrated by the
Siemens Research Laboratories in Erlangen more than 10 years ago.

5.2.3. Japan

The interest in Japan is strongly oriented toward small passenger cars,
probably because of the heavy pollution and the high gasoline prices. However, trucks and buses have also found industrial attention.

Similar to Germany, the industrial approach to electric vehicles is
organized into an intercompany program and receives help from utility companies who are interested in load leveling. Additionally it has the strong
support of the government through the National Research and Development

FIG. 85. Solo Electra bike: 1 hp, 36 V, 37-mile range (company photo).

Program directed by the Ministry of International Trade and Industry.
This government-funded program ($20 million) is a 5-year task to be com-
pleted in 1976, at which time vehicle manufacturers are supposed to make
suggestions for production. Table 28 shows the goals of the 5-year plan
[149].

a. Daihatsu Kogyo Co., Ltd. About 270 units of the EBC-70 (EM 38) people
transporter (six persons) were used at the Osaka Expo '70. Speed was maximum 10 mph; range was about 90 miles. Motor: 3 kW. Weight: 1,700 lb.
Daihatsu produced the small car listed in Table 28, column 3. The electric
van, ES 38 V, was used first as a transporter at Expo '70. In 1973, 20 vehicles were produced; presently about 10 are made per month. The price
is $5,000 now, but in production quantities it should decrease to $3,000.
Figure 86 shows a picture of this vehicle. Its speed is 55 mph (maximum)
and its range is 40 to 50 miles. Motor: 6 kW, dc, recently up to 10 kW,
with battery increase from 72 to 84 V.

TABLE 28

Design Performance of 1973 Prototypes [149]

	Cargo		Passenger cars (and vans)		Buses
	Lightweight	Compact	Lightweight	Compact	
Passenger + loading capacity, kg (lb)	2 persons + 200 kg (440 lb)	2 persons + 1,000 kg (2,200 lb)	4 persons or 2 persons + 100 kg (220 lb)	5 persons or 3 persons + 300 kg (660 lb)	60–80 persons
Gross vehicular weight, kg (lb)	Approx. 1,100 kg (2,425 lb)	Approx. 3,500 kg (7,720 lb)	Approx. 1,000 kg (2,200 lb)	Approx. 2,000 kg (4,400 lb)	Approx. 15,000 kg (33,100 lb)
Maximum speed, km/h (mph)	More than 70 (43)	More than 70 (43)	More than 80 (50)	More than 80 (50)	More than 60 (37)
Mileage per one recharge, km (miles)	130–150 (80–93)	180–200 (112–124)	130–150 (80–93)	180–200 (112–124)	230–250 (143–155)
Acceleration 0–30 km/h (0–19 mph), in sec	Less than 5	Less than 5	Less than 4	Less than 3	Less than 8
Climbing ability (speed of climbing an inclination of 6°), km/h (mph)	More than 40 (25)	More than 40 (25)	More than 40 (25)	More than 40 (25)	More than 25 (16)
Company responsible for development	Toyo Kogyo	Nissan	Daihatsu	Toyota	Mitsubishi

Remarks: 1. The weight of batteries shall be less than 30% of the gross vehicular weight.

2. The energy density of a lead-acid storage battery shall be 60 Wh/kg (27 Wh/lb). This is based on constant output for 5 h.

3. The mileage per one recharge is based on a value in continuous running at a constant speed of 40 km/h (25 mph).

FIG. 86. Daihatsu ES 38 V (company photo).

The ES38P is a platform car on the same chassis as the van. The L38V is a small passenger car (2,000 lb, 60 V).

An electric tricycle (a battery-powered three-wheeler) with a swing frame shows excellent maneuverability. Speed: 20 mph. Range: 20 miles. Motor: 24 V, 0.6 kW. Weight: 600 lb.

b. Toyota Motor Co. [150]. Toyota produces a car for five passengers. Weight: 3,600 lb. It has a three-speed transmission and sixteen 12-V batteries (192 V). Speed: 50 to 60 mph; range: 80 miles at 30 mph. It comes close to the car in Table 28, column 4, except for the range. The model EV-2 features a computer control system and is said to have a range of over 100 miles.

c. Nissan Motor Co. [151]. Nissan produced a small electric service van (2,000 lb) for two persons. Motor: 96 V, 13 kW. Maximum speed: 50 mph; range: 47 miles. Except for the range, it is the car listed in Table 28, column 2.

In addition there is a heavier five-passenger vehicle (3,600 lb) with a
120-V, 16-kW motor. Speed: 55 mph; range: 35 miles. A 1-ton electric
truck (5,400 lb) with a 120-V, 27-kW motor is listed as EV-4 in the program. It is said to go for 100 to 125 miles at 45 mph.

d. Toyo Kogyo Co. Toyo Kogyo produced a flatbed van of only 1,700 lb
(2,400 lb gross weight) with a two-speed transmission and eight 12-V batteries. It is claimed to be able to travel for 80 to 90 miles at 45 mph on
one charge.

e. Mitsubishi Co. Mitsubishi built experimental buses for Nagoya and
Osaka. They are capable of loading 60 to 80 persons. Gross weight: 14 to
16 tons. Motor: 60 to 70 kW, 360 V. Speed: 40 mph maximum; range
(with six stops per mile): 50 miles.

Special battery charging and exchange stations, fully automated, have
been constructed. The 4-ton battery is "pushed through" the base of the
car.

An electric-diesel engine hybrid bus is used in Tokyo. The automatic
charge and exchange station is shown in Fig. 87.

f. Japan Storage Battery Co., Ltd. Japan Storage Battery Co. performed
extensive testing of lead batteries and evaluated them for vehicle operation.
Figure 88 shows a L-38V car used in the test fleet of Japan Storage Battery
Co. Details of new batteries are shown in Table 29 [152].

g. Yuasa Battery Co. Yuasa Battery Co. develops improved lead batteries
with thin electrodes; circulating electrolyte is also considered to be a possible improvement (inside cell circulation channels are provided). As a
new battery type, sodium-sulfur cells are being investigated (see Sec.
4.2.2.5).

h. Matsushita Electric Co. Matsushita works on iron-air batteries (Sec.
4.2.2.2.b).

i. Sanyo Electric Co. Sanyo Electric Co. studies zinc-air batteries (Sec.
4.2.2.2.a).

j. Sony Corp. Sony tested a hybrid car operated on zinc-air and Ni-Cd
batteries (Sec. 4.2.2.2.a).

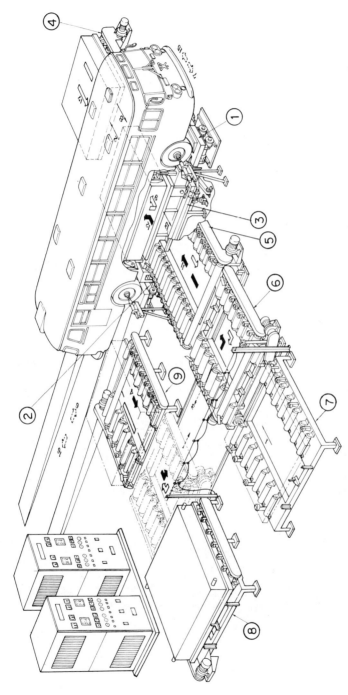

FIG. 87. Battery charge and exchange station for buses in Osaka. 1, 2: Wheel support, positioning device. 3, 4: Battery exchange mechanism. 5, 6, 7: Battery transport roller tables. 8, 9: Charging tables Nos. 1 and 2.

TABLE 29

Details of New Lead-Acid Batteries for Electric Vehicles

	Dimension (mm)				Weight including electrolyte (kg)	Performance			
Type	Total height	Case height	Length	Width		Wh/kg	Nominal capacity (Ah)	1-h capacity (Ah)	Life (cycles)
E75	232	207	304	172	23.5	38	75	49	>500
E100	249	215	410	176	32	38	100	65	>500
E135	257	222	504.5	182	42	38	135	88	>500

	Electric vehicle	Industrial vehicle	Automobile	
Energy density (Wh/kg)	38	30	37 ~ 41	by 5 hR capacity
Life (cycles)	500	1500	150	by industrial battery testing method
Internal resistance (Ω-Ah)	0.12	0.20	0.09	by industrial battery testing method

FIG. 88. The L38-V vehicle used by Japan Storage Battery Co. for fleet operations.

5.2.4. France

a. French Policy. The Inspector General of the Highways Department and
the President of the Interministerial Task Force for Electric Vehicles has
stated French policy as follows:

> Prototypes of electric vehicles have been produced by companies as,
> e.g., the Electricité de France, and the feasibility has been demon-
> strated; to continue this effort and to ensure further development,
> state support for improvements and production in small series has
> been requested. The public and certain regulatory and fiscal authori-
> ties will seek special measures to favor the development of electric
> vehicles [153].

During the past 5 years over $1 million was granted per year for im-
provement of the standard lead battery, for sodium-sulfur, zinc-air, and
fuel cell systems (especially methanol-air batteries). Hybrid electric
sources have also been studied.

Industrial tests: the program for developing small cars covered 100
vehicles at a cost of about $2 million. Vehicles, of the Renault 4 type,
were loaned to specific users for testing.

Financial assistance through the Ministry for the Protection of Nature
and the Environment has been given for the construction of two utility vehicles with a payload of 1,500 lb. Tests were performed with 20- and 50-
passenger buses.

b. Electricité de France [154]. A trial of 90 light utility vehicles was
made under "ordinary working conditions." The design characteristics of
the electric cars were as follows:

> Maximum speed: 40 mph, load capacity: 440 lb
> Overall weight: 1 ton (empty).
> Chassis: Renault R 2108 with R 1126 body.
> Battery: 700 lb, lead-acid, 48 V.
>> Actual performance: 15 Wh/lb at 15 W/lb.
>> Specified: 20 Wh/lb at the 5-h rate.
> Motor: weight: 85 lb, 5.5 kW, 4,500 rpm.

Speed controller: electronic or electromechanical through contactors or switches.

Heating: electric, 650 W, compartmentalized into seat-floor-roof and window heating.

Groups of vehicles were tested in the central city of Paris and in the suburbs. The results were as follows: with three stops per mile, the specific consumption is 0.45 kWh/km (0.72 kWh/mile) and the actual range of operation calculates from 27 to 40 miles.

A result of the tests will be a change to R-5 vehicles with improved features. Ultimately 200 vehicles are planned.

Electricité de France also built buses for 22 and 50 passengers. A trial over 2 years was made [155].

Sorel Electrobus for 22 Passengers. Total weight, loaded: 19,870 lb. Series-wound dc motor: 27 kW, 144 V. Battery: 575-Ah, flat-plate type, or motor: 36-kW, 192-V, with 350-Ah tubular-type battery. Weight: 6,600 lb (each system). Economy: cost with day current is twice as high as with night current. The latter compares to one-half of the gasoline cost per mile (7 cents/mile).

Sorel Bus for 50 Passengers. Weight: 26,000 lb; payload: 3 tons. Motor: 70-kW (662-lb), 240-V, 640-Ah battery (weight 8,830 lb). Six buses are planned. Tests in the cities of Dijon, Paris, and Grenoble (in the Alps).

c. Alsthom-Exxon Enterprises [156]. The cooperation of the French company Alsthom and the Exxon U.S., aimed at the development of highly compact fuel cell power sources, may lead to batteries for electric vehicles (see Sec. 4.2.3.3.d and Fig. 75).

d. Laboratoires de Marcoussis. Centre de Recherches de la Compagnie Générale d'Electricité investigates a zinc-air system with circulating electrolyte. See Sec. 4.2.2.2.

e. The Institut Francais du Petrol. The Institut Francais du Petrol studied the possibility of using a hydrogen-air battery as a vehicle power supply. The hydrogen is obtained from a methanol reforming unit (see Sec. 4.2.3.2.c).

5.2.5. Italy

a. Fiat S. P. A. A small two-seater car was designed to have good handling in city traffic and also in hilly areas [157]. Maximum speed: 55 mph; acceleration: 0 to 35 mph in 9 sec; grade climbing: 20%. Motor: 10 kW (18 kW peak) at 10,000 rpm, 96 V. Battery weight: 440 lb (range 40 miles) or 700 lb (range 60 miles). Figure 89 shows the construction of this car.

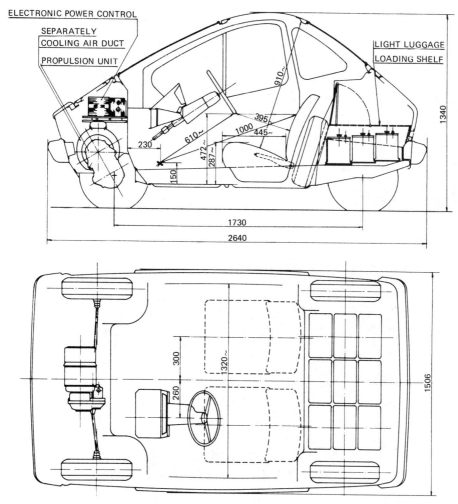

FIG. 89. Construction views of the Fiat two-passenger electric vehicle [157].

A diesel-electric hybrid bus with a total weight of 18 tons, for 100 passengers, was also proposed by Fiat. The specifications were made in view of its use on freeways, requiring speeds up to 70 mph.

b. Ghia S. p. A. and Demasono Automobili. Ghia and Demasono Automobili, subsidiaries of the Rowan Controller Co., Maryland, designed a four- to five-passenger vehicle with a curb weight of 1,300 lb. Speed: 40 mph; range: 50 to 100 miles. Motor: 8 kW, General Electric.

c. Zagato S. p. A.-Elcar Corp. Zegato-Elcar produces a two-seater city car, available in the United States. Model 1000 (1 kW) has a speed of 25 mph and a range up to 50 miles. Model 2000 (2 kW) has speeds up to 35 mph over a 50-mile range. Body: reinforced fiberglass, lightweight. Motor: 48-V dc; controller: 6 speed steps. The price is about $3,000 (1974). Figure 90 shows the car [158].

5.2.6. Sweden

ASEA exhibited an electrically powered Saab. It contained a 120-V Ni-Cd battery weighing 1,000 lb. Top speed was 45 mph, range 25 miles.

FIG. 90. Elcar two-passenger vehicle [158].

The activity in fuel cells at ASEA produced a 240-kW hydrogen-oxygen fuel cell as a demonstration power plant for submarines. Ammonia cracking was used as fuel source (see Sec. 4.3.2.3.f). Smaller units have powered lift trucks and various vehicles [159].

After several years' interruption, activity switched to metal-air cells (iron-air). See Sec. 4.2.2.2.b.

5.2.7. USSR

In the period 1948-1959 several prototypes of electric vehicles were tested in Moscow and Leningrad. A coordinated effort for the development of electric vehicles started in 1968. It involved the Ministry of Automobile Transportation, the Yerevan Polytechnical Institute, and the Scientific Research Institute for Electric Transport [160]. The Soviet prototype vehicle is a 1/2-ton truck, weight: 3,740 lb, speed 27 mph, range 48 miles. Two dc motors are used. An improvement over the EM-0466 has a streamlined body and a more powerful battery.

5.2.8. Australia

Flinders University in South Australia has an electric vehicle with a non-conventional power train. A printed circuit motor running at constant speed is connected to a variable-ratio transmission which operates on a hydrostatic drive principle. Motor: 10 kW, 150 V. Battery: 60 lb. Vehicle weight: 1,200 lb. Range: 50 to 88 miles; maximum speed: 50 mph [161].

6. THE ECONOMY, MARKET POTENTIAL, AND FUTURE OF THE ELECTRIC CAR

6.1. Utility Vehicle Versus Personal Car

The best example for utility vehicles for on-the-road service are the vans ordered by the U.S. Postal Service [129] and perhaps the milk delivery trucks used by the largest electric fleet (4,500 vehicles) in existence in the United Kingdom [162]. These services are characterized by multi-stop/starts and operation over a relatively short distance at low speeds. In a

fleet the maintenance conditions are good; the depreciation of the vehicles
can be stretched over as long a period as 10 years (without need for a
model change). The quietness and cleanliness of the electric trucks are
an environmental advantage, and the low price of night current charging
reduced the operating cost below the gasoline cost. The U. S. Postal Service test reported 8 days downtime in 7, 200 days of testing (30 vehicles).
Yearly operating cost: $384, including 4, 533 kWh of electricity for charg-
ing, etc. Vehicle cost: $10, 000 [129]. There is a good reason to extend
these advantages to utility repair vans, garbage collection trucks, meter
reading vehicles, and so on.

The next best-suited vehicles for electric operation are city buses,
commuter vehicles, school buses (in general, people movers) which operate on a predictable route, in congested traffic, with time available for recharging (or battery exchange) "at the end of the line."

It is therefore no coincidence that trucks and buses are the preferred
objects of electric vehicle developments. Fleet operators and utility companies are the ones who gain from such developments, with the public interest being served in the area of environmental improvement and better
energy distribution.

Surprisingly, battery companies were not the first proponents for developing electric vehicles. It becomes understandable if one considers the
tremendous investment cost in new battery production facilities versus the
uncertainty of recovery of that capital. One-third of all the lead produced
in the United States already goes into batteries, and until the advent of the
catalytic muffler, one-third went into gasoline additives, the rest was used
by other industrial manufacturers. Half of the lead is recycled. The total
yearly lead production in the United States is 1. 5 million tons. The price,
which rose rapidly in 1974, has stabilized and is assumed to be for some
time in the 26 to 30 cent/lb range [163]. It is not to be expected that electric vehicles will rapidly replace gasoline vehicles, but a penetration of
the light utility vehicle field, up to over a million (10% of the truck and bus
population) may happen within a decade. It depends to a large extent on
the attitude of the government, which in turn is strongly influenced by public opinion and industrial lobbies. The public has accepted environmental

measures with unexpected enthusiasm and has reacted to sensible energy-conserving proposals in a very positive sense.

In this respect it may be recalled that the 1967 report about development of electrically powered vehicles, prepared at the request of the Committee on Commerce, U.S. Senate, stated a very negative attitude, referring to performance tests made by the Post Office Department as showing that "the art does not produce a product suitable for daily Postal Service" [11]. The technology has really not changed much, but the situation with respect to energy availability and cost has.

The picture is more complicated in the area of personal vehicles. One argument in favor of small electric cars is the existence of many families with second and third cars. Table 30 shows a statistical method for estimating the market for electric passenger cars [11].

As a result, about one-third of all passenger cars may be electric cars within a decade or two. The sale of small electric cars could be 1 to 2 million per year in 1985 and 3 to 4 million between 1995 and 2000. Why does industry not jump at such an opportunity?

Working against this prediction is the poor performance capability of the lead-acid battery-powered electric passenger vehicle. Even though most of the cars discussed in this chapter show a 30- to 60-mile (50 to 100 km) operating range, sufficient to fulfill the "statistical" needs, the actual experience is different and every honest owner of a present-day electric vehicle with lead batteries will ultimately admit the disappointments in winter time, the unexpected failure of one or more battery units due to nonuniformity, the frustration relating to the low speed, especially at the end of the discharge and during hill climbing.

It will take at least a decade to develop, test, and tool up for mass production of acceptably performing short-distance cars available at a reasonable price. There is no large industrial firm in the automotive field willing to start earnestly until the market is "assured." "Assurance" cannot be given unless the government underwrites the capital expenditure and prevents the still favorable (60 cents/gal) gasoline price from slipping down in future years, perhaps through a political change in the oil-producing countries, or an economical dumping attempt.

TABLE 30

Method for Estimating Potential Market for Electric Passenger Cars

	Total number of passenger cars (millions)	Family-owned passenger cars* (millions)	Number of families owning cars (millions)	Second, third, etc., cars owned by families† (millions)	Market for battery-powered passenger cars	
					Total‡ (millions)	Percent of total passenger cars
1960	62	50	41	9	11	18
1965	76	61	46	15	18	24
1970	87	70	51	19	23	26
1975	99	79	56	23	27	27
1980	110	88	61	27	31	28
1985	120	96	66	30	35	29

*Total number of passenger cars less 20% of all passenger cars which are owned by all governments, Army personnel residing on base, individuals living in institutions, and cars owned by businesses.

†Family owned cars less number of families owning cars.

‡Second, third, etc., family passenger cars plus 20% of the passenger cars owned by all governments, Army personnel residing on base, individuals living in institutions, and cars owned by businesses.

It will take more than a decade to develop rechargeable batteries which can power a car over longer distances at 60 mph (100 km/h). Such batteries exist and are being tested, but mass production is far in the future.

Fuel cells, the ultimate power sources, which will have a high conversion efficiency and no charging delay, are definitely capable of fulfilling all power and range requirements, but again, system development has not begun anywhere and the cost picture is not very bright on the basis of today's catalyst technology.

In a long-range forecast the fuel cell-powered car has better chances than the rechargeable battery car for reasons of far greater efficiency and easy refueling. As pointed out already (Sec. 4.2.1), the incentives to produce "superrechargeable batteries" may not match the return on investment to be expected; besides, a motor generator/battery-hybrid vehicle would outperform it even with present-day technology.

Fuel cells must wait for a new "fuel economy," perhaps based on hydrogen or methanol, providing the necessary widely distributed and cheap fuel [123].

The market estimate made before, that "one-third of all passenger cars may be electric cars in the year 2000," must therefore be qualified with the expectations of the future development of electrochemical power sources. The move to electric vehicles may be described as occurring in three stages:

1. The increased use of lead-acid-powered cars in existing application areas—the present stage.

2. The extension to new application, depending on improvement of secondary cells and their attractiveness in an economic view—next decade.

3. The use of battery- or fuel cell-powered vehicles for general transportation, as alternatives to public transportation, when the availability of petroleum-based fuels becomes seriously restricted—end of the century [164].

6.2. Social, Economic, and Related Electric Power Aspects

The questions of public acceptance of low-powered vehicles, environmental specifications, and fashion trends in a group of the population are imponderable and the course of developments may suddenly change if not prescribed by Government action. However, once a well-designed small electric car enters the competitive market, it will stimulate and advance the technology again. Such a bootstrap process may be initiated by a small company and accelerate fast; it may be the result of a progressive, far-sighted single-person effort. Acceptance and public opinion cannot be legislated directly, but it can be directed by such means as declaring center city areas off-limits for gasoline vehicles.

One example of this is a "metrocab system" for lower and mid-Manhattan, New York City [165]. It was proposed to develop a low-pollution, off-grade right-of-way system using existing electric vehicle types. A small sedan or van which will comfortably seat two with luggage (possibly four-passenger capacity) was considered. This proposal was rejected, but it is only a question of time that it will be repeated.

Another example is the electric "minibus system" for mid-Manhattan [166]. A 20-passenger bus (Electrobus) could serve to relieve congestion by competing with the taxi; it carries more passengers and would reduce the number of vehicles. Special routes should accommodate the transportation needs in three areas: between railroad stations, to the main station, and to the shopping centers. For three loops, 16 buses would be needed. Frequency would be 5 min in rush hours. Fares would be competitive with transit fares of present taxis (minimum 50 cents, maximum $1.00). In 1974 the cost estimates were as follows:

The Electrobus sells for $28,000 without batteries. Battery installation and replacement when needed (hopefully not more frequently than every other year) amounts to $3,400. Cost of electricity is 2 to 3 cents/mile; cost of operation and maintenance is 3.5 cents/mile. The total yearly budget for 17 buses was estimated at $600,000 with yearly revenue of $990,000. The purchase of the vehicles and facilities (garages, charging stations, etc.) would have to be on a bond basis or with the help of subsidies (R&D grants).

There exists much skepticism among planners and city organizations that such systems would work. Unfortunately, only a demonstration would convince people, and the demonstration needs funds (a vicious cycle). Legislation to close midtown Manhattan—or any large city center—to private gasoline vehicles would certainly help the electric vehicle projects. It would be a great testing opportunity for gas/battery hybrid cars!

Similar projects aimed at the operation of electric vehicles on mall-type right-of-ways have been initiated for Lansing, Michigan, and Long Beach, California. The different climates may have determined the outcome of the projects (negative and positive results).

A feasibility study was also made for Honolulu, Hawaii [167]. A 50-passenger bus was considered in this case.

The availability of electric power is very much related to electric vehicle transportation schemes. Certain is the need of the electric utility companies for a load-leveling scheme, storing power in off-peak consumption periods. It sounds quite natural to use electric vehicles with rechargeable batteries for that purpose, perhaps more reasonable than building huge load-leveling batteries for their own sake.

Studies on this subject have been made. Night-time charging could change the load curve of the electricity supply, and substantial savings could be achieved. In most networks there is also an afternoon slack, which can be used for intermediate (parking lot) charging. Figure 91 shows typical load curves and the improvements offered by electric vehicle (time-controlled) recharging [168].

Up to 1972 the electric utilities predicted a huge increase in the need for electric power. It was argued that the changeover to electric cars would make the situation much worse because the combined needs of an increased population and an expanding industry could be satisfied only with difficulties in the years to come.

How wrong such a "sure" prediction may turn out is shown in Fig. 92 and Table 31, which explains the curves.

Since zero population growth is nearly reached now (in the Western countries) and the cost of fuel has gone up at least twice, it is not surprising to find that the choice is certainly between the lower curves. If this is true, and increased generating power is available (atomic plants), then it

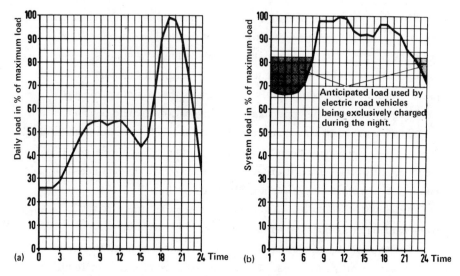

FIG. 91. Load curves of power companies. (a) Load curve of a
poorly electrified residential area. (b) Anticipated load curve of the public
electricity supply in winter 1985.

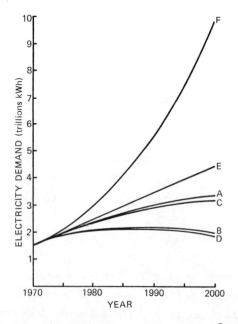

FIG. 92. Electricity demand projections [169].

TABLE 31

Electricity Demand Growth and Alternative Assumptions

Case	Population assumption	Electricity price assumption	Electricity demand (TkWh)			
			1975	1980	1990	2000
A	BEA	FPC	1.98	2.38	3.01	3.45
B	BEA	Double by 2000	1.88	2.07	2.11	2.01
C	ZPG 2035	FPC	1.98	2.37	2.95	3.29
D	ZPG 2035	Double by 2000	1.88	2.05	2.07	1.91
E	BEA	Constant	2.02	2.54	3.56	4.56
F	BEA	*	2.14	3.05	5.66	9.89

BEA, Bureau of Economic Analysis; FPC, Federal Power Commission; ZPG 2035, zero population growth reached in 2035. In the constant price assumption 1970 prices are maintained in each region. In the "double by 2000" assumption, the average price in each region increases annually by 3.33% of its 1970 value for 30 years. In case F, the FPC demand projection and the BEA population projections were used. A total of 1.53 TkWh of electricity was generated in 1970.

*Average prices decline 24% from 1970 to 1980, and 12% each 10 years thereafter until 2000.

will be very advantageous to switch to electric vehicle transportation. With the obvious disenchantment of the U.S. public with mass transportation (electric railroads are well established in Europe!), the personal electric car may become more easily accepted.

6.3. The "Next Best Solution": A Hybrid Car

The concept of hybrid vehicles is not new (see Sec. 2). Right from the beginning of automobile history, some people thought of coupling the low-torque gasoline engine (mechanically) with a Dynamotor in a parallel hybrid system [170]. Attempts to eliminate the clutch and transmission lead to the series hybrid systems in which an engine-generator is used to charge the batteries, which alone operate the electric motor. Such systems are used today for rail vehicles, buses, road machinery, trucks, and military ve-

hicles with multiwheel drives; the speed of the motor can be made completely independent from the wheels. The battery only acts as a "load
leveler," similar to a flywheel-hybrid system.

The parallel hybrid obtains higher efficiency by using the motor power
directly (without double conversion); the electric motor-generator (dynamotor) can be smaller. A series-parallel hybrid was developed by TRW
which had the advantages of both systems [171]. The arguments against
"hybrid systems" (of all kinds) were the double cost of (at least) two drive
systems.

The antipollution efforts brought the hybrid system into the foreground
because an optimized engine can be more easily cleaned up. The energy
crisis put more emphasis on the fuel-saving aspects.

A hybrid would solve the problems of battery cars with respect to
range, speed, and recharging facilities. Obviously the cost of generating
power on board with a heat engine is greater than that of charging from a
central power station and the conversion efficiency is lower, but the overall result for the vehicle user is better than to drive a gasoline engine in a
very inefficient mode and use high-priced gasoline or succumb to the drawbacks of a limited electric vehicle.

In a study by the Aerospace Corporation, four classes of vehicles were
considered: the full-size (4,000-lb) family car, the two-passenger (1,700-
lb) commuter car, the delivery/postal van, and a large 50-passenger highway bus [172]. This study for the Environmental Protection Agency was
aimed at pollution reduction for street vehicles, not at power saving. For
that reason the emphasis was on duplicating the performance of present
gasoline cars. The achievements in cleaning up the combustion engines between 1971 and 1974 made this viewpoint largely obsolete, especially since
the advantages were not really convincing. Large-horsepower engines
were chosen; the batteries were far too small to do much good. The setting
of limits for the combined weight of the power train led to that unfortunate
outcome.

Volkswagenwerk, Germany built a city taxi prototype for exhibition at
the Museum of Modern Art in New York, also showed the vehicle at the
"Convergence 76" in Dearborn, Michigan, 1976. A detailed description
can be found in Ref. 173. In principle it uses the parallel hybrid principle,

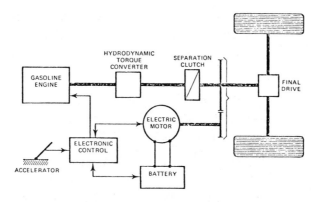

FIG. 93. Volkswagen city taxi hybrid power plant [173].

utilizing a separation clutch to power the rear wheels from the engine and/
or the electric motor. A diagram of the arrangement is shown in Fig. 93.

The gasoline engine develops 50 Din horsepower (1600 cm^3 displacement); the electric motor is a Bosch 130 V, 16 kW dc shunt motor with
32 kW peak power capability.

The vehicle can drive 43 mph (70 km/h) in electric mode, 65 mph
(104 km/h) in hybrid mode. The acceleration behavior is good: 0 to 60
mph in 30 seconds. Total weight: 4740 lb (2133 kg). Gasoline consumption: 20 miles per gallon (8.5 km/liter or 12 liter/100 km). The exhaust
specifications strictly follow Federal standards, which may be the reason
for the rather high fuel consumption.

A hybrid car of less weight and moderate power makes more sense if
general usefulness and gas savings are the main concern: a city vehicle,
maximum 2,200 lb (1000 kg) for two to four passengers with good acceleration, reasonable personal comfort (also in winter time), unlimited range
in the city, highway capability (55 mph, 90 km/h), if needed for short
trips, and considerable gasoline savings (over 40 mpg or less than 6 liter/
100 km) with low pollution levels. On the basis of my experience with the
fuel cell/lead battery hybrid vehicle (Sec. 4.2.3.2.b), I have built the following hybrid combination (which is also aimed at lowest possible cost):

Motor Generator-Battery Hybrid Vehicle:

Vehicle curb weight: 2,200 to 2,500 lb (1,000 to 1,200 kg)
Battery weight: 500 lb (230 kg), eight 12-V batteries, 100 Ah

Engine generator: 200 lb (90 kg), 16 hp, approx. 8 kW, 120 V

Motor: dc series, 10 hp continuous rating, 20 hp peak output

Control system: at least a two-gear transmission, SCR control or field

The car is able to drive at a maximum speed of 55 mph for 25 min to negotiate high-speed stretches of city throughways, and can go at 45 mph for any distance. Acceleration and hill climbing is comparable to any four-passenger small car. It can be heated. The retail battery cost (at 1976 prices) is $350—if SLI (starter) batteries of highest quality are chosen. They are needed for peak power, are not to be deep discharged. The economy of battery usage under such circumstances may be better than using golf cart batteries, which have a poor high-current capability. At this time a golf cart battery costs twice that of an SLI-battery. However, this price mark up is mainly due to the marketing (distribution) situation.

A suitable commercial engine generator (intermittent duty) uses gasoline at the rate about 3/4 gal/hr (2.8 liter/h) on full load, equal to 1 h driving in a mixed city-suburban driving mode.

If the generator is a 120 V ac machine, it needs only a full wave (4 diodes) rectifier and can also be used for topping or equalizing the battery when needed. In the garage the house current can be used more economically. The auxiliary system is a 12-V system, charged from a separate 8A engine generator circuit.

At the International Electric Vehicle Exposition in Chicago, April 26-29, 1977 this hybrid car has been demonstrated under the auspices of the Electric Vehicle Association of Ohio [174]. The drive train and the chassis of the vehicle are the same as described in Sec. 4.2.3.2.b.

The only problem which is not satisfactorily solved at the present time is the noise and exhaust pollution created by the single cylinder air-cooled engines which can be bought from the shelf as trailer or farm power generators; a water-cooled motorcycle engine is probably better suited. There is hope that some modifications will remedy this situation in the near future.

Figure 94 shows the mounting of the 7 kW motor generator (alternator) in the rear section of the hybrid vehicle.

FIG. 94. The 7-kW motor generator set used in the hybrid car built by Kordesch.

Actual road tests of this car demonstrated that the use of an auxiliary engine is of utmost importance, especially in winter driving. The straight driving range of the car—on batteries only—drops from 30 miles (about 50 km) to less than 10 miles (16 km) at -6°F (-21°C). In city driving, with many stops, the situation is even worse. With the engine generator operating the batteries retain their peak power capability—like any SLI battery— and the car accelerates satisfactorily. On the level road the gasoline consumption was measured to be above 50 mpg (5 liter/100 km)—with a clear possibility to further optimize the system, a gas consumption of 60 mpg (4 liter/100 km) seems feasible.

6.4. The Electric Vehicle Amateur

The hybrid car mentioned in the previous section is a worthwhile subject for talented electric vehicle builders or, better said, vehicle converters, since no private person can afford to design a vehicle from the ground up. "Since electric cars are not available at a reasonable price, the only solution is to build one." This statement is taken from a "build-your-own-electric car guide," a small booklet, also distributed by the Electric Vehicle Association of Ohio [174].

A few cautioning remarks are in order:

The cost of converting a gasoline car into an electric car still is in the neighborhood of $2,000 to $4,000—and the task should only be undertaken by people with mechanical expertise and with the availability of proper tools and occasional help of a welding and machine shop. The choice of a suitable vehicle to convert is the first most important step. It should be a rather new car, not a rusted, abandoned vehicle.

The result of at least several months of hard work should bring satisfaction. It is guaranteed if one does not do much highway driving or does not wish to transport more than two persons plus moderate luggage. For city driving a lightweight car is better than a "battery on four wheels," which wastes energy during accelerating its own weight at each stop sign.

Section 3 of this chapter will be helpful in judging the layout of the car and determining the size of the battery needed. The choice of the motor is important too. Rather larger than too small, the weight is not so decisive compared with other components. The tires should be radials. The speed control depends on the character of the builder. A manual four-speed transmission is of great service. A simple automatic transmission is recommended for people who hate shifting gears. The efficiency loss is not too great. SCR chopper circuits are available on the market ($500 to $1,000), but can be built by an "amateur" with electrical-electronic talents.

The most expensive parts of the project will be the engine generator (in case of a hybrid) about $1,000 and the electric motor. A new 10/25-kW dc motor retails for nearly $2,000. There are lower-cost motors ($900) around—some used, some imported. Also military surplus (aircraft starter) motors are available. Alternating current motors are less costly and relatively lightweight, but the control system is more complicated and expensive.

The batteries can be purchased locally from several reputable companies. The choice between SLI batteries and motive power (golf cart) batteries is one of economy: the latter are presently far more expensive and the increase in life is perhaps not worth the money (see Sec. 4).

The following discussion of vehicle conversion is reprinted from Ref. 174 and somewhat extended. There are three ways of building a four-wheel electric vehicle:

1. Convert an existing auto to electrical propulsion.
2. Use the chassis of an existing auto, convert to electrical operation, and provide a plastic body to complete the vehicle.
3. Build one from the ground up, using special fabricated parts, plastic body, etc., based on an original design.

The first method is possibly the fastest and the easiest to perform; it is also least expensive. The choice of the vehicle for conversion is quite wide from many foreign compacts available. Most have rear engines, some have front engines.

6.4.1. Popular Compacts for Conversion

1,250 to 1,400 lb: NSU-Prinz, Fiat 600, 850, 1000, BMW, Renault Dauphine
1,400 to 1,700 lb: DKW, Opel-Kadett, Toyota-Corolla, Datsun, Sunbeam, Austin, Simca, Ford-Anglia, Volkswagen, Saab
Over 1,700 lb: Volvo, Audi, VW-Bus, and most of the four-door imported cars.

Most of the two-door lightweight cars will still be in the original weight class after conversion, depending on the number of batteries on board, and with 2 passengers included they will have 2,000 to 2,500 lbs total weight.

6.4.2. Small U.S. Compacts for Conversion

Practically every small car of the three main American auto makers is suitable. Vegas, Pintos, Gremlins, and Pacers have been successfully converted. They are heavier than the European or Japanese small cars, and after conversion their weight will be over 3,500 lb. For hybrids they are better suited than the lightweight cars.

The conversion is started by removing the engine from the car and leaving the transmission in for a practical reason: it is handy for matching the speeds to the motor in climbing hills, and selecting most economical current levels; also, the reverse gear is useful in place of an electrical switching circuit. The clutch, however, is removed in most cases, unless the effect of flywheel energy storage is to be used [146].

The accelerator pedal is used in similar fashion as in the conventional auto—to start and accelerate the vehicle by activating various control circuits. In one case it may be connected to a variable resistor in a resistance-control system, or it may activate step-switching circuits, depending on the choice of the control system.

The installation of the electric motor as replacement of the engine requires an adapter plate, fabricated from aluminum plate of sufficient thickness to provide threaded holes for fastening the motor and through holes matching transmission flange holes.

Also, a coupling for connecting the transmission shaft to the electric motor shaft must be fabricated from the existing clutch coupling and another coupling matching the motor shaft spline welded in a unit.

The motor installation in a front-engine car is sometimes simpler, but otherwise will be quite similar to that of the rear-engine car, with additional support of the motor to the frame desirable to stabilize it.

All instruments connected to the engine may be removed; the speedometer and turn-signal flasher shall remain. We will need the following electrical instruments: one voltmeter indicating total battery voltage and one which can be switched by a rotary switch to check each battery and the accessory battery (this can be done with one instrument if the range is switched accordingly), one ammeter, and one state-of-charge indicator. A hybrid vehicle needs two ammeters (to check the charge current also). The ampere-hour meter may be an electronic integrating meter, or a remote acid-gravity indicator (see Sec. 4.1.4.3). If there is an on-board charger, a separate ammeter is recommended; it could be a center-zero meter to indicate also the charging current into the accessory battery.

Conversions have been successfully done by many enthusiastic individuals in small and large companies. They are too many to mention, but a few examples may be mentioned.

The Mars Electric Car was a converted Renault [175]. Several models were distributed to electric utility companies and to technical schools for studies (weight over 4,000 lb).

A Kharman-Ghia was converted to electric propulsion with a very exhaustive testing following the building of the car [176]. Weight: 3,500 lb with 1,500 lb of batteries (44%).

A Pinto was converted into an electric car by a team of Normandy High School students [25] and received wide recognition at the exhibition of the Third International Electric Vehicle Symposium in Washington, D. C., in 1974.

The second approach to building an electric car differs from the first one described above only by the fact that we replace the existing body of the car with a plastic body. These are available in a variety of designs from a dune buggy to a slick-looking rally car, in a wide price range.

The third method mentioned is based on total design approach, where all components can be controlled to give the best possible performance and efficiency.

The first consideration is the lowest possible weight of the total vehicle; because power consumption is a direct function of weight, it makes good sense to keep the battery weight to a minimum, which should be not more than 750 lb (340 kg).

The body and chassis construction would provide significant weight savings if it were formed from plastic in two sections—the upper body and the lower body chassis with independent wheel suspensions. The chances of a small business enterprise to produce electric vehicles are only good if government or nonprofit funding is obtainable. This viewpoint and marketing price considerations are given in [177].

An electric car design which comes close to the ideal one is that of the Sundancer [178] and of a two-seater sports car as shown in Fig. 95. However, the cost of those cars (hand production) is prohibitive for any amateur ($30,000 to $50,000).

6.4.3. Participation of the Electric Components Industry

Encouraging for the "electric vehicle amateur" is the growing number of companies engaged in the production of components for electric vehicles, mostly for existing golf carts, personnel carriers, etc., but some of the industrial suppliers are probably interested in selling their product also on the amateur-hobbyist market. A yearly "Directory of the Electric Vehicle Industry" is usually printed in the spring issues of Electric Vehicle News [179]. It also lists electric vehicle-related associations, clubs, societies,

FIG. 95.   This prototype Electric Town Car has a top speed of 55 mph (90 km/h) and can cruise at 40 mph (65 km/h) for 100 miles (160 km).   (Courtesy of Copper Development Association, Inc., New York.)

etc., which may be of particular help to private electric car builders who want information about specific problems.

At the present time there is no single company or organization which sells or recommends a complete (conversion) electric car "kit" with instruction manual, etc., similar to electronic hobby kits, but it is hoped that in the not too distant future an entrepreneur will bring this needed service to a probably very large group of mechanically interested people.

In this connection it is worthwhile to point to a recent (computer program) "Electric Vehicle Study" from the Ford Motor Co. in which the components of an electric car are evaluated from a mass production point of view [180]. It also shows that electric vehicles will cost more because parts are more expensive than gasoline-vehicle parts. A city car (Pinto), a metropolitan car (Maverick), and a delivery van or city bus are discussed.

The small city car has the best chance to be competitive in cost, even with present lead-acid batteries. The delivery van will most need improved batteries to approach general vehicle performance (but can easily make up for higher cost in specialized route applications).

Perhaps this was the reasoning of the Linear Alpha, Inc., designers who presented their 1975 Seneca series in the body styles of the Pinto, Pinto Wagon, and Mustang Ghia [181].

6.4.4. Mechanical and Electrical Specifications, Performance Tests

For the development of battery electric vehicles in England, the Electricity Council set up specifications based on the analysis of small van and car uses [139]. The document worked out is probably the most comprehensive list of electric vehicle requirements published today. It is essentially written for the testing of Enfield passenger cars (see Fig. 84) to assure compliance with safety and reliability concepts. The price set for quantity production including batteries, spare wheel, and tool kit was $1,500 (1976). It requires also a handbook with driving and maintenance instructions.

The vehicle specifications are listed in the Appendix.

6.4.5. The Selection of Batteries (A recurring problem, see Sec. 4.1)

For the amateur electric car builder (or converter), it is not easy to decide on the right battery for his car. Battery manufacturers [182, 183, 184] have

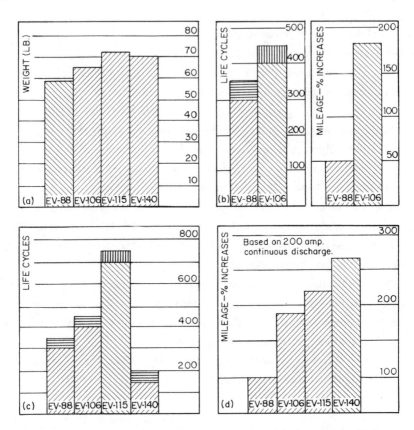

FIG. 96. Comparison of battery types for use in electric vehicles
[182]. (a) Lowest weight battery. Of the four Exide electric vehicle batteries, the EV-88 is the lightest—about 17% lighter than the heaviest of the
four. Yet the battery has an energy density that gives it a power capacity
roughly 17% higher than the industry standard for electric vehicles. This
battery may be your best choice for use in certain small vehicles where the
load capacity is limited and the desired range per cycle is modest.
(b) Higher capacity production battery. For many moderate-range vehicle
applications, the EV-106 may prove to be your best all-round performance
battery. It offers a power capacity roughly 41% higher than the industry
standard. Yet its weight is only 8% greater than the EV-88. At the same
time, this battery offers substantially greater life potential—400-450 cycles
compared with 300-350 cycles for the EV-88. (c) Longest life battery. In
laboratory tests, the EV-115 has shown a life capability of 700-750 cycles—
better than twice that of the EV-88. Also, it has a power capacity roughly
118% higher in typical electric vehicle service (200 A discharge). The EV-
115 features a new-design positive plate that more effectively retains its
power-producing materials. This battery is ideal for most heavy duty applications where long battery life is desired and the user prefers to keep replacements to a minimum. (d) Greatest mileage battery. The EV-140 is

TABLE 32

Example of Battery Specifications [182]

Battery type	Volts	Plates per cell	Dimensions	Weight (lb) Dry	Wet	Quan. electro. (qt)	Capacity at 77°, 75 A to 5.25 V (min)	Life (cycles) (lab.)
EV-88	6	19		40.4	59.8	7.1	88	300-350
EV-106	6	19	L: 10-3/8	47.2	65.1	6.6	106	300-400
EV-115	6	17	W: 7-3/16	56.5	72.5	6.3	115	700-750
EV-140	6	23	H: 11-7/32	52.7	70.5	6.7	140	150-200

recently published a list of battery specifications which at least tries to clear up the confusion between claims of "high power" and "long cycle life" (Table 32). The batteries of other major manufacturers are probably similar in their properties. Figure 96 shows the comparison of the four batteries listed in Table 32 in a graphical way.

6.5. Summary and Outlook for the Future

An attempt has been made throughout this chapter to assess realistically the chances of electric vehicles to become a part of our future life style. The existing and probable future technology has been evaluated—sometimes tinted by wishful thinking—but this is justified in cases of real possibilities based on facts as we know them today. In the end the economic and sociological demands will prevail and determine the future, and the technology which can supply what is best will then be utilized.

FIG. 96 (Cont.). the most advanced electric vehicle battery yet developed for energy density. It enables an electric vehicle to be driven further or at higher speeds for a longer distance than any of the other batteries. The specific range capability of any electric vehicle battery depends also, of course, on the vehicle's speed, the hardness of the tires, the type of terrain and certain other considerations. It should also be noted that the EV-140 has the lowest predicted life of any of the batteries. The designer must decide, therefore, which battery performance characteristics he wants most.

As an electrochemist it is even more obvious to the author that the development of better power systems will govern the wide use of electric vehicles for personal transportation. The responsibility of electrochemical science and technology for the progress of mankind must therefore be considered very high. Unfortunately, the educational base for a great electrochemistry of the future is not too good. There are very few universities and technical institutes where the practical side of galvanic systems is studied. The students going off to industry are usually not well prepared to cope with the still relatively new field. If the government wants to assure the existence of electric vehicles in the future, support is extremely urgent at the ground level of progress in independent research and development with long-range planning.

The most promising present outlook for electric vehicles was created by Public Law 94-413, enacted 1976. As a suitable finish to this chapter a recent announcement of the Energy Research and Development Administration [185] shall be reprinted here.

> Public Law 94-413, the Electric and Hybrid Research, Development, and Demonstration Act of 1976, dated September 17, 1976, requires that "Within twelve months after the date of this Act, the Administrator (of ERDA) shall develop data characterizing the present state-of-the-art with respect to electric and hybrid vehicles. The data so developed shall serve as baseline data to be utilized in order (1) to compare improvements in electric and hybrid vehicle technologies; (2) to assist in establishing the performance standards under subsection (b) (1); and (3) to otherwise assist in carrying out the purposes of this section."

ACKNOWLEDGMENT

The author is indebted to the members of the Electric Vehicle Association of Ohio for checking the accuracy of data. The help of S. J. Cieszewski in assembling the manuscript, proofreading and literature searching is very much appreciated.

APPENDIX A-1

The Electricity Council Battery Electric Passenger Vehicle
Working Party "Outline for Provisional Specification" (1974)

2-Seater, 2-Door, 4-Wheeled Saloon

Body

    Lightweight, possibly plastic or fiberglass construction.

Dimensions

Overall length	Approx. 6 ft, 6 in.	Overall width	4 ft, 4 in.
Height	4 ft, 3 in.	Wheel base	4 ft, 4 in.

Fittings

    Headlights, sidelight, windscreen, wiper, trafficators, horn, heating
and demisting equipment.

Instruments

    Speedometer, odometer, state-of-charge and mains "on" indicator.

Controls

    Accelerator and brake foot pedals, forward/reverse switch, hand
brake, light switches, horn button, heating/demisting switch.

Performance

Maximum speed	40 mph
Cruising speed	35 mph
Acceleration	0 to 40 mph in 10 sec
Hill climbing	To maintain a speed of 25 mph on a gradient of 1 in 12-1/2 and to start on a gradient of 1 in 5.
Ground clearance	5 in. fully laden
Range per charge	30 to 40 miles depending on terrain and other conditions
Battery charger	Provision for charger integral with car
Turning circle	Max. 18 ft

Production Cost

    The car should be designed on the basis of a quantity production cost
of £350 maximum, including batteries.

APPENDIX A-2

The Electricity Council Specification for Small City Car

1.0   Basic Construction

1.1   Chassis, space frame constructed in 1 in. square steel tube
           (surmounted by steel body skeleton), clad in 16 swg commercial-
           quality aluminum panels.
           Car undersealed.
           Front and rear battery bays will be vented to the outside of the
           car and sealed as far as possible from passenger compartment,
           and treated against battery acid.

1.2   Windscreen—triplex safety glass, electrically heated.
           Side and rear windows safety glass.

1.3   Hinged doors with sliding windows and antiburst locks.

1.4   Finished:
           Three colors of the ICI range will be available:
               1. Colorado orange
               2. White
               3. Bright red

2.0   Dimensions

           Overall length   112.0 in. (2.845 m)
           Overall width   56.0 in. (1.420 m)
           Overall height   56.0 in. (1.420 m)
           Wheelbase   67.0 in. (1.700 m)
           Track—Front   48.0 in. (1.219 m)
                  Rear   48.0 in. (1.219 m)

           Ground clearance:
               Body: Driver only   5.00 in. (12.7 cm)
               Differential: Curb weight   4.5 in.   (11.4 cm)

           Curb weight: Front axle   1,000 lb   (453.0 kg)
                              Rear axle   1,150 lb   (522.0 kg)
                                            2,250 lb   (975.0 kg)

Payload 350 lb (161 kg) (including driver).

A tolerance of $\pm 3\%$ shall be acceptable on figures given in this section.

3.0   Suspension and Steering

3.1   Suspension

Front: wide-based wishbones acting as swing axle, with coil spring damper units.

Castor angle: $5°$

Camber angle: $1/2°$ positive (mid-laden)

Wheel alignment: $1/2°$ toe-in

Rear: Live axle, combined coil spring damper units located by twin parallel radium arms and A bracket

3.2   Steering

Rack-and-pinion   2-1/2 turns lock to lock

Turning circle   Not greater than 27 ft

3.3   Brakes

Front: twin leading shoe

Rear: single leading/single trailing shoe

3.4   Tires

Radial 145 SR 10 in. -3-1/2 J rims

3.5   Towing eyes

Provision to be made for fixing towing eyes at both front and rear of the vehicle.

4.0   Transmission

4.1   Drive shaft: with sliding spline, or resilient coupling

4.2   Drive gear: spiral bevel

5.0   Motor and Control System

5.1   Motor

5.1.1   48 V dc series four-pole with ventilated enclosure and with internal cooling fan

5.2   Controller

5.2.1   Manufacturer to specify type of controller

5.3   Arc suppression.

5.3.1   Arc suppression diodes in traction and control circuits.

5.4   Cables

5.4.1   Cable for traction circuit: 70 mm^2 CSP

6.0   Instrumentation and Controls

6.1   Instruments

6.1.1   (a) Speedometer 0 to 60 mph, odometer, trip setting
           (b) Traction battery state-of-charge indicator—current-
                compensated voltmeter type
           (c) Auxiliary battery state-of-charge indicator

6.2   Controls

6.2.1   Control panel

6.2.1.1   (a) Forward/reverse switch
           (b) Lights switches
           (c) Heater switch
           (d) Hazard warning switch
           (e) Screen demist switch
           (f) Emergency cutout button

6.2.2   Steering column

6.2.2.1   Left-hand switch—Wipers (two-speed)
                            Windscreen washer
                            Headlamps on

6.2.2.2   Right-hand switch—Direction indicators
                             Full/dip beam
                             Headlamp flasher
                             Horn

6.3   Steering column lock—operates "ignition" which is also inter-
         locked with handbrake and charger. The headlamps shall also be
         supplied through this switch.

7.0   Electrical Equipment

7.1   Batteries

7.1.1   Traction batteries: 8 lead-acid, 6 cell, 12 V
         Rating: 92 Ah at 5-h rate
         Total weight: 700 lb (318.0 kg)

7.1.2   Auxiliary battery: 1 lead-acid, 6 cell, 12 V
         Rating: 47 Ah at 5-h rate

7.2   Electrical safety

7.2.1   Each pair of batteries fused

7.2.2   Traction circuit "cut" by manual emergency button, handbrake,
         and door microswitches.

7.3   Battery charging

7.3.1   Main battery
         Charger integral with vehicle and capable of giving an 80% dis-
         charged battery a full charge in 8 h.

7.3.2   The auxiliary battery shall be charged by a separate battery
         charge.

7.3.3   A lockable hinged flap to be provided in the normal filler cap po-
         sition for the mains lead for charging. The key for this flap to
         be if possible the same as the door key. The lead to have a 13-A
         plug on one end and a socket for attaching to the vehicle, during
         the time of charging, at the other end. The lead to be 16 ft long
         (5 m).

7.4   Lighting: 12 V

7.4.1   (a) Headlamps: rectangular sealed beam: $2 \times 60/75$ W
         (b) Side lights: in headlamp units 2×6 W
         (c) Rear lights 2×6 W

(d) Indicators	4 × 21 W
(e) Brake lights	2 × 21 W
(f) Number plate lights	2 × 5 W
(g) Courtesy light	1 × 6 W

7.5   Fusing

7.5.1   Traction batteries fused in pairs

7.5.2   Auxiliary battery fused

7.5.3   Each of the following circuits will be individually fused with 35-A cartridge-type automotive fuses.

7.5.3.1   (a) Control circuit

(b) Headlights

(c) Side lights

(d) Horn, stop lights, radio—if fitted

(e) Windscreen wipers, windscreen washer, heater

(f) Windscreen heater

(g) Indicators

8.0   Interior Fittings

8.1   White or black leathergrain upholstery, with black facia.

8.2   White head lining.

8.3   Black leather grain parcels shelf.

8.4   Black or dark gray carpet, black rubber mat.

8.5   Seatbelts, lap and diagonal to BS. AU 48.

8.6   One ashtray.

8.7   Rear-view mirror.

8.8   Glove box.

9.0   Optional Equipment

9.1   Car heater—48 V 600 W with 12-V recirculating fan

Note: Although an extra to the basic price, the Working Party recommend this to be installed in all cars.

9.2   Wing mirrors

9.3   Rapid charger

10.0   Performance

         Figures with driver and 144 lb (65 kg) ballast

10.1   Maximum speed of at least 40 mph on flat and level road

10.2   Distance traveled per battery charge

         Nonstop flat and level road with dry calm conditions at maximum
         speed terminating at 37 mph at least 40 miles

10.3   Acceleration:
         From standing start 0 to 30 mph: not more than 13 sec
         Repeated accelerations: three accelerations of 0 to 30 mph each
         in not more than 13 sec and one in not more than 16 sec—with
         9 miles at 30 to 35 mph between each acceleration

10.4   Hill climbing
         From standing start on 1 in 12.5 will attain 25 mph
         Will restart on 1 in 5

11.0   Braking

11.1   Foot brake
         0.8 g deceleration with a pedal pressure of not more than 65 lb

11.2   Parking brake
         The parking brake must be capable of holding the vehicle, facing
         in either direction, stationary on a gradient of at least 1 in 4

12.0   Reliability

12.1   Suspension, steering, and body construction to be checked by
         MIRA

13.0   Warranty

13.1   Vehicle:
         Free 500-mile service provided
         Six months labor and parts warranty

13.2   Battery (charged with built-in charger) 9 months or 10,000 miles,
            whichever is earlier

13.2.1   In order to determine whether or not a battery satisfies the war-
            ranty, tests shall be carried out in accordance with BS2550
            (1971), Sections 10.1 to 10.1.3., 10.2 to 10.2.8

13.2.2   If any 12-V battery is tested at the end of the manufacturer's
            warranty period, then it shall deliver when fully charged not less
            than 70% of the declared ampere-hour capacity when new
            See BS2550(1971) para. 10.3.8 except read 70 to 80% in this par-
            agraph.

14.0   Accessories

14.1   Each vehicle must be supplied with spare wheel, basic tool kit
            (jack, wheel brace with flattened section for removal of wheel
            embellishers, insulated spanner for battery terminals, box
            spanner either L-shape or with captivated bar for removing rear
            bumper bar, and special battery filler) and an Owner's Handbook
            giving driving instructions, routine maintenance instructions,
            and details of simple fault diagnosis.

APPENDIX A-3

The Electricity Council Battery Electric Vehicle Working Party
Mechanical Test Schedule for Battery Electric Car

1.   Introduction

     1.1 The test to be carried out at the Motor Industry Research Associ-
     ation (MIRA) establishment, Nuneaton.

     1.2 All tests, including driving the vehicle for the performance and
     endurance testing, to be carried out by MIRA Staff.

     1.3 The Electricity Council and the car developer to have observers
     present if so desired.

1.4 On completion of the tests, or before if any major failure occurs, the vehicle to be inspected by a representative of the Department of the Environment (Vehicle Engineering Division).

2.   Vehicle Construction

2.1 All tests, with the exception of 3.8, shall be carried out as a two-seater car with the MIRA standard 300-lb load (including the driver).

2.2 Both dimensions and weight to be recorded in accordance with MIRA standard practice.

2.3 Measurements of camber, castor, and toe-in be recorded for both laden and unladen vehicle.

2.4 Turning circle on full lock to be recorded in both directions.

2.5 The speedometer and odometer to be calibrated.

2.6 Six runs to be made at 20 and 30 mph through 50-ft water trough, the water depth to be 3 in. Water seepage into the car noted and the effects on electrical components to be examined.

2.7 In addition, MIRA will carry out a general examination of the mechanical construction of the vehicle—this to include also fitting of windows, doors, bonnet, and boot lid; tire wear, etc.

3.   Performance

3.1 Hydraulic and parking brakes tests to be carried out in accordance with MIRA standard practice, this to include hill holding with handbrake.

3.2 Brake fade to be tested also the effects of water ingress into brake drums after test 2.6.

3.3 A test to be carried out to produce a speed/time graph which will enable figures of acceleration to be obtained at any speed up to the maximum.

3.4 The vehicle shall be accelerated from 0 to 30 mph and the time recorded. The vehicle shall continue to be driven at 30 mph for

9 miles[*] and then stopped. The time taken for the journey to this point shall also be noted. The vehicle shall be kept at rest until 1 h has elapsed from the starting of the test. This test procedure shall be continually repeated until the vehicle can no longer be accelerated to 30 mph.

When the vehicle can no longer be accelerated to 30 mph the test procedure as outlined above shall continue except that the vehicle shall be accelerated to the maximum speed possible. When the vehicle can no longer be accelerated to 20 mph or the speed during the 9-mile run falls below 20 mph, the test shall be discontinued.

3.5 The speed which the vehicle could ultimately achieve, on a 1 in 12.5 gradient, from rest to be recorded (this to be simulated on the dynamometer).

3.6 A test to be carried out of the ability to pull away from rest of hills of 1 in 4, 1 in 5, and 1 in 6.

3.7 The maximum speed of the vehicle to be recorded. The test to be carried out on the level twin track.

3.8 The distance traveled to be recorded from rest with the full power applied the whole time. The distance to be noted when the specified maximum speed can no longer be maintained and the test to be discontinued when the speed drops to 20 mph. A note should be made also of the total distance traveled during the test. During this test a full speed history should be obtained.

3.9 The distance traveled to be recorded from rest with the driver only (no further weight to make up to the standard 300 lb) the full power to be applied the whole time. The distance to be noted when the specified maximum speed can no longer be maintained and the test to be discontinued when the speed drops to 20 mph. A note should be made also of the total distance traveled during this test. During this test a full speed history should be obtained.

---

[*]Nine miles has been chosen because it is a convenient distance on the MIRA track.

3.10 The distance traveled to be recorded from rest with the vehicle driven at a steady 30 mph. The distance to be noted when 30 mph can no longer be maintained and the test to be discontinued when the speed drops to 20 mph. A note should be made also of the total distance traveled.

3.11 The distance traveled to be recorded from rest with the vehicle driven at a steady 20 mph. The distance to be noted when 20 mph can no longer be maintained.

3.12 The distance traveled to be recorded with a maximum speed of 30 mph with four stops and starts per mile.

Maximum power should be applied to accelerate to 30 mph and this speed maintained until moderate braking is applied to stop the vehicle at the 440-yard mark. The vehicle should be kept at rest for 30 sec, then the procedure repeated. The distance shall be noted when the vehicle cannot be accelerated up to 30 mph. The procedure should be continued even when the vehicle cannot be accelerated at 30 mph but should be discontinued when the vehicle cannot be accelerated to 20 mph and the final distance noted.

Note: For tests covered by 3.2 to 3.11 the battery must be in a fully charged condition (According to the maker's specification) at the commencement of the tests, tires correctly inflated, the roads to be dry, and there to be negligible wind. Battery specific gravity should be recorded before and at the completion of each test.

The battery on the vehicle must be always charged overnight through the vehicle built-in charger. Wherever possible any standby batteries shall be charged with a charger identical to that built into the vehicle.

4.   Endurance Tests

4.1 The vehicle shall be driven for 100 miles at 25 mph over the Pavé circuit. The vehicle shall be fully examined after not more than each two circuits of the test track (approximately 3 miles) and any defects rectified by the car developer as they occur, and the test restarted.

4.2 The vehicle shall be driven for 500 miles over the Ride and Handling circuit. The vehicle to be full, examined in the early part of the test after each 25 miles and if no failures occur then at each battery change or battery charge (if the batteries are not changed) and any defects rectified by the car developer as they occur.

Note: If failure of any part of the vehicle or component occurs during the Pavé or ride and handling tests, then the repair—or if necessary modification—shall be carried out by the manufacturer and, as far as this part or component is concerned, the test shall be restarted. If, however, failure occurs in any part or component after it has successfully completed the specified distances (i.e., during the extended running) on both the Pavé and ride and handling circuit tests, then this latter failure shall not be taken into account when assessing the test results.

5. Noise Test

5.1 A noise test shall be carried out on the vehicle to B.S. 3425 Specification.

6. Tests with Alternative Motors

6.1 If the vehicle is to be offered with alternative motors, then the modified vehicles shall be tested as paragraphs 3.3, 3.4, 3.5, 3.6, 3.7, and 3.12.

7. Report of Test

7.1 MIRA to prepare a detailed report on the results of the tests. This report to cover factual data obtained, safety, and reliability, also a subjective assessment of the ride and handling characteristics.

APPENDIX A-4

The Electricity Council Battery Electric Vehicle Working
Party Electrical Test Schedule for Battery Electric Cars

1. Introduction

   1.1 The test to be carried out at the Electricity Council Research
   Centre at Capenhurst.

   1.2 The car developer can have an observer present during the period
   of the tests if so desired.

   1.3 The vehicle to be fitted with heater but not with radio.

2. Vehicle Examination

   2.1 A check shall be carried out of all electrical equipment to ensure
   that it conforms with the manufacturers' specification. This to cover
   cable sizes, fuse ratings, contactor ratings, etc.

   2.2 Check all auxiliaries—all lights, direction indicators, screen
   wipers, heater.

3. Vehicle—Electrical Tests

   3.1 Install instrumentation to record histograms of battery current,
   speed, and time spent in each controller step. Install thermocouples
   to record motor, battery, and transmission temperatures.

   3.2 Install the vehicle on the rolling road and carry out the following
   tests:

   (a) Simulated urban and city runs
   (b) During (a) above, record total range and from the current-
        time product obtain effective ampere-hour capacity of the bat-
        tery under working conditions.

   3.3 While the vehicle is operating on the rolling road the following
   will be checked:

   (a) Peak transient voltages in the system to assess the suitability
        of the insulation.

(b) Contact resistances of contactors. (Determine whether or not this value rises continuously.)

(c) Cable temperatures. (If high, check with manufacturers.)

(d) Effects of auxiliary equipment on range.

(e) Equality of discharge rates of each battery.

4. Charger—Electrical Tests

4.1 Test charger for satisfactory functioning. When charging, record current and voltage and compare with makers' specification.

4.2 Record the temperature of the charger components during the charging cycle.

4.3 Since the charger can be considered as a piece of "mains" operated domestic equipment, an examination and test should be carried out at the Appliance Testing Laboratories.

5. Battery

5.1 The batteries shall be tested under BS. 2550 (1971).

6. Vehicle Tests and Assessment

6.1 Measure the rolling resistance

6.2 Measure the level road speed corresponding to each step of the controller.

Note: 1. During the above tests the batteries shall be charged by means of the vehicle built-in charger.

APPENDIX B

Electric Vehicle Test Procedure—SAE J227a (Abstracted)
SAE Recommended Practice,
Last Revised February 1976
Society of Automotive Engineers, Inc.

B-1. Purpose and Scope—This SAE recommended practice establishes uniform procedures for testing electric battery powered vehicles which are capable of being operated on public and private roads. It is the intent of these recommended practices to provide standard tests which will allow various performance characteristics of electric vehicles to be cross-compared on a common basis in specifications, technical papers, and engineering discussions. The tests concern attributes of the total vehicle system rather than those of its subsystems and components. Tests of components such as batteries are the subject of separate procedures.

The road tests specified in this standard practice are recommended for use whenever possible particularly to establish vehicle performance specifications. The dynamometer procedures are included primarily to facilitate development testing.

Section 2 provides definitions of terminology used in this document. Section 3 specifies test conditions and instrumentation which are to be used for all the tests specified in this recommended practice while Section 4 identifies the data which are to be recorded for all tests. The specific tests covered by this document are:

Range at Steady Speed (Sec. 5)
Vehicle Range When Operated In A Selected Driving Pattern (Sec. 6)
Acceleration Characteristics On A Level Road (Sec. 7)
Gradeability Limit (Sec. 8)
Gradeability at Speed (Sec. 9)
Vehicle Road Energy Consumption (Sec. 10)
Vehicle Energy Economy (Sec. 11)
Deceleration (Sec. 12)

B-2. Terminology

2.1 Curb Weight—The total weight of the vehicle including batteries, lubricants, and other expendable supplies but excluding the driver, passengers, and other payloads.

2.2 Drive Line Ratio—The motor shaft rpm divided by the rpm of the traction wheels of the vehicle.

2.3 Gradeability—The maximum percent grade which the vehicle can traverse for a specified time at specified speed. The gradeability limit is the grade upon which the vehicle can just move forward.

2.4 Initial State of Charge (of Battery)—The amount of energy stored in the battery. When practical, the initial state of charge should be expressed as a percent of the capacity obtainable from a fully charged battery when discharged at a rate equivalent to the vehicle maximum cruise speed discharge rate.

2.5 Projected Frontal Area—The total frontal area of the vehicle obtained by projecting its image on a vertical plane normal to its direction of travel.

2.6 Tractive Force—The force available from the driving wheels at the driving wheel ground interface.

2.7 Tire Rolling Radius—The effective radius of a tire when it is deformed by the weight of the vehicle ballasted to its rated gross vehicle weight (SAE J670c).

2.8 Maximum Cruise Speed—The highest vehicle speed sustainable for at least one hour under specified environmental road test conditions starting with a fully charged battery; or such other maximum cruise speed as may be recommended by the vehicle manufacturer.

B-3. Test Conditions and Instrumentation Common To All Tests—The following conditions shall apply to all tests defined in this recommended practice unless otherwise stated in specific test procedures.

3.1 Condition of Vehicle

3.1.1 Vehicle shall be tested in its normal configuration with normal appendages (mirrors, bumpers, hub caps, etc.).

3.1.2 The vehicle shall be tested at manufacturer's rated gross vehicle weight.

3.1.3 Manufacturer's recommended tires shall be used. Tire pressure shall not exceed pressures recommended by Tire and Rim Association (TRA). Tire tread shall not be worn to the point where the tread wear indicators are exposed.

3.1.4 Normal manufacturer's recommended lubricants shall be employed.

3.1.5 The vehicle shall be stored for a minimum 8 h soak at ambient temperature (Sec. 3.3.1.1) before tests which start with a fully charged battery.

3.2.3 Full charge is to be established using manufacturer's recommended charging procedure and equipment.

3.2.4 For tests requiring an X% discharged battery at the start (for example, gradeability tests), the required initial state-of-charge will be established as follows. A Range at Steady Speed test (Sec. 5) shall be performed at recommended maximum cruise speed, and the end-point time and watt-hours consumed to the end-point of range determined. To achieve X% discharge of a fully charged battery, the battery will be discharged for X% of the end-point time either by driving the vehicle at recommended maximum cruise speed or by discharging the battery through a load at an equivalent constant power. Tests conducted with the battery partially discharged at the start must be initiated no longer than 10 min after the desired initial state-of-discharge is reached.

3.2.5 For tests in which the effects of battery initial state-of-charge are to be investigated, tests should be conducted with the propulsion batteries 0%, 40%, and 80% discharged.

3.3 Environmental Conditions

3.3.1 General

3.3.1.1 Temperature during vehicle and battery ambient soak period shall be within the range of $16°$-$32°$C ($60°$-$90°$F). Ambient temperature during road testing shall be in the range of $5°$-$32°$C ($40°$-$90°$F).

3.3.2 Road Tests

3.3.2.1 Road tests are to be performed on a road which is level to within $\pm 1\%$ and having a hard, dry surface. Tests shall be run in opposite direction when they are performed on a road test route. The direction of travel need not be reversed when operating on a closed test track.

3.3.2.2 The recorded wind speed at the test site during test shall not exceed 16 km/h (10 mph).

3.3.3 Dynamometer Tests

3.3.3.1 Dynamometer load must be programmable at various vehicle speeds to simulate vehicle road load versus speed characteristics.

3.3.3.2 Dynamometer road load power settings shall be made using either of the following procedures. Data from the coast-down tests described in Section 10 which establish the power required at various vehicle speeds to overcome aerodynamic drag and rolling resistance may be directly used to

program the power absorbed by the dynamometer. Alternatively the dynamometer road load points can be set to require the same power output from
the battery as is required to propel the vehicle at constant speed on a level
roadway. Battery power required to maintain various steady road speeds
shall be measured using the vehicle testing procedures described in Section 5 of this document.

3.3.3.3 Dynamometer flywheel shall be engaged with the nearest available
inertial weight which equals or exceeds the rated gross vehicle weight.

3.4 Test Instrumentation—This section provides a list of instruments
which are required to perform the tests specified in this recommended
practice. The overall error in recording or indicating instruments shall
be no worse than ±2% of the maximum value of the variable to be measured
(not including reading errors). Periodic calibration shall be performed
and documented to insure compliance with this requirement.

3.4.1 General Instrumentation—The following classes of instruments are
required for the purpose of tests outlined in this procedure.

> DC watt-hour meter or watt-time recorder
> Vehicle speed versus time recorder
> Distance versus time recorder
> Tire pressure gauge
> Ambient temperature versus time indicator
> DC watt meter
> Battery temperature indicator
> Electrolyte hydrometer (for vehicles with lead-acid batteries)
> AC kilowatt-hour meter

3.4.2 Road Tests

> Wind speed and direction measurement versus time
> Means for determining grades of test route segments
> Fifth wheel for measuring vehicle speed and distance

B-4. Data to be Recorded for all Tests

4.1 General

4.1.1 Vehicle Identification

4.1.2 Overall maximum dimensions (including projected frontal area)

4.1.3 Weight: curb weight and test weight to within ±2%.

4.1.4 Battery

Manufacturer.

Type and normal rating at specified discharge rate.

Previous history of the battery including chronological age, number and nature of charge-discharge cycles, description of the last discharge and recharge processes, and a brief description of known adverse usage conditions.

State of initial charge using the definition of percent charge presented in Sections 3.2.3 and 3.2.4. Where meaningful, other parameters such as open circuit voltage, electrolyte specific gravity, etc., shall also be stated.

Watt-hours discharged during test.

Temperature at start and end of test (either within electrolyte or at cell terminal, as appropriate).

4.1.5 Motor type and rating.

4.1.6 Overall drive train ratio(s) available, and those used during test, plus vehicle speeds at shift points if manual transmission.

4.1.7 Tires: manufacturer, design, size, rolling radius as specified by tire manufacturer, and pressure at start and end of test.

4.1.8 Power consumption of individual accessories, and times when each accessory was on during the test.

4.1.9 Environmental Conditions:

Range of ambient temperature during test

Range of wind velocities during test

Range of wind direction during test

Presence of any precipitation during test

Mean test site altitude relative to sea level

4.1.10 Running surface (road surface or dynamometer wheel).

4.1.11 Description of test route or dynamometer load program—road class, road surface type and condition (Table 9 of SAE J688), and lengths and grades of test route.

4.1.12 Date and starting and ending times of test.

4.1.13 List of all instrumentation used in test (manufacturer, model no., serial no.) and their last calibration date.

4.1.14 Any deviation from test procedure and reason for deviation.

4.2 Road Tests

4.2.1 Data shall be recorded and averaged for tests in opposite directions when tests are run on a road test route. The data reported shall be the average of at least two test runs in each direction. The range of test results and the number of test runs also shall be reported.

4.3 Dynamometer Tests

4.3.1 Description of dynamometer used (including drum or roll diameter and number of tire contact points).

4.3.2 Road load set points.

4.3.3 Equivalent inertial weight used.

4.3.4 Vehicle speed from dynamometer roll.

B-5. Range at Steady Speed

5.1 Purpose of Test—The purpose of this test is to determine the maximum range an electric road vehicle can achieve on a level road at steady speed.

5.2 Test Procedure—These road or dynamometer tests are to be conducted subject to the test conditions and data requirements described in Sections 3 and 4. Range tests are to be conducted at a minimum of three different test speeds including one test at the recommended maximum cruise speed of the vehicle. Individual tests shall be started with the vehicle propulsion battery in a full state of charge.

5.2.1 Road Tests—The vehicle shall be operated in a normal manner and be accelerated under its own power to the preselected test speed. The range test shall be continued without interruption at the preselected speed which is to be maintained to within ±5% until the vehicle reaches its end of range as defined in Section 5.3. The vehicle range shall be determined as the average of several tests made around a closed test track or in opposite directions over a road test route. The steady speed reported is to be the average speed maintained over the distance traveled.

5.2.2 Dynamometer Tests—The vehicle shall be brought to the preselected test speed under its own power and operated without interruption at within ±5% of this speed until the end of range is reached as defined in Section 5.3. Dynamometer test conditions are defined in Section 3.3.3. Range shall be determined as the average of several tests and the reported speed shall be the average speed maintained during the testing.

5.3 Definition of End of Range—The end of driving range is reached when the vehicle speed falls below 95% of the initially programmed steady speed or when such other vehicle performance limitation is reached as may be specified by the vehicle manufacturer. For example, if continuing the range test might result in deleterious operation of the battery, the vehicle manufacturer may relate the end of driving range to some characteristic of the battery such as terminal voltage under load.

5.4 Special Data Recording—In addition to recording the data specified in Section 4 the following special data shall be reported.

5.4.1 The test data shall be plotted as a curve showing range as a function of vehicle speed. The actual test points shall be indicated on this curve. When reporting these data, it shall be specified whether they are based upon road test or dynamometer test results.

5.4.2 The factor(s) involved in determining the end of range as defined in Section 5.3 shall be reported.

B-6. Vehicle Range When Operated In A Selected Driving Pattern

6.1 Purpose of Test—The purpose of this test is to determine the maximum range traveled and energy consumed by a test vehicle when operated on a level surface in a definite repeatable driving cycle. The driving cycles defined in this procedure are not necessarily intended to simulate a particular vehicle use pattern. Rather it is the intent of this section to provide standard procedures for testing electric road vehicles so that their performance can be cross-compared when operated over a fixed driving pattern.

6.2 Definition of Test Cycles—Four test cycles are defined to allow the vehicle to be tested under conditions which best match its intended use.

where: V = vehicle cruise speed—km, h (mph)

t_a = acceleration time—s

Table 1—Test Schedule for Repeatable Driving Pattern

Schedule	A	B	C	D
V	16 ± 1.5 km/h (10 ± 1 mph)	32 ± 1.5 km/h (20 ± 1 mph)	48 ± 1.5 km/h (30 ± 1 mph)	72 ± 1.5 km/h (45 ± 1 mph)
t_a	4 ± 1	19 ± 1	18 ± 2	28 ± 2
t_{cr}	0	19 ± 1	20 ± 1	50 ± 2
t_{co}	2 ± 1	4 ± 1	8 ± 1	10 ± 1
t_b	3 ± 1	5 ± 1	9 ± 1	9 ± 1
t_i	30 ± 2	25 ± 2	25 ± 2	25 ± 2
T	39 ± 2	72 ± 2	80 ± 2	122 ± 2

Note: All times shown are in seconds.

t_{cr} = cruise time at speed V—s

t_{co} = coast time—s

t_b = braking time to zero speed—s

t_i = idle time at zero speed—s

T = total cycle time—s

Values for the parameters of the four test cycles are presented in Table 1.

6.2.1 Driving Schedule A—Schedule A is characterized by a peak speed of 16 km/h (10 mph) and is intended for use in testing a vehicle designed for use on a fixed route with high frequency stop and go operation (for example, residential postal delivery van, milk truck, etc.).

6.2.2 Driving Schedule B—Schedule B is characterized by a cruise speed of 32 km/h (20 mph) and is intended for use in testing a vehicle designed for use on a fixed route with medium frequency stop and go operation (for example, bakery truck, shuttle bus, etc.).

6.2.3 Driving Schedule C—Schedule C is characterized by a cruise speed of 48 km/h (30 mph) and is intended for use in testing a vehicle designed to be used over a variable route with medium frequency stop and go operation (for example, parcel post delivery van, retail store delivery truck, etc.).

6.2.4 Driving Schedule D—Schedule D is characterized by a cruise speed of 72 km/h (45 mph) and is intended for use in testing a vehicle designed to be used over a variable route in stop and go driving typical of suburban areas (for example, commuter car, etc.).

6.3 Test Procedures—The road or dynamometer tests defined in this procedure are to be conducted subject to the test conditions and data requirements of Sections 3 and 4. The tests are to be started with the battery fully charged using the vehicle manufacturer's standard procedures.

6.3.1 Road Tests—The test vehicle shall be operated repeatedly and without interruption over the selected driving schedule on a level road or test track until it reaches its end of range as defined in Section 6.3.3. The vehicle range shall be determined as the average of at least three tests made around a closed test track or in opposite directions over a road test route. The steady speed reported is to be the distance traveled divided by the total elapsed time.

6.3.2 Dynamometer Tests—The test vehicle shall be operated repeatedly and without interruption over the selected driving schedule on a dynamometer until it reaches its end of range as defined in Section 6.3.3. Dynamometer test conditions are defined in Section 3.3.3.

6.3.3 End of Range—The end of driving range is defined as the end of the driving cycle immediately preceding the cycle in which the vehicle either ceases to meet the requirements of the selected driving schedule or reaches some other vehicle performance limitation specified by the vehicle manufacturer. For example, if continuing the test might result in deleterious operation of the battery, the vehicle manufacturer may relate the end of range to some battery characteristic such as its voltage under load.

6.4 Special Data Recording—In addition to recording the data specified in Section 4 the following special data shall be reported.

6.4.1 The range achieved, the number of test cycles successfully completed, and the test schedule used shall be recorded for each range test. The range reported shall be the average range achieved over at least three tests. The number of tests and the spread of the data also shall be reported. When reporting these data, it shall be specified whether they are based upon road test or dynamometer test results.

6.4.2 The factor used to define the end of range in Section 6.3.3 shall be identified and reported.

6.4.3 When dynamometer tests are run the road load set points used in the dynamometer shall be specified.

B-7. Acceleration Characteristics On A Level Road

7.1 Purpose of Test—The purpose of this test is to determine the maximum acceleration the vehicle can achieve on a level road with the propulsion battery at various initial states-of-charge.

7.2 Test Procedure—The road and dynamometer tests defined in this section are to be conducted subject to the test conditions, instrumentation, and data recording requirements of Sections 3 and 4.

7.3 Road Test Procedure

7.3.1 A suitable, straight, paved test route shall be selected upon which the vehicle can be safely accelerated to speeds near its peak speed.

7.3.2 The test vehicle is to be accelerated from a standing start at its maximum attainable, or permissible, acceleration rate until either the vehicle's peak speed is reached or until a safe limit speed is attained.

7.3.3 At least two successive runs shall be made in opposite directions over the test course to establish the vehicle's maximum acceleration characteristics at each of the three battery states-of-charge specified in Section 3.2.5. The time interval from the start of coast-down to the beginning of the next successive acceleration run at each battery state-of-charge shall not exceed five minutes.

7.4 Dynamometer Tests—Dynamometer test conditions are defined in Section 3.3.3. Vehicle speed shall be determined for these tests from measurements of the dynamometer drum or roller speed.

7.5 Special Data Recording—In addition to recording the data specified in Section 4, the following special data shall be reported.

7.5.1 The vehicle's acceleration characteristics shall be plotted as speed versus time for each of the initial states-of-charge. For each state-of-charge, the data to be plotted shall be the average results of at least two runs for that initial state-of-charge. When reporting these data, it shall be specified whether they are based upon road test or dynamometer test results.

B-8. Gradeability Limit

8.1 Purpose of Test—The purpose of this test is to determine the maximum grade on which the test vehicle can just move forward.

8.2 Test Procedure—Direct measurement of gradeability limit on steep test grades generally is impractical. Therefore, the gradeability limit is to be calculated from the manufacturer's recommended gross vehicle weight and the measured tractive force delivered by the vehicle at a speed near zero.

8.2.1 The tractive force shall be measured on a suitable horizontal surface and is the maximum force which can be maintained by the vehicle propulsion system for a period of 20 s while moving the vehicle at a minimum speed of 1.5 km/h (1 mph).

8.2.2 The tractive force shall be determined for various battery states-of-charge where the latter are defined in Section 3.2.5.

8.2.3 Because the high-rate discharge capability of batteries is time dependent, two tractive force tests are to be made for each battery state-of-charge. The lower of the two tractive force measurements shall be used to determine the gradeability limit.

8.3 Calculation of Gradeability Limit—The percent gradeability limit is to be determined.

8.4 Special Data Requirements—The procedures just defined establish the gradeability limit of the test vehicle as a function of the battery state-of-charge. If the traction force is limited by slippage between the vehicle's drive wheels and the road surface this fact should be recorded.

B-9. Gradeability At Speed

9.1 Purpose of Test—The purpose of this test is to determine the maximum vehicle speed which can be maintained on roads having different grades. The effect of battery state-of-charge on this vehicle capability is to be brought out in these tests. Two alternate procedures are described. An analytical method using data collected in Section 7, "Acceleration Characteristics On A Level Road," is described along with a direct dynamometer procedure.

9.2 Analytical Method

9.2.1 Using the speed-time data from the road tests of Section 7, the vehicle's acceleration characteristics shall be plotted for each state-of-charge. Data for successive time intervals then are to be used to determine the vehicle's average acceleration.

The data derived from these calculations shall be plotted as average acceleration versus vehicle speed and a smooth curve shall be drawn through the calculated points for each state-of-charge. If the test vehicle is equipped with a recording accelerometer as well as speedometer during the test of Section 7, the information is obtained directly and can be plotted as illustrated. The percent grade the vehicle is able to traverse at any selected speed is now to be calculated.

9.3 Dynamometer Tests—A chassis dynamometer also can be used to determine gradeability at speed providing that the total road power loss of the vehicle has been established for various vehicle speeds using the procedure of Section 10.

9.4 Special Data Recording

9.4.1 The calculated percent gradeability of the vehicle shall be recorded for each test speed and for the three battery initial states-of-charge specified in Section 3.2.5.

9.4.2 When reporting these gradeability data, it shall be specified whether they are based upon road test or dynamometer test results.

B-10. Vehicle Road Energy Consumption

10.1 Purpose of Test—The purpose of this procedure is to determine the power and energy consumed at varying vehicle speeds to overcome aerodynamic drag and rolling resistance.

10.2 Test Procedure—Vehicle road power and energy consumption at various steady speeds are to be determined from a coast-down test which shall be performed in the following way.

10.2.1 Accelerate the test vehicle under its own power on a level road or test track to its maximum safe speed.

10.2.2 Disconnect the drive motor(s) and allow the vehicle to coast freely to zero speed while recording vehicle speed versus time.

10.2.3 Repeat the coast-down test at least three times in opposite directions over the road or track to compensate for the effects of wind and grade.

10.2.4 During the coast-down tests, the power train loads which are coupled to the wheels shall be minimized or removed. If the vehicle has a transmission then the motor shall be isolated from the drive line by placing it in neutral. If the motor cannot be mechanically isolated then both the armature and field of the motor shall be electrically open circuited. A correction factor described in paragraph 10.4.4 shall be used to compensate for those power train loads which cannot be removed easily.

10.3 Data To Be Recorded

10.3.1 The speed or deceleration of the vehicle shall be recorded as a function of time during the coast-down tests.

10.3.2 In addition to the general information to be recorded which is specified in Section 3, any special modifications of the vehicle which were made to minimize power train loads during the coast-down tests as described in the previous paragraph shall be recorded.

10.4 Data Reduction—The vehicle speed versus time data obtained during the coast-down tests shall be processed to establish an average coast-down characteristic for the test vehicle. This characteristic curve shall be plotted.

10.4.1 Vehicle Road Load Power—The vehicle propulsion power required to overcome aerodynamic and rolling resistance is to be determined.

10.4.2 Vehicle Road Energy—The road energy consumed per kilometer in propelling the vehicle at steady speed also can be determined from the coast-down characteristics previously plotted.

Again using the vehicle speed, V_n, at successive time intervals, t_n, the road energy consumed at the average speed shall be determined.

The calculated road energy consumption per kilometer of travel for each calculated average value of road speed shall be plotted.

10.4.3 Alternate Procedures—If instantaneous values of vehicle acceleration are available from instruments which record acceleration directly, then instantaneous values of dissipated energy and power can be determined using appropriate equations to produce the relationships illustrated in these figures.

10.4.4 Correction for Power Train Loads—The values determined using this procedure are to be the energy and power dissipated external to the vehicle by aerodynamic drag and rolling losses. Corrections therefore must be made to the values of energy and power determined in paragraphs 10.4.1 and 10.4.2 to compensate for those power train loads which could not be eliminated during the coast-down tests. Specifically the windage and friction losses in the motor and drive line may be significant if they cannot be decoupled. In this case data describing the windage and friction losses of each component are to be obtained from the component manufacturer. Gear ratios in the drive line then are to be used to relate component speeds to vehicle speed and the energy and power dissipated in the drive line established as a function of vehicle speed. These data can be used to correct values of road energy and power previously obtained and the corrected energy and power curves plotted.

B-11. Vehicle Energy Economy

11.1 Purpose of Test—The purpose of this test is to define a measure of the overall energy economy of an electric vehicle and to define procedures for its evaluation.

11.2 Definition of Energy Economy—Energy usage involves the process of charging the battery as well as the consumption of this stored energy for vehicle propulsion. Because the vehicle user pays for charging energy, vehicle energy economy is defined as the vehicle range in various operating modes divided by the AC energy required to return the battery to its original state-of-charge. The vehicle energy economy therefore is defined as follows:

$$\text{Energy Economy} = \frac{\text{Range in prescribed driving mode}}{\text{AC energy required to recharge battery}}$$

$$\text{km/kWh (miles/kWh)}$$

11.3 Test Procedure—Tests for determining vehicle range at steady speed and vehicle range over a definite repeated driving cycle have been defined in Sections 5 and 6, respectively. Either of these range tests can be used to establish values for vehicle energy economy.

11.3.1 Vehicle manufacturer's recommended procedures shall be used to charge the battery to full capacity both before and after the selected range tests.

11.3.2 A watt-hour meter shall be installed across the AC energy source source used to charge the battery and the total energy consumed to return the battery to full charge shall be measured following the range test.

11.4 Data Recording—The range of the vehicle at steady speed or its range over a repeated driving cycle is to be divided by the AC energy required to return the battery to its initial state-of-charge as defined above. This quotient shall be reported as the energy economy of the electric vehicle under the particular conditions of the range test.

In the case of the continuous speed driving mode, the energy economy data can be presented in the form of a curve.

In the case of the repeatable driving pattern, the results should be presented in tabular form for the driving modes tested.

B-12. Deceleration

12.1 The vehicle shall be subjected to the same braking tests as other road vehicles and existing SAE standard practices shall apply.

APPENDIX C

List of the electric vehicles described in details in the Proceedings of the International Conference on Automotive Electronics and Electric Vehicles, CONVERGENCE 1976 [186].

Car name	Company name and address
AM General DJ-5E	TM General Corporation 32500 Van Born Road Wayne, Mich. 48184
Battronic Minivan	Battronic Truck Corp., Boyertown, Pa.
Braun Elec	Braunlich-Roessle Co. 3117 Penn Avenue Pittsburgh, Pa. 15201

Car name	Company name and address
CitiCar	Sebring Vanguard Inc. Sebring Air Terminal P.O. Box 1479 Sebring, Fla. 33870
Copper Electric Town Car	Copper Development Association, Inc. 405 Lexington Ave. New York, N.Y. 10017
Daihatsu B-20 - Motor Tricycle Daihatsu EH S40 Daihatsu EH-S40VM	Daihatsu Motor Sales Co., Ltd. 1, Daihatsu-Cho, Ikeda-City Osaka, Japan
Electra King PFS 125 Electra King PFS 135	B & Z Electric Car 1418 W. 17th Street Long Beach, Calif. 90813
Electra Van	Jet Industries, Ltd. 1141 West 6th Street Austin, Texas 78703
ES-512-Pb, Range: 80 km ES-512-Zn/Ni, Range: 148 km	General Motors Engineering Staff Warren, Mich.
EVA Electric Luxury Sedan	Electric Vehicle Associates, Inc. P.O. Box 9803 Brook Park, O. 44142
Lucas 1 tonne payload van (fixed battery pack) (Prototype) Lucas 1 tonne payload (Slide-in battery pack) (Prototype) Lucas Electric CF1 Pullman Lucas Electric Midi-Bus Lucas Electric Taxi (Prototype)	Lucas Industries Limited Great King Street Birmingham B19 2XF, England
MAZDA Electric Bongo Van MAZDA Electric Family Van MAZDA Electric Porter Cab Truck	Toyo-Kogyo Co., Ltd. 3-1 Shinchi, Fuchu-cho, Aki-gun Hiroshima, Japan
Silent Karrier Silent Rider	Chloride Technical Limited Wynne Avenue Swinton Manchester, England, M27 2 HB

Car name	Company name and address
THEV No. 2	National Tsing Hua University Hsinchu, Taiwan, ROC
Townobile 120	Elroy Engineering Pty, Ltd. 20 Lutanda Close Pennant Hills. 2120. Australia Tel. (02) 848-9385
Transformer I	Electric Fuel Propulsion Corp. 2237 Elliott Avenue Troy, Mich. 48084
Hybrid City Taxi	Volkswagen of America, Inc. 818 Sylvan Ave. Englewood Cliffs, N.J. 07632
Windmobile (Electric-Wind Hybrid)	J. L. Amick 1464 Cedar Bend Dr. Ann Arbor, Mich. 48105

Note: Since this last vehicle is really unique, its data are reprinted here:

Performance data
    Maximum speed: 65 mph
    Cruise speed: 50 mph
    Range (cruise): 220 miles

Propulsion battery
    Type: Lead acid
    No. of cells: 18
    Capacity: 35 A for 2 h
    Size: 20" × 12" × 9"
    Weight: 174 lb

Controller
    Type: EVC-23648
    Size: 4" × 6" × 2"
    Weight: 1 lb

Vehicle size
    Wheelbase: 6.5 ft
    Length: 11.5 ft
    Width: 8.0 ft
    Height: 8.0 ft
    No. of pass: 1

Vehicle weight:
    Curb weight: 750 lb
    GVW: 900 lb
    Payload: 150 lb

Propulsion motor (2)
    Type: permanent magnet
    Size: 4.5" diam. × 11"
    Weight: 23 lb
    Power: 1.5 hp

REFERENCES

1. The Automobile and Air Pollution, Parts I and II, PB-176884/85, U.S. Department of Commerce, Washington, D.C., December 1967.

2. "Progress in Areas of Public Concern," Conference Proceedings, General Motors Proving Ground, Milford, Mich., February 1971, pp. 5-13.

3. J. O'M. Bockris, ed., Electrochemistry of Cleaner Environments, Plenum Press, New York, 1972.

4. "Research and Development Opportunities for Improved Transportation Energy Usage," Dept. of Transportation, DOT-TSC-OST-73-14 (1972).

5. Shell Oil Company, "The National Energy Position," Houston, 1973.

6. S. K. Meucher and H. M. Ellis, "The Comparative Environmental Impact in 1980 of GMV vs. EMV," Gordian Associates, Inc., 1971.

7. G. J. Huebner, Jr., and D. J. Gasser, in: Energy and the Automobile, SP-383 Society of Automotive Engineers, paper No. 730518 (1973).

8. Energy flow patterns in the U. S. A., courtesy of Lawrence Livermore Labs.

9. "Automobile Facts and Figures," Automobile Manufacturer's Association (1970).

10. "A Review of Electric Vehicle Technology," a brochure published by the Applied Res. Lab. Scientific Res. Staff of Ford Motor Co. (about 1966).

11. "The Development of Electrically Powered Vehicles," Bureau of Power, Federal Power Commission, prepared at the request of the Committee on Commerce, U. S. Senate, 1967.

12. M. Pöhler, VDI-Berichte, 149, pp. 68-84 (1970).

13. G. Wilke, Akkumulator-Triebwagen, R. Oldenburg Verlag, Munich, 1954.

14. M. Barak, in: Power Systems for Electric Vehicles, U. S. Dept. of Health, Education and Welfare, Washington, D. C., 1967, pp. 105-119.

15. H. J. Schwartz, NASA Technical Memorandum TM X-71471 Electric Vehicle Battery Research and Development; also: Electrochemical Society Meeting, Boston, 1973.

16. Tire Performance Data, courtesy of Goodyear Tire and Rubber Co., Akron, Ohio.

17. D. M. Tenniswood, H. A. Graetzel, Automotive Engineering Congress, Detroit, Michigan, January 1967, Society of Automotive Engineers, paper no. 670177.

18. A Study of Low-Pollution-Potential Vehicles-Electric, for the Dept. of Health, Education and Welfare, Arthur D. Little, Inc., Contract No. PH86-67-108 (1968).

19. Electric Highway Vehicles—A System Approach, Anderson Power Products, Inc., Power Equipment Div., Boston, Mass., 1973.

20. W. E. Goldman, Proceedings of the Symposium on Batteries for Traction and Propulsion, March 7-8, 1972, pp. 242-254; the Columbus section of the Electrochemical Society, Columbus, Ohio, 1972.

21. H. Schmidt, Elektro-Transporter Bölkow, Varta A. G. Mitteilungen, ZV 5A Nr 649, March 10, 1971.

22. A. Kusko, Solid State D. C. Motor Drives, M. I. T. Press, Cambridge, Mass., 1969.

23. E. Kröhling, Varta A. G. , personal communication.

24. V. Wouk, "Electronic Circuits for Speed Control and Braking," in Power Systems for Electric Vehicles, U. S. Dept. of Health, Education and Welfare, Symposium at Polytechnic Institute of Brooklyn, April 6-8, 1967.

25. Electric Vehicles, a series of articles in Machine Design, October 17, 1974, issue, Penton Publication, Cleveland, Ohio.

26. General Electric Co. , SCR Manual, 5th ed. , 1972, pp. 369-387.

27. Thornton Power Systems, Waltham, Mass. , 1973.

28. H. Domann and S. Renner, Robert Bosch GmbH, "Experiences with Electric Drives for Vehicles," Third Electric Vehicle Symposium, Washington, D. C. , 1974, Paper No. 7450 A.

29. E. A. Rishavy, W. D. Bond, and T. A. Zechin, General Motors Corp. , "Electrovair," Automotive Engineering Congress, Detroit, Mich. , 1967, Society of Automotive Engineers, Paper No. 670175.

30. P. D. Agarwal and I. M. Levy, "AC Electric Drive System," Automotive Engineering Congress, Detroit, Mich. , 1967, Society of Automotive Engineers, paper no. 670178.

31. Ch. Bader, H. G. Plust, Deutsche Automobilges mbH, "Electric Propulsion Systems for Road Vehicles, State of the Art and Present Day Problems." Third Electric Vehicle Symposium, Washington, D. C. , 1974, paper no. 7478.

32. A. H. Walz, "Theoretical Approach to Electric Battery Vehicle Performance," Third Electric Vehicle Symposium, Washington, D. C. , 1974, paper no. 7418.

33. S. Sandelowsky, "How to Convert Petrol Cars into Electric Cars," Third Electric Vehicle Symposium, Washington, D. C. , 1974, paper no. 7421.

34. Akiya and Shinya Kozawa, "An Electric Car and Lead Acid Batteries," U. S. Branch of the Electrochemical Society of Japan, Cleveland, Ohio, (Yayoi Publishing Co. , Tokyo, 1971).

35. Toyo Kogyo, "Wankel-Engine Hybrid," Automobiltechn. Z. , 72, 454 (1970).

36. J. J. Gumbelton, D. L. Frank, S. L. Genslak, and A. G. Lucas, General Motors Corp. , Society of Automotive Engineers Meeting, May 1969, Chicago, paper no. 690461.

37. J. O'M. Bockris and D. M. Drazic, Electrochemical Science, Taylor & Francis, London, 1972.

38. Sears, Roebuck, and Co. , Die Hard battery, size 24F, manufactured by Globe-Union, Inc. Sears, Roebuck sales catalogs (1972-1974).

39. G. A. Mueller, ed. , Gould Battery Handbook, Gould, Inc. , Mendota Heights, Minn. , 1973.

40. Data provided by Furukawa Battery Co. , Japan.

41. Instructions, Maintenance and Service Manual, Gould Industrial Battery Div., Gould Natl. Batteries, Inc., St. Paul, Minn. 1972.

42. Electric Vehicle Power Systems, Gould, Inc., Automotive Battery Div., St. Paul, Minn., 1972.

43. K. V. Kordesch, J. Electrochem. Soc., 119, 1053 (1972).

44. C. C. Christianson, Third Electric Vehicle Symposium, Washington, D.C., 1974, paper no. 7469.

45. K. V. Kordesch and F. Kornfeil, U.S. Pat. 2864055 (1958).

46. M. Lurie, H. N. Seiger, and R. C. Shair, Power Sources Conference, 17, 110 (1963).

47. G. Lander, Varta Batterie A.G., Battery Council International, Mexico City, Mexico, 1975.

48. Yuasa Battery Co., Japan (1971).

49. G. S. News, Technical Reports, Japan Storage Battery Co. [Vol. 28, No. 1 (1970) and following years].

50. T. Ishikawa, K. Shimizu, Denki Kagaku, 39, 605-612 (1971).

51. J. F. Macholl and A. G. Koch, Battery Symposium, Cleveland 1956, AABM, 1956. Similar results of Russian work: E. I. Krepakowa and B. N. Kabanov (Akkumulatoren), also unpublished work by J. J. Lander, Naval Research Laboratory (1958).

52. S. U. Falk and A. J. Salkind, Alkaline Storage Batteries, Wiley, New York, 1969.

53. U.S. Pat. 3,853,624 (1974), Westinghouse.

54. Varta Batterie A.G., "The Fenox Battery," 4th International Vehicle Symp., Düsseldorf, Germany, 1976.

55. J. J. Lander and A. Fleischer, Zinc-Silver Oxide Batteries, Wiley, New York, 1971.

56. K. V. Kordesch, "Alkaline Manganese Dioxide-Zinc Cells," in Batteries, Vol. 1, Manganese Dioxide, K. V. Kordesch, ed., Marcel Dekker, New York, 1974.

57. A. Charkey, Proceedings of the 7th Intersociety Energy Conversion Engineering Conference, San Diego, Calif., American Chemical Society, Washington, D.C., 1972, pp. 110-113.

58. G. C. Kugler, Second International Electric Vehicle Symposium, Atlantic City, N.J., Nov. 1971.

59. Ni-Volt Batteries, Energy Res. Corp., Bethel, Conn.; Contract DAAB07-72-C-0114, Ft. Belvoir, Nov. 1973.

60. E. Luksha, Gould, Inc., NASA CR-134658, Lewis Res. Center, 1974.

61. G. Caprioglio and J. T. Porter, II, Advances in Battery Technology Symposium, Vol. 3, pp. 86-100, The Electrochemical Society, Southern Calif.-Nev. Section, 1967.

62. A. J. Appelby and J. P. Gabano, Proceedings of the Symposium and Workshop on Advanced Battery Res. & Design, Argonne Natl. Lab., March 1976. The Electrochem. Soc., 1976.

63. R. R. Witherspoon, Society of Automotive Engineers, Detroit, Mich., Jan. 1969, paper no. 690204.

64. Hideo Baba, Society of Automotive Engineers, Detroit, Mich., Jan. 1971, paper no. 710237.

65. P. Lefant, Proceedings of the Symposium on Batteries for Traction and Propulsion, March 7-8, 1972. The Electrochemical Society, Columbus Section, Columbus, Ohio, 1972.

66. F. Kober and M. Yarish, Westinghouse, 132 Meeting of the Electrochemical Society, Chicago, October 1967.

67. G. Kraemer and V. A. Oliapuram, Varta A.G., Meeting of the International Society of Electrochemistry, Battery Section, Marcoussis, France, May 1975.

68. H. Cnobloch, D. Gröppel, D. Kühl, W. Nippe, and Gisela Siemsen, Siemens Forschungslaboratorium, Brighton Power Source Conference, 1974, paper no. 17.

69. O. Lindström, Swedish National Development Co., Brighton Power Source Conference, 1974, paper no. 18.

70. S. Zaromb, Power Systems for Electric Vehicles, Symposium at Columbia University and Polytechnic Institute of Brooklyn, April 6-8, 1967, U.S. Dept. of Health, Education and Welfare, pp. 255-267.

71. L. A. Heredy, H. L. Recht, D. E. McKenzie, Power Systems for Electric Vehicles, Symposium at Columbia University and Polytechnic Institute of Brooklyn, April 6-8, 1967, U.S. Dept. of Health, Education and Welfare, pp. 245-253.

72. M. Kelin, Proceedings of the 6th Intersoc. Energy Conv. Eng. Conf., Boston, 1971, Society of Automotive Engineers, New York, 1971, pp. 79-83.

73. J. Dunlop, J. Stockel, and G. van Ommering, (COMSAT), Power Sources Symposium, Brighton, 1974, paper no. 20.

74. C. J. Amato, Society of Automotive Engineers Meeting, Detroit, Mich., Jan. 1973, paper no. 730248.

75. P. C. Symons, Energy Development Associates, Proceedings of the Third International Electric Vehicle Symposium, Washington, D.C., 1974, paper no. 7432.

76. R. Zito and D. L. Maricle, General Electronic Laboratory, Proceedings of the Second International Electric Vehicle Symposium, Atlantic City, N.J., 1972.

77. G. Glerici, M. deRossi and M. Marchetto, Power Sources 5 (1974), D. M. Collins, ed., pp. 167-181, Academic Press, N.Y., 1975.

78. S. Gratch, J. V. Petrocelli, R. P. Tischer, R. W. Minck, T. J. Wahlen, Proceedings of the 7th Intersociety Energy Conversion Engi-

neering Conference, San Diego, Calif., 1972, American Chemical
Society, Washington, D.C., 1972.

79. C. A. Levine, G. G. Heitz, and W. E. Brown, Proceedings of the 7th
     Intersociety Energy Conversion Engineering Conference, American
     Chemical Society, Washington, D.C., 1972.

80. E. T. Seo, R. R. Sayano, M. L. McClanahan and H. P. Silverman,
     26th Power Sources Symposium, May 1974, PSC Publication Comm.,
     Red Bank, N.J.

81. H. A. Wilcox, 21st Annual Power Sources Conference, 1967, PSC Pub-
     lication Comm., Red Bank, N.J., pp. 39-42.

82. E. C. Gay, W. J. Walsh, J. D. Arntzen, and E. J. Cairns, Proceed-
     ings of the 7th Intersociety Energy Conversion Engineering Conference,
     San Diego, Calif., 1972, American Chemical Society, Washington,
     D.C., 1972, pp. 54-70.

83. P. A. Nelson, A. A. Chilenskas, and R. K. Steunenberg, Argonne
     National Laboratory Report ANL-8075 (1974). See also ANL-8058
     (1974) and ANL-8039 (1973).

84. L. R. McCoy, S. Lai, R. C. Saunders, and L. A. Heredy, 26th Power
     Sources Symposium, May 1974, and EPRI Report, 1975.

85. J. O'M. Bockris and S. Srinivasan, Fuel Cells: Their Electrochem-
     istry, McGraw-Hill, New York, 1969.

86. C. Berger, ed., Handbook of Fuel Cell Technology, Prentice-Hall,
     Englewood Cliffs, N.J., 1968, chapter on low-pressure, low-
     temperature fuel cells by K. V. Kordesch.

87. C. Marks, E. A. Rishavy, and F. A. Wyczalek, Automotive Engineers
     Congress, Detroit, Mich., Jan. 9-13, 1967, SAE paper no. 670176.

88. C. E. Winters and W. L. Morgan, Automotive Engineers Congress,
     Detroit, Mich., Jan. 9-13, 1967, SAE paper no. 670182.

89. K. V. Kordesch, in Fuel Cells, W. Mitchell, ed., Academic Press,
     New York, 1963.

90. M. B. Clark, W. G. Darland, and K. V. Kordesch, Electrochem.
     Technol., 3, 166-171 (May-June 1965).

91. K. V. Kordesch, J. Electrochem. Soc., 118, 815 (1971).

92. K. V. Kordesch, Proceedings of 6th Intersociety Energy Conversion
     Engineering Conference, Society of Automotive Engineers, 1971,
     paper no. 719015, pp. 103-111.

93. K. V. Kordesch, Power Sources for Electric Vehicles, Modern As-
     pects of Electrochemistry, Vol. 10, J. O'M. Bockris, ed., Plenum
     Publishing Co., New York, 1975.

94. A. Michel and W. Frie, Third International Electric Vehicle Sympo-
     sium, Washington, D.C., 1974, paper no. 7452.

95. H. Cnobloch, M. Marchetto, H. Nischik, G. Richter, and F. von
     Sturm, Siemens A.G., Third International Symposium on Fuel Cells,
     Brussels, June 16, 1969, pp. 203-209.

96. Y. Breele, Report of the Institut Francais du Petrole, 1972.

97. Fuel Cells, A Review of Government Sponsored Research, NASA SP-120, National Aeronautics and Space Administration, Washington, D. C., Contract to Dept. of Chem. Engr., North Carolina State Univ., L. G. Austin, ed., John I. Thompson & Co., 1967.

98. H. A. Liebhafsky and E. J. Cairns, Fuel Cells and Fuel Cell Batteries, Wiley, New York, 1968.

99. A. M. Adams, F. T. Bacon, and R. G. Watson, in Fuel Cells, W. Mitchell, ed., Academic Press, New York, 1963.

100. "Powercels," A Pratt & Whitney Aircraft Co. Publication (1967/1968), describes ground power and also Apollo Mission Performances of fuel cell.

101. Acid Electrolyte Fuel Cell Technology Program, Final Report, SPR-113, Contract NAS9-12332, General Electric Co., NASA-Lyndon B. Johnson Space Center, R&D Branch, Houston, Texas, Oct. 1973.

102. Fuel Cell Technology Program, Final Report, PWA-4756, NASA Contract No. NAS9-11034, Exhibit G, Pratt & Whitney Aircraft to L. B. Johnson Space Center, Houston, Texas, June 1973.

103. Fuel Cells, A Technology Survey, B. J. Crowe, Computer Sciences Corp., NASA SP-5115, Contract NASW-2173, Technology Utilization Office, Washington, D. C., 1973.

104. Code of Federal Regulations, Title 49, Parts 170-190 (Rev. Jan. 1969), U. S. Government Printing Office, Washington, D. C., 1969. See also Ref. 123.

105. F. A. Martin, Seventh Intersociety Energy Conversion Engineering Conference, San Diego, Calif., American Chemical Society, Washington, D. C., 1972, paper no. 729209, pp. 1335-1341.

106. J. H. N. van Vucht, F. A. Kuijpers, and H. C. A. M. Bruning, Philips Research Reports, 25, 133-140 (1970).

107. K. Beccu, U. S. Patent 3,824,131 (1974), Battelle Institute, Geneva.

108. O. J. Adlhart, Seventh Intersociety Energy Conversion Engineering Conference, San Diego, Calif., American Chemical Society, Washington, D. C., 1972, paper no. 729163, pp. 1097-1102.

109. G. Evans and K. V. Kordesch, Science, 158, 1148-1152 (1967).

110. P. Dantowitz, Power Systems for Electric Vehicles, Symposium, U. S. Dept. of Health, Education and Welfare, Columbia University and Polytechnic Institute of Brooklyn, April 6-8, Public Health Service Publication No. 999-Ap-37, 1967, pp. 297-306.

111. M. R. Andrew, W. J. Gressler, J. K. Johnson, R. T. Short, and K. R. Williams, Automotive Engineering Congress, Jan. 1972, Society of Automotive Engineers, New York, paper no. 720191.

112. B. Warszawski, B. Verger, and J. C. Dumas, Marine Technol. Soc., 5, (1), 28-41 (Jan./Feb. 1971).

113. G. Ciprios, Proc. 20th Power Sources Conference, 1966, pp. 46-49.

114. K. R. Williams, "Hydrocarbon and Methanol Low Temperature Fuel Cell Systems," 6th International Power Sources Symposium, Brighton 1968, paper no. 38.

115. E. H. Okrent and C. E. Heath, Fuel Cell Systems II, Advances in Chemistry Series, Vol. 90, American Chemical Society, Washington, D.C., 1969, pp. 328-340.

116. B. S. Baker, ed., Hydrocarbon Fuel Cell Technology, Academic Press, New York, 1965 (contains several publications on molten salt fuel cells).

117. M. V. Burlingame, Fuel Cell Systems II, Advances in Chemistry series, Vol. 90, American Chemical Society, Washington, D.C., 1969, pp. 377-382.

118. W. Hausz, G. Leeth, and C. Meyer, ECO-Energy, Proceedings of the 7th Intersociety Energy Conversion Engineering Conference, San Diego, Calif., American Chemical Society, Washington, D.C., paper no. 729206, pp. 1316-1322.

119. C. Marchetti, Eurospectra, European Communities Information Service, Washington, D.C., 1971, pp. 117-129.

120. W. Vogel, J. Lundquist, and A. Bradford, Electrochim. Acta, 17, 1735-1744 (1972).

121. K. V. Kordesch and R. F. Scarr, "Thin Carbon Electrodes for Acidic Fuel Cells," Proc. 7th Intersociety Energy Conversion Engineering Conference, San Diego, Calif., American Chemical Society, Washington, D.C., 1972, paper no. 729003, pp. 12-19.

122. H. J. Halberstadt, "The Lockheed Power Cell," 8th Intersociety Energy Conversion Engineering Conference, Philadelphia, 1973, paper no. 739008.

123. The Hydrogen Economy, Miami Energy Conference, 18-20 March 1974. ("THEME"). Papers on all aspects of hydrogen production, transport, storage, and usage for energy distribution, load leveling, and automobiles. Proceedings published by Univ. of Miami, Coral Gables, Florida (1974).

124. G. A. Dalin, "Performance and Economics of Silver-Zinc Batteries in Electric Vehicles," Symposium on Power Systems for Electric Vehicles, April 6-8, 1967, Yardeney Electric Corp.

125. L. Martland, A. E. Lynes, and L. R. Foote, Automotive Engineering Congress, May 1968, Detroit, Society of Automotive Engineers, paper no. 680428.

126. L. R. Foote, J. F. Hough, Automotive Engineering Congress, Detroit, Mich., Jan. 12-16, 1970, Society of Automotive Engineers, paper no. 700024.

127. H. N. Seiger, S. Charlip, A. E. Lyall, and R. C. Shair, 21st Power Sources Conference, Atlantic City, N.J., May 1967.

128. Electric Vehicle News, May 1974, p. 7.

129. D. P. Crane, presentation at the Battery Council International (BCI) meeting in Hollywood, Fla., April 1975.

130. Electric Vehicle News, Nov. 1972, pp. 6-9.

131. Electric Vehicle News, Aug. 1974, pp. 10-13.

132. Electric Vehicle News, Nov. 1973, pp. 20-22.

133. D. Gates and M. Pocobello, Automotive Engineering Congress, Detroit, Mich., Jan. 8-12, 1973, Society of Automotive Engineers, paper no. 730249.

134. R. G. Beaumont, Third International Electric Vehicle Symposium, Washington, D. C., 1974, paper no. 7413.

135. Electric Vehicle News, Nov. 1974, p. 3.

136. R. R. Aronson, Third International Electric Vehicle Symposium, Washington, D. C., 1974, paper no. 7420.

137. Electric Vehicle News, Nov. 1974, pp. 15-17.

138. N. A. Cook, Advances in Battery Technology Symposium, Southern Calif.-Nev. Section, The Electrochemical Society, 1967, pp. 1-29.

139. D. H. Smart, Mechanical & Electrical Specifications, Performance and Testing of a Small Battery Electric Car, The Electricity Council, United Kingdom, Paper 7481, Third International Electric Vehicle Symposium, Washington, D. C., 1974.

140. "GES" Gesellschaft für elektrischen Strassenverkehr, mbH, Dusseldorf, Germany.

141. H. Hagen, "The MAN Electrobus," Third International Electric Vehicle Symposium, Washington, D. C., 1974, paper no. 7460.

142. A. M. Muller-Berner, P. Striffer, Automobiltechn. Z., 72, 78-84 (1970).

143. P. Strifler, Automobiltechn. Z., 74, 244-248 (1972).

144. A. Kalberlah, Third International Electric Vehicle Symposium, Washington, D. C., 1974, paper no. 7445.

145. H. G. Raschbichler, Third International Electric Vehicle Symposium, Washington, D. C., 1974, paper no. 7443.

146. J. Hellig, H. Schreck, and B. Giera, Third International Electric Vehicle Symposium, Washington, D. C., 1974, paper no. 7453.

147. Electric Vehicle News, Feb. 1975, pp. 9-10.

148. G. Kucera, H. G. Plust, and C. Schneider, Automotive Engineering Congress, Detroit, Mich., Feb. 1975, SAE paper no. 750147.

149. G. E. Smith, Second International Electric Vehicle Symposium, Atlantic City, N. J., 1973, paper no. 7473.

150. Electric Vehicle News, Aug. 1974, pp. 15-17.

151. H. Chinc, Third International Electric Vehicle Symposium, Washington, D.C., 1974, paper no. 7474.

152. K. Kamada, Third International Electric Vehicle Symposium, Washington, D.C., 1974, paper no. 7429; also presentation at Battery Council International Meeting in Florida, Feb. 1975.

153. L. Saulgeot, Third International Electric Vehicle Symposium, Washington, D.C., 1974, paper no. 7479.

154. M. M. Cochat-Heurin-Wolf, Third International Electric Vehicle Symposium, Washington, D.C., 1974, paper no. 7480.

155. L. Lefort and G. M. Martin, Third Electric Vehicle Symposium, Washington, D.C., 1974, paper no. 7462.

156. Power Applications, Inc., Valley Stream, N.Y.

157. G. Brusaglino, Third International Electric Vehicle Symposium, Washington, D.C., 1974, paper no. 7415.

158. Electric Vehicle News, Aug. 1974, p. 24.

159. O. Lindstrom, "Fuel Cells for Traction Purposes," Electric Review, Aug. 12, 1966.

160. J. D. Busi and L. R. Turner, U.S. Army Foreign Science and Information Center, Charlottesville, Va., J. Electrochem. Soc., 121, 183C-190C (1974).

161. J. O'M. Bockris, Flinders University, South Australia, private communication; Electric Vehicle News, Aug. 1974, p. 9.

162. G. E. Pearson, Express Dairy Co., South Ruislip, Middx., England; Electric Vehicle News, Aug. 1974, pp. 19-20.

163. R. J. Kenny, E.S.B., Inc., presentation at Battery Council International Meeting, Florida, 1975.

164. J. H. B. George, Arthur D. Little, Inc., Third International Electric Vehicle Symposium, Washington, D.C., 1974, paper no. 7428.

165. R. Adams and A. J. Mautner, Third International Electric Vehicle Symposium, Washington, D.C., 1974, paper no. 7463.

166. Assessment of the Potential for Electric Minibus Systems for Midtown Manhattan, The Electric Vehicle Council, New York, 1973.

167. H. H. Wang and R. T. Okawa, Third International Electric Vehicle Symposium, Washington, D.C., 1974, paper no. 7459.

168. K. J. Oehms and H. Busch, Third International Electric Vehicle Symposium, Washington, D.C., 1974, paper no. 7411.

169. D. Chapman, T. Tyrell, and T. Mount, Science, Nov. 1972.

170. H. Pieper, U.S. Patent No. 913,846 (1909).

171. G. H. Gelb, N. A. Richardson, T. C. Wang, B. Berman, Meeting, Society of Automotive Engineers, Jan. 1969, paper no. 690071.

172. "Final Report, Hybrid Engine/Electric System Study," Report Nos. TOR-0059 (6769-01)-2, Vol. 1 and 11. The Aerospace Corp., 1971.

173. "The City-Taxi Project," Electric Vehicle News, 5 (3), pp. 14-15 (1976).

174. Electric Vehicle Association of Ohio, 9135 Fernwood Drive, Olmsted Falls, Ohio 44138.

175. R. R. Aronson, Society of Automotive Engineers, May 1968, paper no. 680429.

176. T. N. Thiele and D. L. Moore, "Case History and Economic Study of an Electric Commuter Vehicle," Advanced Technology Center, Allis Chalmers Corp., 1971.

177. G. M. Naidu, G. Tesar, and G. G. Udell, The Electric Car, Publishing Sciences Group, Inc., Acton, Mass., 1975.

178. R. S. McKee, B. Borisoff, F. Lawn, and J. F. Nosberg, "A Test Bed Electric Vehicle," Society of Automotive Engineers, Jan. 1972, paper no. 720188.

179. Electric Vehicle News, P.O. Box 533, Westport, Conn. 06880.

180. L. E. Unnewehr, Third International Electric Vehicle Symposium, Washington, D.C., 1974, paper no. 7414.

181. Linear Alpha, Inc., P.O. Box 591, Skokie, Ill., 60076.

182. Exide, E.S.B., Inc., "Electric Vehicle Batteries," 1975/1976.

183. Globe Union, Inc., "Electric Vehicle Batteries," 1976.

184. Gould, Inc., "Electric Vehicle Batteries," 1976.

185. Electric Vehicle News, 6 (2), pp. 7-21, May 1977.

186. Society of Automotive Engineers, SAE-Catalog No. P-68, IEEE No. 76CH1146-OVT.

Chapter 3

THE RECENT STATE OF THE TECHNOLOGY

I. LEAD STORAGE BATTERIES IN GERMANY

Dietrich Berndt

Varta Batterie A.G.
Research and Development Center
Kelkheim
Federal Republic of Germany

1. INTRODUCTION

Lead storage batteries always have in principle the same construction, but different applications result in a multitude of design variables accommodating specific uses. The development of different products was favored in Europe because many separate market areas existed, which only recently (after World War II) overlapped.

A similar situation exists with respect to standardization. At the present time, British, German, and French standards exist in parallel, but an attempt to unify the standards, also including U. S. standards, is being made with the help of the International Electrochemical Commission (IEC).

No attempt is made here to give a complete summary; however, in many respects the situation in Germany seems to be representative for Europe.

2. COMPONENTS

It should be recognized that the preference for certain battery types has to a large extent historic and geographic reasons. The lead/calcium alloys, for example, play a far smaller role in Europe than in the United States. Regarding plate types, the positive tubular plate is used much more in Europe. There are also many different separator types and methods for the manufacture of dry-charged cells.

2.1. Alloys

The grid in the plates serves not only as an electric conductor but also as the support for the active material. With the exception of the Planté plates, which consist of pure lead, all grids are made from alloys which have a greater hardness than lead.

The preferred alloying element was always antimony, which has an undesired influence on the hydrogen overvoltage of the negative electrode, but is beneficial for the cycling behavior of the positive plate. For that reason even today 9% antimony is added by many manufacturers to the positive grids of batteries which should have a long cycle life, as for instance traction batteries. In starter batteries (SLI type), the antimony content was reduced to 4 to 5%. Special alloys have been developed for maintenance-free batteries [1] containing only 2% Sb; some stationary batteries may have even less antimony.

In Europe, lead/calcium alloys are presently used only in small portable maintenance-free cells. Further, an antimony-free alloy is known under the name Astag [2]. It contains additives under 0.1% (0.009% As, 0.065% Te, 0.008% Ag) and shows a low hardness value compared with the other alloys mentioned. Presently it is used for special applications and to a small extent for stationary batteries.

2.2. Electrode Designs

The principal electrode constructions correspond to those in the United States; however, in many details there are differences.

2.2.1. Grid Plates

Two types used in industrial cells are shown in Fig. 1. The diagonal grid shown on the right side is the older design; today an attempt is made to use a fine gridwork and large mass fields, especially with negative plates, in order to save lead. The thickness of the grids in conventional stationary and traction batteries is about 4 mm.

2.2.2. Tubular Plates (ironclad plates, "Panzerplatten," Pz.)

Not only the grid plates but also the tubular plates reached wide utilization for the construction of stationary batteries. Figure 2 shows two types of tubular designs. Figure 2b shows a braided glass fiber tube supported by a perforated polyvinylchloride (PVC) sheet. Figure 2a shows a woven tube which is used without any further coating. It is produced as a continuous tape from which the necessary number of tubes is cut. The material is a pure plastic fiber (polyester) or a mixture of polyester and glass fiber [3]. The tube diameter is usually 8 mm; in batteries for electric vehicles the tube diameter is reduced to 6 mm in order to increase the high-current output.

The tubular plate is mechanically very stable because the active material is enclosed by the tube. For the same reason the mass itself can be very porous, which causes excellent mass utilization. In addition, the cen-

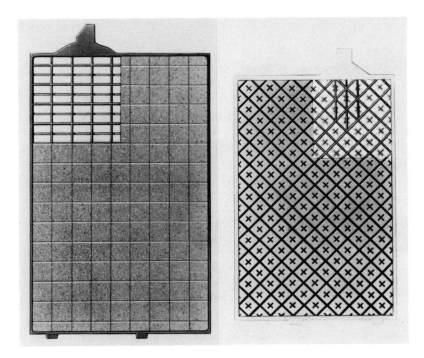

FIG. 1.   Two different types of grid plates for industrial batteries.

trally located collector rod assures a very uniform current distribution over the active material [4]. In tubular plates the mass utilization figure is up to 80 Ah/kg. However, at high currents the tubular plates are at a disadvantage due to the relative thick geometry, if compared with thin-grid plates or Planté plates. This can be seen by comparing the performance curves in Sec. 3.1.

2.2.3.   Planté Plates ("Grossoberflaechenplatte, Gro") and Negative
         Box Plates

Figure 3a shows a cut through the positive Planté plate in which the fine lamellas are very noticeable, causing an increase of the geometric surface by a factor of 8 to 12. Older designs of Planté plates are 12 mm thick, newer types only 8 mm.

    Planté plates have recently been used increasingly in emergency batteries because of their suitability for short-time high-current drain. In

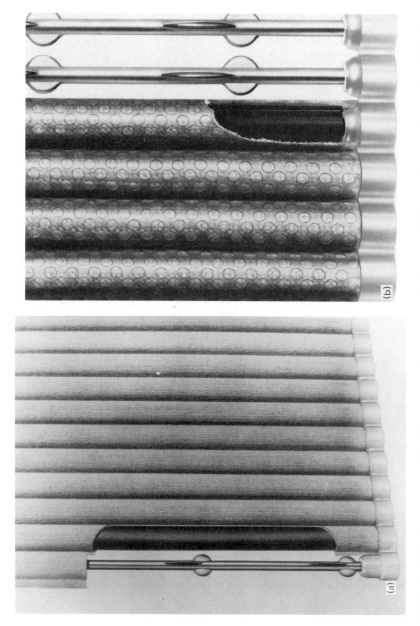

FIG. 2. Tubular plates. (a) Tubes woven from pure plastic material or polyester-glass mixed fabric. (b) PG tubes: braided tubes of glass fibers, with perforated PVC coating.

FIG. 3. Sectional cuts through electrodes. (a) Positive Planté plates. (b) Negative box plate.

this plate the thin layer of active material gives good mass utilization at
high loads. The disadvantage is the large amount of lead employed as conductive material.

The negative box plate (Fig. 3b) consists of a hard lead grid which is
enclosed on both sides by perforated lead sheets. The life of this plate is
excellent, but its electrical output, especially at high currents, is poor because a large part of the surface is covered by the lead.

Planté plates are used in combination with box plates for batteries with
lower current requirements. It is to be expected that these plates will be
replaced by newer constructions soon (see Sec. 3).

2.3. Separators

For starter batteries (SLI), separators made from sintered PVC or cellulose are used. The cellulose is usually coated with Phenoplast (phenolic
formaldehyde resin) to impart a better mechanical and chemical durability.
A list of separators which are used in stationary and traction batteries is
given in Table 1. "Open porosity" is the porosity which is accessible to
the electrolyte; it is measured by determining the water uptake of the material. Easiest to produce are the PVC-sinter separators, which can be manufactured in small plants. These separators have a relative low porosity
and a coarse pore structure. The feared splitting off of chlorine, leading
to corrosion, depends largely on the starting material and on the sintering
process. It can be kept very low. Even precipitated PVC (Ameri-Sil in
Table 1) can split off chlorine, but the silicate coating produced by the co-
precipitated silicate is said to protect the PVC to a large extent.

Among separators which are not based on PVC, the rubber separator
has proven its suitability for 50 years. The Darak 5000 separator combines good mechanical properties with high porosity due to its fabric (non-
woven material) inlay. The chemically very stable phenol-resorcinol resin
makes it possible to place this separator directly adjacent to the positive
plate without destruction by oxidation being observed.

TABLE 1

Separators Used for Stationary and Traction Batteries

Commercial name	Material	Production method	Average pore diameter (μm)	Open porosity (%)
Mipor B	Rubber	Mixing of natural rubber with volatile substances. Pores are formed during vulcanization.	1.8-0.06	55
Darak 5000	Phenoplast on polyester	Dispersion of phenol-resorcinol formaldehyde resin and water is rolled into polyester felt and crosslinked. The water is forming pores after removal [5]	0.8	60
Sinter PVC	PVC	Sintering of PVC powder	22-16	35-47
Porvic I	PVC	PVC/starch mixture is sintered starch is then dissolved.	0.6	87
Amer-Sil	PVC	Precipitation of PVC from organic solvents together with sodium silicate	0.1	65

2.4. Drying Methods for Negative Electrodes

For the production of dry-charged cells it is necessary to preserve the reduced state of the negative electrode and prevent the oxidation of the active lead. Besides the well-known methods of drying in overheated steam or in vacuum, two other processes have been used economically in Europe: the boric acid method [6] and the petroleum process [7]. In the first process, which is used mainly for SLI batteries, a dual immersion of the washed plates in boric acid solution prevents the oxidation of the lead, so that the plates can be dried in an air stream and keep their charge for several years. Compared to the preservation with organic substances, this method has the advantage of freedom of side effects, e.g., the wetproofing of the plates, which can influence the initial behavior of the battery [6].

In the petroleum process the plates are dipped into $130°$C petroleum after the washing; remaining solvent is then removed by air drying. The protective action of the petroleum can be increased by the addition of paraf-

fin [8]; excellent storage life is thereby obtained. Using improper expanders may lead to excessive protection, which can impair the charging
processes so that the battery can only be charged with small currents after
filling.

3. CELL TYPES

A whole spectrum of cell types can be produced by combining different electrodes. On the basis of specific properties and application requirements,
certain cell types have been developed.

Figure 4 presents a survey of the situation in Germany. It must be
realized, however, that the ever-moving market may change such a simplified tabulation; besides that, the picture as it appears for Germany may be
only conditionally true for the rest of Europe.

3.1. Stationary Batteries

As can be seen from Fig. 4, stationary batteries are built either with positive tubular plates or with Planté plates. Modern installations which require several hours discharge, for instance, telephone installations, are
served exclusively by tubular plate batteries. Increasing demand for stationary high-load batteries has lead to the type Gro-E, to be used for emergency supply batteries in computer centers, hospitals, nuclear power
plants, and other places.

3.1.1. Stationary Batteries with Tubular Plates

Cells with positive tubular electrodes in transparent containers are shown
in Fig. 5. Single cells and batteries (block battery) are pictured. Between
the tubular plates (see Fig. 2a) and the grid plates (see Fig. 1) the microporous separators are located. The electrolyte is sulfuric acid of 1.24 g/
ml density.

The discharge curves of one cell from this production series are shown
in Fig. 6. The current is related to one positive plate (100 Ah at the 10-h
rate). Multiplying by the number of plates gives the total cell current.

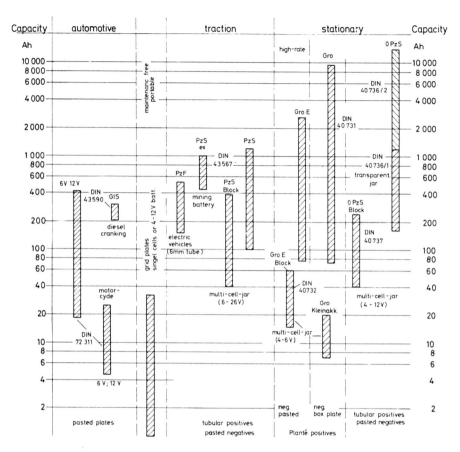

FIG. 4. Simplified schematic of the battery types standardized in Germany. The capacity and the application of the lead battery series is shown [9].

A rapid drop of cell voltage and capacity is noticeable at high current drains. The best efficiency of this cell type is achieved at discharge times between 1 and 10 h. The application in telephone installations, where at least 4 h emergency service is needed in case of a line failure, suits this battery type and it is used in large numbers for this purpose. Large cells of this type are mounted in hard rubber containers. The largest cell of the standard series which is commercially available is the Varta 13,750-Ah cell. The cell contains 55 positive tubular plates with 250 Ah nominal ca-

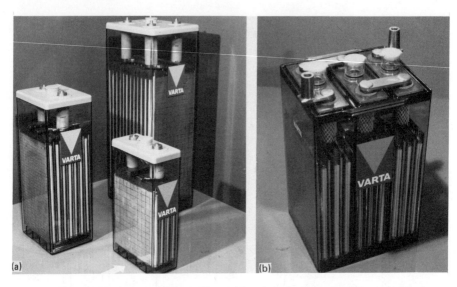

FIG. 5. Stationary cells with positive tubular electrodes. (a) Single cells. (b) Three-cell block battery.

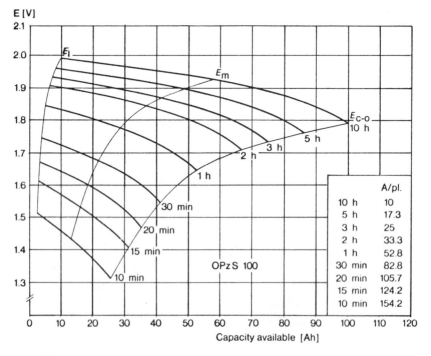

FIG. 6. Discharge curves of a stationary tubular plate cell (with reference to the 100-Ah positive electrode).

pacity. The cell is not sealed completely because an exchange between the gas space in the cell and the battery room is desired in order to prevent the accumulation of explosive gas mixtures. However, the tendency to construct closed cells is noticeable in the industry, even for that size; for that reason the development of cell vents with explosion-proof features has been pushed forward.

3.1.2. Stationary Batteries with Planté Plates

The older types (in Fig. 4 listed with the symbol Gro) of large size are built in hard rubber enclosures, or—smaller sizes—in plastic housings. In these cell types, positive Planté plates are combined with negative box plates. In more recently designed cell types, positive Planté-type plates are combined with negative grid plates (Gro E).

Figure 7 shows as an example a cell of the series Gro-E Varta, 1,400 Ah). The positive plates are suspended in the cell. One can recognize the

FIG. 7. Stationary cell with Planté plates. Type 14 Gro E 1400; Varta.

enlarged top of the plastic container on which the extended plate yokes are resting. The negative grid plates (replacing the older box plates) which are four parallel terminals brought out through the top cover.

The discharge curves shown in Fig. 8 indicate the good performance at high currents. The behavior during the first 10 min characterizes the short time-peak load capabilities (Fig. 8b).

Gro-E batteries are also used as small multicell block batteries for small emergency installations (e. g., operating room surgical lamps). The positive plates consist of pure lead and therefore the water loss during continuous charge at a "floating voltage" of 2.2 to 2.3 V is so small (similar to batteries with antimony-free grids) that service intervals of 5 years are sufficient at normal temperatures.

3.1.3. Stationary Cells with Grid Plates

Grid plate cells for stationary installations are not listed in Fig. 4, although they are produced in small numbers in England, France, and Germany. They play a larger role in some other European countries, but this cell type has not found the wide distribution for stationary installations as is realized in the United States.

3.2. Traction Batteries

For common use as traction batteries, the cells with tubular plates have found wide application. Figure 4 lists only manufacturing series with this plate type combined with negative grid plates. Only very recent developments aimed at electric vehicle batteries with extremely high energy densities (40 Wh/kg) for street vehicles have gone back to grid plate cells with reduced life expectancies.

3.2.1. Traction Batteries with Tubular Plates

The traction batteries of the series PzS listed in Fig. 4 use the same plate types as the corresponding stationary cells, but they are built in a far more compact way since the volume of vehicle batteries must be kept as small as possible. The design is shown in Fig. 9.

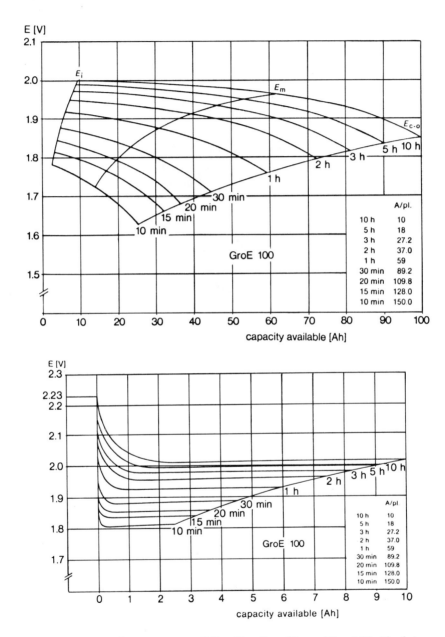

FIG. 8. Discharge curves of Gro E cells with positive 100–Ah plates. Above - Voltage as a function of the dischargeable capacity at different discharge currents. Below - Voltage during the first 10% of the discharge time (Varta).

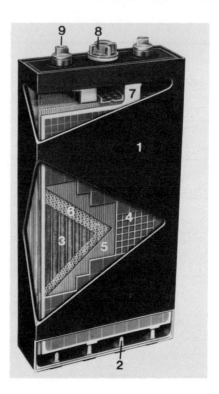

FIG. 9.   Traction cell with tubular plates. 1, jar; 2, sediment bridge;
3, positive electrode (tubular); 4, negative electrode (pasted); 5, microporous separator; 6, punched PVC separator; 7, negative bridge.

Adjacent to the negative grid plate one finds a microporous separator
which is separated from the positive plate by a distance holder made from
perforated plastic. At the bottom of the cell are prisms on which the plates
rest.

The free volume is kept so small that the amount of acid is just enough
to satisfy acceptable maintenance requirements. Automatic water filling
systems are presently tested to facilitate this procedure.

The cell cover is often sealed against the container by melted thermoplastic material (bitumen); however, new designs use welded cover container seals. A disadvantage of the welded cells is the impossibility of

making repairs by opening the defective cells, adjusting plates, or exchanging separators.

Traction batteries are operated with acid of a density of 1.27 g/ml. The volume of the acid should be sufficient so that the density can drop to 1.1 g/ml during discharge. The higher acid density results in a higher initial voltage, compared to stationary cells, as can be seen by comparing Fig. 6 with Fig. 10. The limited amount of acid, on the other hand, causes a steeper decay of the voltage at the end of discharge as a result of the volume restriction.

Smaller traction batteries are sold in multicell block arrangements. Figure 11 shows as an example a 6-V (three-cell) battery with a nominal capacity of 108 Ah (10-h discharge rate) produced in Sweden. Therefore, the designation differs from the German system used in Fig. 4. Such batteries are used in small transport carts. They are available between 40 and 400 Ah (see Fig. 4).

A special field of application is mining vehicle operation. Long distances and elevated temperatures are considerable handicaps. In addition there is the requirement for explosion protection, which is a factor to be considered in cell cooling and ventilating systems. In Germany most of the

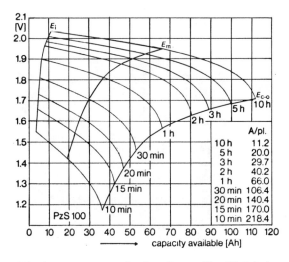

FIG. 10. Discharge curves of a traction cell with tubular plates (current with reference to a 100-Ah positive plate).

FIG. 11. Block battery for traction applications, 3 TR 108 (Noack AB).

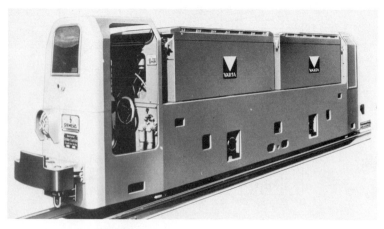

FIG. 12. Mine locomotive with tubular plate battery, 2 54 cells of the type 8 PzS 800 (170 k Wh).

batteries used in mines are standard traction batteries; only in a few older installations are cells with positive grid plates still in use.

Figure 12 shows a mine locomotive which uses (in two compartments) two sets of 54 cells of the type 8 PzS 800 so that a total of 170 kWh are available. The battery carriers (trays) are built with special plug connectors which are explosion-proof and allow a safe exchange of the batteries when taken off for recharge. The ventilation of the closed trays is achieved by leading the air through an arrangement of sheet metal lamellas which prevent a penetration of flames. To assure the absence of sparks in the battery enclosure, the cells are either welded together at the terminals or a special, very strong type of connection with conically shaped terminals is used. In spite of all precautions, some (weak) explosions occur occasionally. For that reason new developments have the goal of isolating all current-carrying parts on the battery surface. The battery tray may then be ventilated directly. Figure 13 shows a picture of such isolated connectors.

The battery-operated train is a further special application for cells with tubular plates. Figure 14 shows one of 230 battery-powered engine

FIG. 13. Sectional view of a battery for mine locomotives; all current-carrying parts are completely isolated.

FIG. 14.   Battery-operated engine car of the German Federal Railroad.

cars presently used by the German Federal Railroads (Bundesbahn) [11].
Each engine car has up to 630 kWh of battery capacity installed.

Depending on the schedule, such trains can operate over distances
ranging from 400 km (local commuter) to 500 km (transit traffic) [11].

Battery-powered boats have been operated for many years on some
lakes where the use of combustion engines is prohibited, as on the
Koenigssee and the Maschsee. Traction batteries with tubular plates are
used and the ships need to be charged only after their daily routes are completed.

3.2.2.   Traction Batteries with Grid Plates

Traction batteries with positive grid plates are used to a small extent only,
as an example in the Fulmen batteries (France). Small transport vehicles

FIG. 15. Spillproof, low-maintenance, miniature lead cell [14].

are operated with them. Grid block batteries of that type look like the battery shown in Fig. 11. These batteries are partially derived from starter batteries (SLI) and adapted with special separators and change of the mass formulation to serve better in deeper cycling applications. New developments have produced batteries with extremely high current output for electric street vehicles. These are discussed in Part 2 of this chapter (K. Salamon).

3.3. Maintenance-free Small Lead Batteries

The maintenance-free miniature lead-acid battery has a long history in Germany [13]. Figure 15 shows a version which has been available commercially since 1949. In this model the acid still can be refilled, a special vent design making the battery spillproof, but it cannot be operated upside down. The state of charge can be recognized from the position of three plastic balls of different specific weight, each starting to sink at a certain acid density. These cells use antimony-free alloys (Pb/Ca) to lengthen the intervals of maintenance. The poorer cycle life could be tolerated in some applications (photoflash units).

After 1950, the development of truly maintenance-free miniature batteries started. Figure 16a shows cells containing gelled electrolyte,
offered in different capacities up to 36 Ah [14]. In the batteries shown in
Fig. 16b the electrolyte is immobilized by a special separator similar to
the construction of gas tight alkaline batteries [15]. The hermetical sealing of lead acid batteries is not considered possible unless the electrodes
are provided with devices capable of catalytically reacting the gases evolving from the electrodes [10,20]. Lacking such means the pictured cells
all have vents of some kind. Using antimony-free alloys reduces the water
requirements so much that filling with water need not be done during the
life time of the battery.

3.4. Starter Battery (SLI)

Starter batteries are usually built as 12-V block batteries; only very large
types are constructed in 6-V modules. The most frequent types are, corresponding to the vehicle size, between 36- and 70-Ah batteries. Such batteries are now usually produced in polypropylene containers.

The nominal capacity of starter batteries is related to the 20-h discharge, but more significant data specify the cold starting and charge acceptance capabilities. In Germany, the standards are given in DIN 72 311
as follows:

Cold start test: the battery is cooled down to -18°C and then discharged with the specified cold-test current which can be calculated approximately by multiplying the 20-h capacity figure (e.g., 70 Ah) with a factor
of 4 to 4.5 expressed in amperes (e.g., 280 A). After 30 sec discharge
time the battery voltage must still be 9 V and is allowed to drop to 6 V only
after 150 sec.

The charge acceptance test requires that a battery, discharged 50%,
must be able to accept the current calculated by the formula: 0.2 × nominal
capacity, expressed in amperes at 0°C, 10 min after a voltage of 14.4 V
(2.4 V/cell) is applied to the battery. Car manufacturers in Scandinavia
consider this test formula too low with respect to current acceptance and
want to provide a higher acceptance capability even if reducing the cold
starting figure would be the consequence.

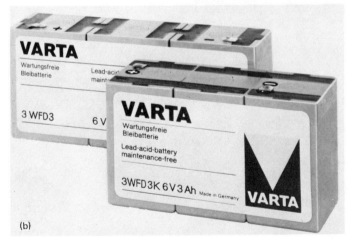

FIG. 16. Maintenance–free miniature lead storage batteries. (a) With gelled electrolyte [13]. (b) Electrolyte soaked up in the separator [15].

As a test for life expectancy, which is determined mainly by grid corrosion, the <u>overcharge test</u> has shown merit. The battery is overcharged
at $40°C$ for a period of 100 h with the 10-h rate charge current and after-
wards a cold start test is performed. The battery voltage must stay for
30 sec above 6 V in the course of six such overcharge cycles.

The inside structure of a modern starter battery is shown in Fig. 17.
The connections between the single cells are processed before the assemblies are inserted into the housing [16]. The upper part of the partition is
then formed later, in a process which also seals the cell interconnectors
[17]. Frequently, the new (U.S.) Globe-Union Process [18,19] is practiced in Europe also.

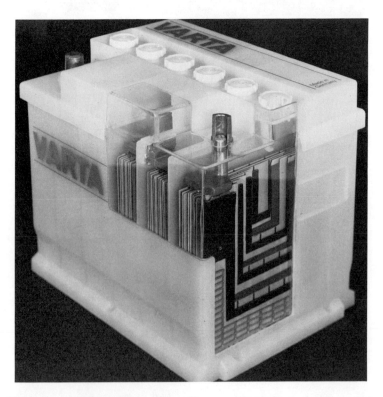

FIG. 17. Starter battery with cell connections through the partition
plastic-embedded by extrusion molding [17].

3.4.1. Maintenance-free Starter Batteries

Several ways to arrive at maintenance-free starter batteries have been followed. The goal is to avoid maintenance costs and to be able to place the
battery away from the motor compartment, somewhere in an available but
inaccessible location in the car.

Figure 18a shows a conventional battery with recombination plugs on
top of the cell openings [20]. Water decomposed during overcharge is reformed and fed back into the electrolyte. This system is limited by the
speed at which recombination occurs and by the amount of nonstoichiometric gassing.

Figure 18b shows a battery containing low-antimony plates in which
plate gassing is decreased to such a degree that water loss is negligible
during the whole life time, when the battery is working under normal conditions. To avoid increased overcharge the voltage regulation of the generator has to be correct.

3.5. Charging Procedures

3.5.1. Vehicle Batteries

Charging procedures are defined more strictly in Europe than in the United
States. In the German DIN 46 772, the different charging methods are described in detail, and the characteristics of the charging process are prescribed with small tolerances.

Battery manufacturers recommend for intermediate charging a constant voltage of 2.4 V per cell, and for complete charging the continuation
with a smaller current (about the 20-h rate). Some believe that constant-
voltage charging should be limited to 2.35 V per cell. These procedures
allow fast and safe charging, but the cost of providing well-regulated charging equipment is high.

For that reason, most charging installations for traction batteries pre-
fer simpler devices. If the charging current is allowed to drop with rising
cell voltage, a tapered charging is achieved. The disadvantage of this
method is that the charging time for the battery must be adjusted for different types and that variations in the line voltage may give rise to excessive or insufficient charging of the battery.

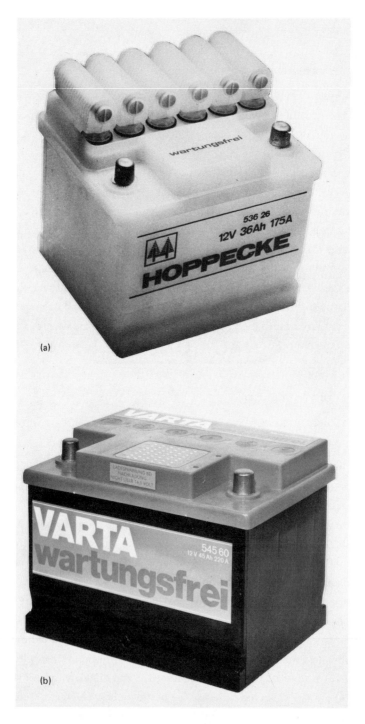

FIG. 18. Maintenance-free starter batteries. (a) Standard battery with catalytic gas recombination devices (Hoppecke). (b) Battery with low-antimony alloy grids (Varta).

These difficulties can be avoided by using additional circuit elements. The Poehler switch [21] activates a timer which provides a predetermined overcharge. The Spegel Charger [22] causes a voltage-controlled on and off switching of the charge current whereby the voltage of the cells during the off time is used to regulate the charging process. Other methods use the gradually decreasing voltage differentials (per time unit) to influence the charger.

3.5.2. Stationary Installations

With stationary batteries a continuous charge with a voltage of 2.23 V per cell is commonly practiced. Some manufacturers recommend a constant-voltage charge with 2.4 V per cell to achieve a faster recharge after discharge periods. Once a year an "equilibration charge" with voltages up to 2.7 V/cell is also proposed.

REFERENCES

1. H. Borchers, S. C. Nijhawan, and W. Scharfenberger, Metall., 28, 863 (1974).

2. Aktiebolaget Tudor, Brit. Patent 1,105,548 (1968).

3. D. Evers et al., German A. S. 1,162,895 (1964).

4. J. Euler and T. Horn, Electrochim. Acta, 10, 1057 (1965).

5. E. Decker, German Patent 1,279,795 (1968).

6. H. Haebler, German Patent 1,088,118 (1960); U. S. Patent 2,996,563 (1961).

7. G. E. Sundberg et al., German A. S. 1,016,334 (1957).

8. F. Sundman, German Patent 1,192,718 (1965).

9. H. A. Kiehne, Elektrotechn. Z., B27, 95 (1975).

10. A. Winsel, German A. S. 2,340,945 (1975).

11. G. Wilke, Akkumulator-Triebwagen, München, R. Oldenbourg Verlag, Munich, 1956.

12. W. Rappenglück, Die Bundesbahn, 17/18, 1 (1968).

13. K. Eberts, Elektrotechn. Z., B21, 297 (1969).

14. K. Eberts, Elektrotechn. Z., B27, 104 (1975).

15. H. Niklas et al., U. S. Patent 3,836,401 (1974).

16. F. Bronstert, German Patent 1,067,899 (1959).

17. H. G. Lindenberg, German Patent 1,804,800 (1970).

18. A. Sabationo et al., U.S. Patent 3,313,658 (1967).

19. A. Sabationo, U.S. Patent 3,476,611 (1969).

20. G. Sassmannshausen, German A. S. 2,008,218 (1970).

21. E. Witte, Blei- und Stahlakkumulatoren, Mainz, 1969, p. 120f.

22. D. A. Clayton, Brit. Patent 1,370,301 (1974); Brit. Patent 1,370,361 (1974); Brit. Patent 1,371,047 (1974).

II. LEAD-ACID BATTERY-OPERATED ELECTRIC STREET VEHICLES: NEW CONSTRUCTIONS FROM GERMANY

Klaus Salamon

Varta Batterie A. G.
Research and Development Center
Kelkheim
Federal Republic of Germany

1. CELL CONSTRUCTIONS FROM THE VIEWPOINT OF THE ELECTRIC VEHICLE DESIGNER

The development of the battery for electric road vehicles has a different history in Europe, as compared with the United States or Japan. In Europe the long-cycle-life industrial cell was the basis of the evolutionary process, while in America and Japan the starter battery and later the golf cart battery were the initial design models, in general, three-cell batteries of standardized dimensions, with high current capabilities, but less good

cycling characteristics. The manifold and new requirements of a road vehicle battery with respect to very high specific energy and power density
combined with reliability and long life were easier to fulfill with the improved industrial-type cell, not least because the development was based
on the more flexible individual cell construction than on the rigid designs
of automobile starter batteries. Street vehicle batteries have presently
reached energy densities of 32 to 40 Wh/kg (used at the 5-h rate), while industrial cells of conventional design have values between 24 and 28 Wh/kg
at the same discharge rate. However, even these improved characteristics are still below the energy densities required by vehicle designers. It
is, therefore, important to consider how much the lead-acid storage battery can be further improved with respect to energy density and still be
economical to produce.

In a recent survey [1] when this question was discussed, two cases
were distinguished. First, a lightweight cell made with positive grid plates
may have an ultimate energy density of about 45 Wh/kg but a life of only
700 cycles (80% depth of discharge at the 5-h rate). The life limitation is
in this case determined by corrosion of the positive grids and by the resulting sludge formation. Second, there may be a much more robust type of
cells with positive tubular plates characterized by an energy density of
about 36 Wh/kg at the 5-h rate, but with a cycle life of at least 1,500
charges and discharges at an 80% depth of discharge. At the present time
such a cycle life is reached with the standard industrial cell at 24 to 28 Wh/
kg. Improvements will not be based on better designs of the electrochemical system but on the development of new cell constructions, utilizing new
materials and new technologies in manufacturing batteries.

1.1.   Plates and Post Straps

About 30% of the mass lead is presently used for the grids and current collectors inside and the posts of the cells. A large portion of this lead serves
as mass carrier and is needed for the electrochemical functioning of the
system, but another considerable quantity used in the conductor, as in plate
lugs, straps, and posts, can be saved with suitable design changes. The

requirement for high current output makes it mandatory that the cross sections of the conductors be sufficient for currents of 200 to 600 A, depending
on the type of vehicle and the need for acceleration and hill-climbing performance. Any compromise must therefore be carefully evaluated.

A good example for a successful reduction of the amount of lead without
performance loss can be seen in the construction of starter batteries with
polypropylene housing and through partition connectors. In these cells the
terminal straps of one cell are connected directly to the straps of the next
cell, immediately above the plate lugs, through the cell wall [2,3].

At the same time a modern manufacturing method, the "cast-on strap"
process, allows a direct casting of the plate connectors [4]. An advantage
of this highly automated process is not only the saving of manufacturing
time but also the reduction in lead required for the lead straps. In the old
welding-on process, the cross section of the plate-connecting "bridge" was
uniform; in the new design it is tapered in accordance with the increasing
current flow toward the ends. The differences are shown in Fig. 1. In recent times this construction has also been used for traction batteries and
single cells.

Figure 2 shows a construction which was especially developed for electric street vehicles [5]. The cells are connected through the wall to form
three- or multicell batteries with no connections above the cell covers, resulting in a considerable saving of lead due to the lack of posts and outside
connectors. A special problem was created by the fact that the through-
the-wall connections are below the electrolyte level, requiring tight seals
in spite of the increased mechanical stress produced by the closeness of
the plate assemblies.

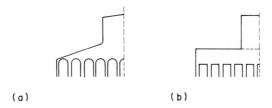

(a)   (b)

FIG. 1. Construction of terminal straps. (a) Cast-on strap.
(b) Welded-on strap.

FIG. 2. Fulmen Monoblock battery P 12180 (12 V, 180 Ah, 5-h rate).

Another construction [6] allows a saving of 5 to 10% of the lead and increases the high current capability of the vehicle battery. The lugs of the
plates are carried directly through the cover (without an inside connecting
strap) and are connected with a lead-coated copper bar, which in the case
of batteries, connects the cells over the shortest possible distance. Each
plate is therefore connected directly to the bus bar, and resistance losses
are minimized. In order to avoid a reduction of the necessary gas space
above the electrolyte, it was found necessary to design a dome-shaped
cover. A further advantage is the excellent heat dissipation due to the
large-surface-area metal connectors. A problem is created by the many
seals required. For that reason wide application of this cell type has not
yet been established.

1.2. Cell Assemblies

The requirements of low weight and small volume are important for a vehicle battery. From this viewpoint it is useful to consider the cell con-

FIG. 3. Vehicle battery with blown polypropylene housing shown during pressure test (overpressure: 3 atm, approx. 45 psi).

tainer as a candidate for reduction of weight and volume. The container must fulfill essentially two requirements: to be leaktight under the most adverse conditions and to assure perfect isolation between cells and to the outside. The function of supporting the cells is taken care of by the battery container or trough.

Blow-molding is a simple and economical manufacturing process for all cell sizes up to 1,000 Ah; relatively thin walls for cell containers can be produced without large tooling expenses.

The average wall thickness of cell containers made that way is about 1 mm, shown in the example of Fig. 3. Hard rubber cases have a wall thickness of about 4 mm. The polypropylene used is very light (weight-saving), and the volume gained allows an increase in active components of

FIG. 4. Three sizes of modern vehicle cells with positive tubular
plates in blown polypropylene housings (150-, 230-, and 450-Ah Cells).

the battery. In spite of this thin wall, such a cell can withstand an internal
pressure of 5 atm (75 psi). Such a value is far above the normal operating
condition (see Fig. 3). By using thin wall cases, the weight portion of the
"packaging" of the vehicle cell is reduced to less than 2% of the total weight.
With the usual traction cells the portion of the casing weighs between 5 and
10% of the cells. Welding the polypropylene cover to the case assures a
tight seal.

In Fig. 4 three modern vehicle cells are shown. These types have been
used for several years in tests of road vehicles. The cells pictured in
Fig. 4 are high-power cells of 150, 230, and 450 Ah capacity, with a cycle
life of 1,500 cycles. The average specific energy density is 32 to 33 Wh/kg
at the 5-h rate. The 450-Ah cell is equipped with a cooling loop for very
heavy-duty applications, for instance in buses.

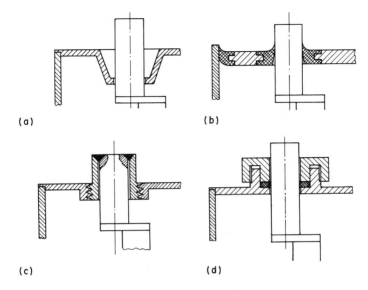

FIG. 5. Design of terminal through-connections. (a) Sealing of post
by means of poured asphalt. (b) Post sealed in with soft rubber gasket.
(c) Seal produced by welding the post to a lead gromet molded first into the
cover. (d) O-ring seal.

1.2.1. Terminal Through-Connectors

The leaktight seal is a special problem with vehicle batteries. Leakage is
very detrimental to the electric isolation properties of the battery trough
or carrier compartment, and any electrolyte contamination leads to increased corrosion of the outside battery components and to the battery environment. The usefulness of vehicle and battery is severely affected by
acid leakage—it is not just an "appearance problem."

Since the welding of cover and case has become commonplace for polypropylene batteries, the most frequent source of leakage has been shifted
to the terminal through-connectors. Several constructions are in use, of
which the most important ones are shown in Fig. 5.

The designs shown in Figs. 5c and 5d are considered to be the most
reliable ones; the constructions shown in Figs. 5a and 5b cannot be used
for road vehicles because of their higher leakage rate. The design Fig. 5c
deserves special consideration: the lead socket is first covered with a
plastic jacket which is allowed to shrink, then the cover is welded to it [7].

In the case of direct molding to the lead socket, the shrinking would be in
the direction away from the center, producing a poor seal. The lead socket
is later welded to the lead post (on top).

1.3. Intercell Connections

The specific energy and power density of a vehicle battery is considerably
influenced by the type of the intercell connectors. The conventional way to
connect cells with each other was to apply a lead bar designed for the average load between the terminals. It could also be a copper rail or a copper
cable with leaded clamps; the connectors were either screwed on or welded
on. The disadvantages of these massive connectors were their considerable
weight, the rigid connection between posts, and trouble with the maintenance
of screw connections or objectionable heating during welding.

Figure 4 shows the two smaller cells equipped with crimp connectors,
two on each terminal. The connecting cables are just inserted and the hollow receptacles are compressed, in the same way as has been proven successful in general techniques for power cable connectors. Of course, this
type of connector must be kept free of acid to avoid corrosive action. If not
possible, a pinhole-free coating is needed. In spite of this difficulty, the
low resistance of unleaded copper cables and their relatively light weight
are very attractive and the improvements in energy and power density of the
complete vehicle battery are easily demonstrated. The flexibility of the
connections result in a stress release between the terminal posts, increasing the safety and life expectancy of the whole system.

1.4. Maintenance Reduction

Once battery-operated vehicles have found wide usage, the problems of
maintenance gain in practical importance. In contrast to the starter (SLI)
battery, which has to serve for short times at a high power level (about
10 Ah are usually removed during heavy cranking) with a following overcharge period, the vehicle battery is regularly discharged to 80% of its

nominal capacity. Lead–calcium alloys or alloys low in antimony [8] provide a solution for reducing the maintenance for starter batteries, but not
for batteries which must be cycled deeply many times. The life expectancy
under these conditions is poor for such batteries [9], and other means had
to be found to reduce maintenance. The replenishment of the water which
is lost during the last phases of charging through electrolysis (hydrogen and
oxygen formation) must be done periodically for each cell and is the biggest
problem. The simple task of making up the lost water by refilling with distilled water becomes a large effort when a multicell vehicular battery must
be serviced.

1.4.1. Hydrogen–Oxygen Recombination

The catalytic recombination of the hydrogen and oxygen to water and its return to the electrolyte reduces maintenance considerably.

The recombination can be done in a central large recombination unit
designed into the gas removal and water filling system [10]. It can also be
achieved by providing an individual recombinator for each cell in the cell
vent plug [11-13]. The efficiency of the recombination poses a problem
too. The total water loss is larger than the water obtained from the recombination device. The reason is to be found in the nonstoichiometric gassing
over a large period of the charge process. With batteries subjected to repeated deep cycling the efficiency of water return is about 70 to 75% [14].

Figure 6 shows the ratio of hydrogen to oxygen during the charging of a
300-Ah cell. It can be seen that the described inefficiency of water return
is the reason that the water fill periods can only be extended by a factor of
two, or at best by three.

1.4.2. Central Water Filling and Gas Removal

The use of a central filling mechanism which connects all cells is therefore
an important feature for maintenance reduction. A level-regulating device
must be included to keep the electrolyte in each cell at the prescribed height.
Many designs suggest the refilling with water via the vents in the cell plugs
[15-19]. Figure 7 shows one example of a relatively simple design of a
"topping up" device. The distilled water provided by the central supply

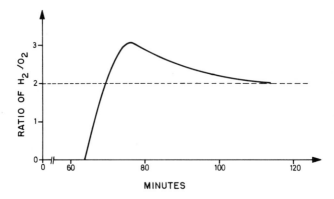

FIG. 6. Composition of the gases evolved during the charging of a 300-Ah cell (the ratio of hydrogen to oxygen as a function of time).

reaches the electrolyte through a thin-wall tube which is sharply bent at a predetermined spot (kink) through the lifting action of a float. The regulating action is not abrupt and is a function of the electrolyte level and the filling time. Figures 8 and 9 show the behavior of the "topping-up" arrangement. The pressure range over which this mechanism works well lies between 40 and 60 cm water column. The material for the tube must withstand the attack of sulfuric acid chemically and must remain elastic for a long time.

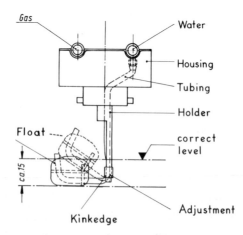

FIG. 7. Cell plug with means to add water automatically to a given level (German Design Pat. 72,149 "Topping-up Device").

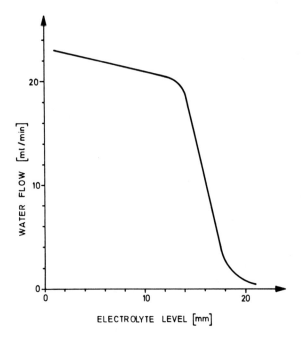

FIG. 8. Water flow versus electrolyte level when using cell plug of Fig. 7.

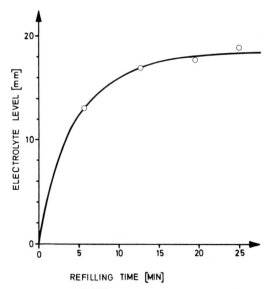

FIG. 9. Electrolyte level versus refilling time, using device shown in Fig. 7.

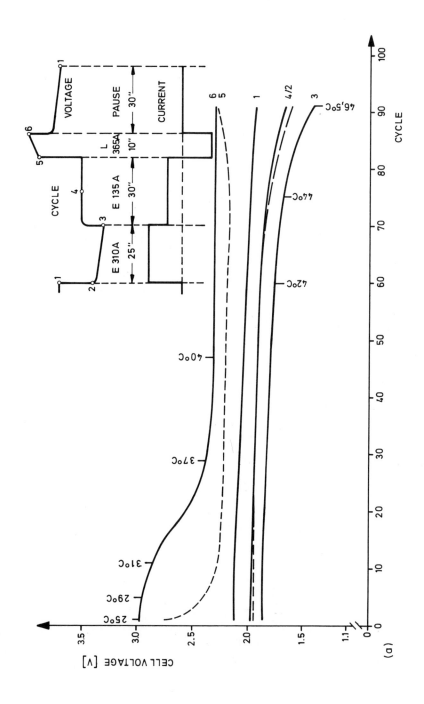

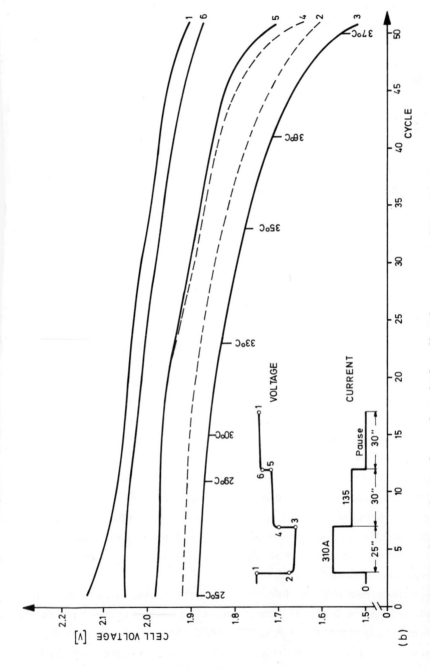

FIG. 10. Working and end voltage of a vehicle cell with a capacity of 250 Ah at the 5 h rate. (a) Simulated cycle without and (b) cycle with regenerative braking.

A central water filling system can easily be combined with a gas removal system. It provides a safe atmosphere in the battery compartment.
The removal of the sulfuric acid vapors to the outside also has the advantage that the corrosion of battery and vehicle components is reduced—resulting in less maintenance.

2. ELECTRIC ROAD VEHICLE PROJECTS IN GERMANY: A SUMMARY

Electric road vehicle projects started in Germany in the 1970s, and it is
noteworthy that the emphasis was put on the driving characteristics of the
vehicles, not on the range. Acceleration values close to those of automobiles with combustion engines (albeit small ones compared to those in the
United States) were a major requirement. This is necessary to avoid traffic blockage by underpowered electric vehicles.

The handicap established by the short range of battery-operated vehicles was reduced by the selection of utility vehicles used in urban traffic
patterns, which are inherently short-distance, multi-stop schedules. In
addition it was realized that the battery exchange method [20] could be used
under certain conditions to multiply the useful range of the vehicle.

The application of modern power electronics made the nearly loss-free
regulation of the current flow to the motor possible; also regenerative braking was advantageously used. Compared to the previous methods of control
by means of resistors, a range improvement of up to 20% was achieved and
driving comfort was much improved.

Figure 10 shows the influence of regenerative braking on the range of a
vehicle which in a cyclic manner goes through the phases of acceleration,
constant-speed driving, braking (with regenerative circuit), and waiting.
Such simulated driving patterns produce current profiles which can then be
compared with the effects of steady current discharges on batteries of different constructions.

With respect to the usefulness of regenerative braking, it can be demonstrated that the lead-acid battery can take up the high currents generated,
except in fully charged condition. The acceleration losses of stop-and-go
traffic can thereby be effectively reduced and the useful range increased.

2.1. Buses

The best-developed road vehicle in Europe is the MAN Electrobus. After
finishing testing two prototypes (1971), 20 Electrobuses of an improved type
were built for the evaluation of technical parameters. This large-scale experiment was funded by the Gesellschaft für Elektrischen Strassenverkehr
and supported by public money. The problems of battery-operated vehicles
shall be studied over a period of 5 years in scheduled bus-line traffic,
mainly in Mönchengladbach and Düsseldorf. The MAN concept suggests
that the additional weight of the battery should be carried by an extra axle
in order to be able to accommodate the full capacity of 100 persons (7,500
kg). With medium ranges of 80 km in typical line operations, acceleration
values of 10 sec for 0 to 30 km/h and 23 sec for 0 to 50 km/h have been accomplished. Due to the fixed-line operation, the exchange of the battery
can be done at special stations. The replacement against a fully charged
battery takes only 4 min. Additional data are given in Table 1. The total
weight of the battery, including the battery enclosure, control and safety
devices, and accessories for cell cooling, for central water addition and
gas removal, is 6,000 kg. The battery assembly is shown in Fig. 11 with
the cover of the trailer open. The maximum temperature of this battery is
set at $55^{\circ}C$, the compact construction making cooling necessary.

A typical operational schedule considers a 2-h discharge followed by a
2-h charge. The 2-h charge period is not sufficient to keep all the cells
uniformly fully charged (no overcharge), and at carefully programmed intervals an equilibrium charge is performed to maintain the capacity of the
cells at their original high level.

Daimler-Benz A.G. has gone through the initial steps of providing a
bus of the type OE 302 with electric propulsion and after testing the prototypes there are now 30 buses of type OE 305 being built (1975). The technical data can also be found in Table 1. The concept of Daimler-Benz includes
a hybrid energy supply including a diesel generator capable of producing
70 kW. This unit charges the battery (series hybrid) but can also provide
some of the current directly to the motor. During stops a fast recharge of
the battery is possible. In this manner it is feasible to achieve daily ranges
of 300 km with one charge. The bus lines are arranged in such a way that

TABLE 1

Electric Road Vehicle Projects in Germany (1974)[*]

Vehicle manufacturer	Daimler-Benz OE 305	MAN Electrobus	Daimler-Benz LE 306	Volkswagen	Messerschmitt-Bölkow-Blohm
Length (m)	11.1	11.1 + 2.7	5.0	4.4	4.9
Width (m)	2.5	2.5	1.8	1.7	2.1
Height (m)	2.9	2.9	2.2	1.9	2.1
Total weight (kg)	18,600	16,000 + 8,000	3,500	2,700	2,900
Load (kg)	6,900	7,600	1,000	750	600
Passengers	92	105	8	2	8
Acceleration, 0–50 km (sec)	20	23	10	10	12
Top speed (km/h)	70	70	80	80	80
Climbing grade (%)	16	12	13	25	21
Range (km) at (km/h)	80/50	140/50	60/50	125/30	60/50
Power of diesel hybrid (kW)	70	–	–	–	–
Spec. energy usage (Wh/t km)	110	140	90	86	120
Cont. power (kW)	115	115	27	16	22
Peak power (kW)	150	176	56	32	44
Revolutions (U min – 1)	4,800	4,000	6,700	6,700	6,700
Motor weight (kg)	530	700	100	85	95
Nominal voltage (V)	360	360	180	144	144
Max. current (A)	600	550	400	300	400
Solid state control with regenerative braking	Bosch Siemens	Bosch Siemens	Bosch Siemens	Bosch Siemens	Bosch Siemens
Battery type (Varta)	6 PzF 230	7 PzF 455	9 GF 180	5 PzF 150	9 GF 180
Capacity, 5–h (Ah)	230	455	180	150	180
Weight (kg)	3,000	6,100	980	860	780
Nominal voltage (V)	360	360	180	144	144
Specific energy (5–h), System	27	27	34	25	32
(Wh/kg) Cell	32	33	39	31	39
Life (cycles)	1,500	1,500	600	1,500	600

[*] An excellent survey of all German contributions to the electric vehicle technology, as practically shown at the IV International Electric Vehicle Symposium (1976), has been collected in the special issue (Sonderheft) of the Elektrotechnische Zeitschrift, January 1977.

FIG. 11. Battery for the MAN Electrobus. View of the cooling aggregate.

the buses operate on the battery during their runs within the inner city and use the additional power from the diesel generator in the suburbs where the exhaust pollution is less serious. The same bus type can be used without the diesel power supply in a pure battery operational mode. For that reason both battery troughs are arranged under the floor of the vehicle and can be easily exchanged with the accessories provided. A battery change can be done at similar automated stations as the MAN bus is using.

2.2. Transport Vehicles

Daimler-Benz (Fig. 12) and Volkswagen (Fig. 13) started the development of transport vehicles on the basis of existing constructions. Messerschmitt-Bölkow-Blohm decided to build an electrically powered transport vehicle from the ground up, using new concepts in plastic material utilization. To

FIG. 12. Daimler–Benz transport vehicle (1 ton). Battery: 180 V, 27/54 kW.

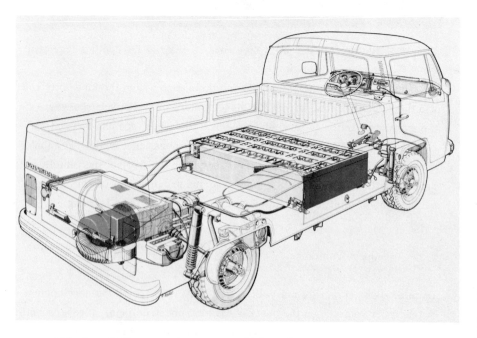

FIG. 13. Volkswagen transport vehicle (3/4 ton). Battery: 144 V, 16/32 kW.

FIG. 14. Messerschmitt-Bölkow-Blohm Vehicle (0.6 ton). Battery: 144 V, 22/44 kW.

date there are 15 prototypes available which include three groups: buses, trucks, and vans. The principal data are also shown in Table 1. Figure 14 shows the vehicle as a personnel mover.

The "cross-exchange technique" used for Volkswagen and Daimler-Benz vehicles allows a battery change within a few minutes, using auto-mated equipment. It works the same way as for the buses: the battery is replaced from the side (crosswise to the driving direction) and needs only one person for supervision. In contrast to this, the exchange of a trailer (uncoupling) needs more manpower and the vehicle cannot move on its own. The design of a permanent coupling is also simpler and results in better tracking characteristics (MAN bus). However, the batteries need not be exchanged if the operational schedule allows recharging overnight. In this sense the systems are flexible and make most economical use of their in-herently limited range.

A thorough collection of all efforts aimed at electric road vehicles has been published (1973) in a special volume of the Elektrotechnische Zeit-schrift [1].

REFERENCFS

1. D. Berndt and H. Niklas, "Physikslische Grenzen heutiger Speicher-batterien und Chancen fúr die Entwicklung neuer Systeme," Electrotech. Z, <u>94</u> (11), 694-699 (1973).

2. H. G. Lindenberg, Varta, German Pat. 1804 800 (1970).

3. A. Sabatino and D. Orlando, Globe-Union, Inc., U.S. Pat. 3,313,658 (1967).

4. F. Bronstert, Varta, German Pat. 1,067,899 (1959).

5. B. Gicquel and A. Meunier, Fulmen, BF 7317 006 (1973).

6. H. Niklas, "State of the Technology of Electric Vehicles and Batteries in Germany," Proceedings of the 85th BCI Meeting, San Francisco, Calif., 1973.

7. K. Salamon, Varta, German Design 75 09 377 (1975).

8. S. Nijhawan, Varta, German Pat. 2151 733 (1973).

9. D. L. Douglas and G. W. Mao, "Laboratory Evaluation of Non-Antimonal Lead Alloys for Lead-Acid Batteries," Proceedings of the 8th International Symposium on Lead-Acid Batteries, 1972.

10. K. Salamon, H. Busch, and E. Zander, Varta, GES, German OS 2350 353 (1975).

11. G. Sassmannshausen and N. Lehme, Hoppecke, German AS 2008 218 (1971).

12. T. Marni and A. Yokogi, Yuasa, German OS 2201 652 (1973).

13. A. Winsel, M. Jung, and E. Voss, Varta, German AS 2340 945 (1975).

14. W. Borger, Varta, personal communication.

15. E. Westberger, AB-Tudor, Schweden, German OS 2415 101 (1975).

16. J. P. Caligary, Fulmen, German OS 2423 412 (1974).

17. J. D. Harris, Lucas, German Pat. 1671 824 (1971).

18. S. Ikeda, Yuasa, German OS 2356 882 (1974).

19. H. Zweighardt, Varta, German Design 7214 990 (1972).

20. G. Hebel and T. Merkle, GES, Voith, German AS 2238 671 (1975).

III. LEAD-ACID BATTERY HISTORY
AND NEW TECHNOLOGY IN JAPAN

Akiya Kozawa

Union Carbide Corporation
Battery Products Division
Technology Laboratory
Parma, Ohio

Tokujiro Takagaki

Japan Storage Battery Co.
Kyoto, Japan

1. INTRODUCTION

The purpose of this report is to outline the historical development of the Japanese lead-acid battery industry and the present status of the technology of the lead-acid battery in Japan. The early history of the lead battery in Japan, industrial standards on various types of batteries, major manufacturers and their brand names are included, since no previous book has published these items in English. The largest number of automobile batteries is produced in the United States; therefore, the Japanese situation will be compared with the U. S. situation, whenever possible, with respect to production volume and technology.

In 1973 Japan consumed 200,000 tons of lead for lead-acid batteries, which was approximately one-third of the U. S. amount. Since 1969 the Japanese production has been in second place worldwide. Automobile battery production in Japan now accounts for 80% of the total lead-acid battery production. As shown in Table 1, the Japanese automobile industry has grown rapidly, particularly in the past 15 years. The total number of vehicles produced including passenger cars, buses, and trucks in 1965 was about 1,880,000 units in Japan, but in 1973 it was 7,080,000 units, which is almost four times the 1965 production. At the end of 1973, the total

TABLE 1

Automobile Production in Japan 1961-1974 (Unit = 1,000 Vehicles)

	Cars registered in Japan			New car production			
	Passenger cars	Trucks and buses	Total	Passenger cars	Trucks and buses	Total	Cars exported
1961	660	2,210	2,870	250	560	810	60
1965	2,180	4,800	6,980	700	1,180	1,880	200
1970	8,780	9,050	17,830	3,180	2,110	5,290	1,220
1971	10,570	9,490	20,060	3,720	2,090	5,810	1,920
1972	12,530	10,050	22,580	4,020	2,270	6,290	1,960
1973	14,470	10,530	25,000	4,470	2,610	7,080	2,070
1974	15,850	11,050	26,900	3,930	2,620	6,550	2,620

number of registered vehicles (including trucks and buses) in Japan was
25 million units, which is roughly one car per every four persons in Japan.
In the United States, the total number of registered vehicles in 1973 was
125 million units, as shown in Table 4, which corresponds to one car per
every 1.68 persons. Since this rapid industrial growth in Japan was such
an unusual period in the history of industry, as much statistical data as
possible has been collected on the production of automobiles and lead-acid
batteries.

Japan played an important role in the early technical development of
the lead-acid battery: that is, the invention of the ball mill method for
manufacturing lead powder by Mr. Genzo Shimazu. Mr. Shimazu obtained
the Japanese patent in 1920 and later obtained a number of patents in the
United States, Germany, France, etc. The ball mill process is very
unique and is still the major industrial method used today.

2. HISTORY OF THE LEAD-ACID BATTERY (1900-1945)

2.1. Early Events

In 1859 G. Planté invented the lead-acid cell in France. His cell consisted
of a wound lead foil soaked in sulfuric acid and formed by electrolysis. In
1881, C. Faure invented the paste-type lead-acid battery. In Japan, S.
Kato of the Government Electricity and Telegraph Institute investigated
lead-acid batteries and prepared 60 batteries in 1886. The batteries were
used for the illumination of the Tokyo telegraph office. In 1895, G.
Shimazu prepared a 1-Ah electrode plate and in 1904 he prepared a 150-Ah
chloride-type battery which was installed as standby power source for his
own factory. The use of central electric power spread throughout Japan
between 1900 and 1905 and electric illumination was started. In those
days power failures were so frequent that the need for standby power in
movie theaters and playhouses was great. Most of the standby power batteries were imported. The first facility for manufacturing lead-acid batteries in Japan was set up by G. Shimazu in his factory in 1910. The facility produced a Tudor-type (formed lead foil electrode) battery. In 1917 the
plant was separately operated as Japan Storage Battery Company, the first
lead-acid battery company in Japan.

G. Shimazu applied for the patent "Method for Manufacturing Reactive Lead Powder" in November 1920. The patent was granted in February 1922 in Japan and later in various countries. The process was essentially a ball mill process in which lead balls (about 3 cm in diameter) crash into each other under a hot air flow (about 250° C). The powdered lead is carried away by the air flow and collected at the outlet. The lead powder thus prepared had unusually superior properties for preparation of the electrode of lead-acid batteries, and no better suited material has been found. This invention can be compared to the discovery of electrolytic manganese dioxide which was first prepared around 1931-1934 and was recognized for having a far superior property than natural MnO_2 for dry cell cathodes. Since then, no better MnO_2 has been found. It is interesting to note that the quality of the active material of both lead-acid batteries and MnO_2-Zn dry cells depends mainly on the manufacturing method.

The invention of the ball mill process had the following historical background: It was recognized that the superior performance of the German submarines of World War I was due mainly to the better lead-acid battery. Japan Storage Battery Company tried to buy the technology from France, but it was too expensive. Under these circumstances Shimazu decided to invent a new process and founded the ball mill process within a year. Figure 1 shows a ball mill for lead powder production (around 1920). Figure 2 shows Dr. Fritz Haber, a well-known German chemist, and Mr. Genzo Shimazu shaking hands with each other in 1924, when Dr. Haber visited Japan Storage Battery Company and saw the ball mill process. Dr. Haber reported the ball mill operation to the German patent office, and the German patent was granted soon thereafter.

Yuasa Battery Company and Furukawa Battery Company were established in 1924.

2.2. Historical Records

1908   Lead-acid battery production started in the Shimazu factory.
1912   First Japanese patent on lead-acid batteries was granted.

FIG. 1. A ball mill for lead powder production in the early days (1920-1925).

1907-1912   Main applications of lead-acid batteries: telephone and telegraph stations, trains, lights for street vehicles, and electric cars.

1914   Production of Tudor-type lead-acid batteries for railroad trains was initiated in Japan. Japanese railroads (Tokaido line) started the service in 1898, and until 1914 all railroad batteries were imported from England. The beginning of the domestic production of Tudor-type batteries in Japan was the consequence of the onset of World War I.

1915   Production of Tudor-type lead-acid batteries reached 1,000 tons, and a 3,600-Ah chloride-type battery was produced and installed at the top of a mountain in Taiwan for telegraph operation.

FIG. 2. Mr. G. Shimazu (left) and Dr. Fritz Haber at the Japan Storage Battery Company in 1924.

1917	Japan Storage Battery Company was established. Within a few years Yuasa Battery Company, Furukawa Battery Company, and Shin Kobe Denki Company were established, since an urgent need of lead-acid batteries arose in World War I.
1918	Yuasa Battery Company introduced the battery production technology obtained from France.
1917–1919	Lead-acid batteries were exported from Japan to China, India, etc.
1919	After World War I ended, the production of lead-acid batteries decreased to one-third.
1925	Use of the lead-acid battery as standby power source had spread considerably in Japan. Mr. S. Toyota recognized the importance of the standby battery for spinning mills and offered $200 million (in today's money) to an inventor of a better battery than the lead-acid battery.

1926   Japanese radio broadcasts started in the Tokyo area and grad-
            ually spread to other parts of Japan. The radio receivers
            used lead-acid batteries.

1931   The Japanese government banned payments in gold to foreign
            countries. Since General Motors and Ford were producing
            automobiles in Japan and could not import batteries, they were
            forced to produce batteries in Japan. Because of this new sit-
            uation, lead-acid battery production in Japan increased.

1932-1945   During the period of the China war and World War II, Japanese
            lead-acid battery production increased rapidly, as shown below:

            1930-1935   3,000 tons/year
            1940   11,339 tons/year
            1944   19,250 tons/year

1945-1975   After World War II, lead-acid battery production increased
            with the increase of automobile production. This period of
            growth will be described in the next section.

3. DEVELOPMENT OF THE JAPANESE LEAD-ACID BATTERY INDUSTRY AFTER WORLD WAR II

1945-1950. The production of lead-acid batteries in the first year after
World War II decreased to less than 50% of the highest production year during the war (1944). The number of employees in 1946 decreased abruptly
to one-third of the 1944 employees even in the case of Japan Storage Battery
Company, which was not damaged by air raids. During the years immediately after World War II, power failures were so frequent that the demand
for lead-acid batteries for standby power was great, and also lead-acid batteries for fishing fleets were in strong demand since fish as food supply was
very important. During the war years, lead-acid batteries were mostly for
military use and had no commercial sales route. Therefore, the companies
were forced to establish commercial marketing after the war. The production recovery was rather rapid, however, since the demand for lead-acid
batteries for railroad trains and telephone and telegraph stations also increased. Battery production of lead in 1949 reached 13,400 tons, which is
about 70% of the maximum production (1944) during the war. The 70% figure indicates an extremely good recovery compared to other industries,
which recovered only to 23%. At the end of 1949 the government tightened

the economy in order to stop inflation. This policy could have influenced
the battery industry considerably.

1950-1955. In June 1950, the Korean War started and the lead-acid battery
industry began to grow because of extensive procurement by the U. S. military force in Japan. Japanese automobiles used lead-acid batteries having
3-mm-thick plates during and after World War II. Since Japanese battery
manufacturers learned that U. S. military car batteries used 2-mm-thick
plates, Japanese battery manufacturers gradually changed to 2-mm-thick
plates around this time.

1955-1975. These 20 years represent a unique growing period for both the
Japanese automobile and the lead-acid battery industry. Table 2 shows the
production figures for this period. We can see that the rapid growth of
automobile batteries began around 1958, and in 1974 it reached 14 times
the production of 1955. The percentage of automobile batteries among the
total batteries increased from 40% in 1945 to 80% in recent years (1965-
1975). Such a rapid growth of automobile batteries enabled Japanese manufacturers to rebuild production facilities to accommodate highly automated
processes. All the major manufacturers built special new plants for auto-

TABLE 2

Lead Consumption for Batteries in Japan (1946-1974) (Unit: Tons)

Year	Total tons	Automobile batteries	
		Tons	Percent
1946	13,217	5,480	41.5
1950	11,980	6,498	54.2
1955	14,826	9,827	66.3
1960	35,213	23,486	66.7
1965	61,434	46,943	76.4
1970	156,812	(127,600)	81.5
1971	160,593	(129,992)	81.5
1972	176,852	(144,164)	82.0
1973	204,430	(166,891)	80.5

mobile batteries during this period. A detailed breakdown of figures from
1969 to 1974 in respect to usage is shown in Table 3. Export is very important for Japanese industries. It should be noted that a high percentage
of export batteries are for motorcycle batteries and automobile batteries.

Figure 3 shows a comparison of lead consumption in battery production
in the United States, Japan, and Great Britain. In the United States,
400,000 tons were used in 1964 and 620,000 tons in 1974, which is 1.5
times the 1964 figure. During the same period Japanese consumption increased three times from 66,640 tons to 220,000 tons. The relatively slow
growth in the United States between 1962 and 1967 is due to the fact that
automobiles were already there in large numbers. The next period of
rapid growth around 1969-1972 is due mainly to the transition from two
cars per family to three cars per family, as seen in Table 4.

These historical facts tell us the following: The lead-acid battery industry depends on other industries such as automobile, telephone and tele-

TABLE 3

Breakdown of Production Types for Japanese

Lead-Acid Batteries with Respect to Lead Consumption (unit: Tons)

Year	1969	1970	1971	1972	1973
1. Automotive batteries:					
For new cars	49,691	57,380	64,172	72,932	82,694
For replacement	48,629	55,750	56,594	62,361	74,180
For export	7,666	7,976	8,206	9,021	7,776
Total of automotive batteries	105,986	121,106	128,972	144,314	164,650
2. Motorcycle batteries					
For new vehicles	2,642	3,552	4,236	4,561	3,655
For replacement	1,851	1,529	1,509	1,421	1,986
For export	3,089	3,329	3,058	3,703	4,550
Total of motorcycle batteries	7,582	8,410	8,803	9,685	10,191
3. Stationary batteries	6,090	6,199	6,895	7,414	8,506
4. Traction batteries	1,344	1,355	1,286	1,377	1,596
5. Batteries for ships	755	683	713	648	569
6. Batteries for electric vehicles	5,215	7,030	7,924	9,166	12,121
7. Batteries for trains	2,282	2,203	2,469	2,135	1,954
8. Others	1,943	1,878	1,914	2,340	1,864

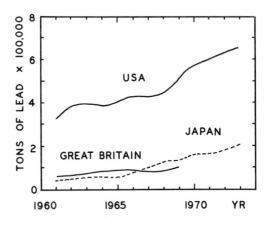

FIG. 3.   Lead consumption for battery production.

graph, railroad, safety systems, etc. Depending on the change in our society, new systems and new equipment will probably need new types of lead-
acid batteries. The new areas would be energy conservation, use of com-
puters, automatic systems, material recycling, pollution reduction, etc.
Development and acceptance of new battery-operated equipment may trigger
a big growth of the lead-acid battery industry. Emergency lighting systems

TABLE 4

Population and Total Number of Motor

Vehicles Registered in the United States and Japan (Unit: Millions)

	Year	Number of cars[*]	Population	Number of people per vehicle
United States	1950	49.3	151.3	3.07
	1960	73.9	180.0	2.44
	1965	90.3	193.5	2.14
	1970	108.4	203.8	1.88
	1973	125.2	209.9	1.68
Japan	1960	2.8	93.5	33.39
	1965	6.9	98.0	14.20
	1970	17.8	103.7	5.83
	1973	25.0	108.7	4.35

[*]Includes passenger cars, buses, and trucks.

TABLE 5

Use of Lead-Acid Batteries in Japan and the United States[*]

	United States (1972)		Japan (1973)			
	$ million	(%)	$ million	(%)	Lead in tons	(%)
1. Automotive SLI batteries	662	(74%)	250	(76%)	164,650	(82%)
2. Motorcycle batteries			20	(6%)	10,191	(5%)
3. Traction batteries	82	(9%)	22	(7%)	12,121	(6%)
4. Stationary batteries	30	(3%)	24	(7%)	8,506	(4%)
5. Other	124	(14%)	14	(4%)	5,983	(3%)
Total	898	(100%)	330	(100%)	201,451	(100%)

[*]Based on estimated data by the authors.

In the United States batteries are being used for golf carts, lawn mowers, snowmobiles, and private airplanes. In Japan, there is no such use. Motorcycle batteries are being produced in considerable numbers. There is no motorcycle battery production in the United States.

for buildings and short-range electric cars would be new developments promising growth in the lead-acid battery industry.

Table 5 shows some difference in the uses of lead-acid batteries in Japan and the United States.

4. COMPARISON OF THE AUTOMOBILE BATTERIES MADE IN JAPAN AND IN THE UNITED STATES

Since automobile batteries account for 70 to 80% of the total lead-acid battery business, production figures and the technology of automobile batteries in Japan and the United States will be compared.

4.1. New Car Batteries Versus Batteries for Replacement

Table 6 shows the ratio (A:B) of the number of batteries produced for new cars (A) versus replacement batteries (B). The ratio is 1:4 to 1:3 in the United States, but it is roughly 1:1 in Japan. The new car batteries are

TABLE 6

Comparison of Automotive Batteries in the

United States and Japan for New Cars and for Replacement[*]

United States	Batteries (in millions)			
	Total batteries	For new cars (A)	For replacement (B)	A:B
1958	30.7	5.9	24.8	1:4.2
1963	41.4	9.2	32.2	1:3.2
1967	40.7	10.4	30.3	1:2.9
1972	50.4	11.8	39.6	1:3.4
1973	53.6	13.0	40.4	1:3.1

Japan		Tons of lead		
	Total batteries	New car batteries	Replacement batteries	A:B
196 1966	64,930	27,272	33,811	1:1.24
1968	97,282	43,739	47,227	1:1.08
1970	121,106	57,380	55,750	1:0.97
1973	164,650	82,694	74,180	1:0.89

[*]Including batteries exported, which is about 5% throughout.

TABLE 7

Total Registered Vehicles and New Cars Produced (Unit: Millions)

		Total number of registered vehicles[*] (T)	Number of new vehicles produced (N)	Percent new vehicles
United States	1965	90.0	9.3	19.3
	1970	108.3	6.6	6.1
	1973	125.2	9.7	7.8
Japan	1965	6.9	1.9	27.6
	1970	17.8	5.3	29.9
	1973	25.0	7.1	28.4

[*]Including cars, buses, and trucks.

sold to automobile manufacturers and the profit is very small for battery manufacturers. The replacement batteries, however, will bring relatively good profit to battery manufacturers, since the batteries are sold to individual users through garages and their own auto parts stores. Why is the ratio A:B in Table 6 so different between Japan and the United States? The following factors may be the main reasons:

1. Thirty percent of the total automobiles produced in Japan are being exported in recent years, as was shown in Table 1. Exported automobiles do not produce a demand for replacement batteries.

2. In Japan, the increase in the number of registered cars is only a recent happening (1967-1974). The percentage of relatively new cars is large. In the United States, the annual new car production is only 6 to 8% of the total of running cars, but nearly 30% in Japan as shown in Table 7.

3. The average life of a car is 5 to 6 years in Japan, but it is 9 to 12 years in the United States. On the other hand the average life of a car battery is roughly 2 years in Japan, but it is 3 years in the United States. Therefore, an average American car uses 3 to 4 batteries during the car life, but a Japanese car uses only 2 to 3 batteries. The average driving distance per car in a year is about 9,000 miles, both in Japan and in the United States.

4.2. Comparison of Typical Automobile Batteries

In Table 8 typical U.S. and Japanese automobile batteries are compared. It can be seen that there is little difference in the engine-starting capability (voltage after 5 sec at 150 A discharge at $0°F$) and the volume energy density (Ah/liter) for the most widely used technology for mass produced batteries is approximately equal in Japan and the United States at this time (1975).

Table 9 gives the specification for the Y 110.5 battery.

Ten years ago, Japanese cars (most of them having a 2,000-cc engine) used a 50-Ah battery (N50), but now cars of this type use a 35-Ah battery (NS40Z). This transition to a smaller battery occurred 10 years ago in

TABLE 8

Comparison of Typical 12-V Automotive
Batteries Made in the United States and Japan

	U. S. specification (BCI)		Japanese specification (JASO)		
Size no.	24	24	24	24	
Type of battery	(9M3B)		(N50)	(NS70)	(NS40Z)
Capacity: Ah at the 20-h rate	50	70	50	65	35
0°F discharge time (min)	3.7	6.0	3.4*	6.2†	3.3*
Voltage at 5 sec at 150 A (V)	8.4	9.3	8.1*	9.3†	8.8*
Charge current (A)	4.	5.25	5.	6.5	3.5
Number of plates per cell	9	13	9	13	11
Volume (liters)	10.1	10.1	10.1	10.1	5.8
Ah/liter at 20-h rate	4.9	6.9	4.9	6.4	6.0
Ah/liter at 0°F at 150 A	0.92	1.48	0.84	1.53	1.42

*These values were extrapolated from those at -15°C.

†These values were obtained from the 300-A data (with temperature effect considered) based on the battery characteristics known.

TABLE 9

Specifications of Y110-5-type Battery

Type of battery	N50	N50Z	NS70	Y110-5	N100Z
Nominal voltage (V)	12	12	12	12	12
20-h rating capacity (Ah)	50	60	65	60	100
Number of plates per battery	54	66	78	90	114
Type of separator	Paper	Paper	Paper	"Yumicron"	Paper
Battery dimensions					
Length (mm)	260	260	260	260	412
Width (mm)	173	173	173	173	176
Height (mm)*	204	204	204	204	213
	(225)	(225)	(225)	(225)	(233)
Battery weight (kg)	17.5	19	19.5	22.5	31.5
Sp gr of electrolyte (20°C)	1.260	1.260	1.260	1.280	1.260

*The number in parentheses indicates total height.

Japan when the alternator replaced the dc generator. On the other hand, American cars changed from a 50-Ah battery to a 70-Ah battery when the charging system changed. This means that in the United States the trend shifted to a higher-capacity and longer-life battery, but in Japan the trend shifted to a smaller-Ah battery, which can maintain about the same life (2 years) and the same engine-starting capability as the old battery. With the old dc generator system, the charge was not sufficient at low engine speeds and the car battery often was left in an undercharged condition, but with the new alternator (ac generator plus a diode) adopted around 1965, the car battery could be maintained in a full charge condition even at low engine speeds. Since American cars began to install air conditioners and other equipment around 1965, the battery capacity had to be increased at the same time as the charging system was changed.

Another difference between the U.S. battery and the Japanese battery is the double-insulation separator system, which is standard in Japanese automobile batteries. The double-insulation system consists of a two-layer (glass mat and treated paper) separator, the glass mat layer being in contact with the positive electrode plate. This separator system was needed because of the bumpy Japanese roads. The glass mat supports the positive active material, acts mechanically as a shock absorber, and prolongs battery life. After most Japanese roads were paved, the double-insulation separator system in Japanese automobile batteries was kept because with it, batteries with similar capabilities and life as U.S. batteries can be made and the lead requirements are reduced.

4.2.1. Plate Thickness

To improve the engine-starting capability the battery should maintain a higher voltage at high currents (150 to 300 A). To accomplish this, one way is to use thinner plates, so that the total number of plates per battery can be increased. In Japan, the plate thickness decreased from 3 mm to 2.0 mm around 1950 and further decreased to 1.5 mm around 1962. U.S. batteries have always used 2.0-mm-thick plates during this period. When the battery case material was changed from hard rubber to polypropylene, the number of plates in U.S. batteries increased and the benefit was a longer life expectancy due to the lower specific load. However, in general, thin plates have a shorter life, particularly when single insulation was used. Another important factor which determines battery life is corrosion of the grid of the positive plate during overcharge. Generally, a thicker plate has a longer life from this (overcharge) standpoint.

4.2.2. Battery Case Material

Until 1960 automobile battery cases had been made from hard rubber (approximately 8 to 10 mm thick). In Japan, the battery case walls were changed to 2 to 3 mm thick AS resin (acrylonitrile-styrene resin) and ABS resin (acrylonitrile-butadiene-styrene resin) after 1960. In Japan as well as in the United States, battery case material for high-volume automobile batteries was changed again to polypropylene around 1968. The AS resin or ABS resin cases used an adhesive to seal the lid and the case, but the polypropylene resin case uses thermal welding instead of an adhesive.

Since the impact resistance of polypropylene is small at low temperatures, a copolymer of polypropylene and polyethylene is being used for today's automobile battery cases. Even today some hard rubber cases are still used for large-size automobile batteries which are not mass produced.

4.2.3. Connection Between Cells

Two systems, connection-over-partition (COP) and connection-through-partition (CTP) have been used. The CTP system requires less lead and exhibits less voltage drop, but the system needs high-technology sealing since the joint portion is immersed in the electrolyte. In the United States the COP system was adopted around 1962 and the CTP system around 1966. In Japan most mass-produced automobile batteries have been using the CTP system since around 1967. The COP system is used mostly for AS and ABS resin case batteries and the CTP system for polypropylene case batteries.

4.2.4. Separator

Since the battery separator must be stable in a $5\underline{M}$ H_2SO_4 (sp gr 1.28) solution, the following materials have been used:

TABLE 10

Characteristics of Various Types of Separators

Item	Type			
	Paper separator	Rubber separator	Sintered PVC separator	Yumicron separator
Material	Linter or craft pulp	Rubber	PVC	Plastics
Thickness (mm)	0.5~1.0	0.4~1.0	0.5~1.0	0.1~0.2
Resistance (Ω/cm^2/sheet)	0.0012~0.003	0.0012~0.003	0.0012~0.003	0.0002~0.0004
Antioxidation (h/sheet)	40~70	100~500	50~100	100~120
Mean pore size (μm)	20~30	0.5~5	20~40	0.1~1
Porosity (%)	60	55	55	60

1. Wood plate (about 1 mm thick)
2. Paper treated with a phenol resin (0.6 to 0.8 mm thick)
3. Microporous rubber (about 0.4 mm thick)
4. Nonwoven cloth made of polypropylene and Teflon
5. Special thin plastic film, e.g., Yumicron (about 0.1 to 0.2 mm thick

Most of today's automobile batteries (80%) in Japan use the treated paper separators. Table 10 shows characteristics of various separators.

4.2.5. Dry-charged Batteries

Dry-charged batteries contain plates which have been formed and dried. The battery can be activated simply by adding the electrolyte, usually without a need for immediate charging. This type of battery appeared on the Japanese market around 1968 and its share gradually increased and reached about 50% of the total replacement batteries on the market in 1975. In the United States, the percentage of dry-charged batteries was about 55% in 1968-1969 and 60% in 1973.

The Japanese climate is extremely high in humidity. Even under such conditions, many Japanese dry-charged batteries (e.g., GS Myca 5) can hold the charge for more than 1 year without special sealing of the electro-

lyte filling hole. This was achieved by an improved manufacturing process.
Since dry-charged batteries are very advantageous from the standpoint of
shipping and maintenance, it is expected that the market share will increase
further in Japan. A disadvantage is the higher production cost.

4.2.6. Maintenance-free Batteries

A 12-V battery consists of six cells, and the water addition must be done
separately for each cell. A new simplified water addition system, which
enables a one-shot addition, distributing the water properly to all the cells,
was incorporated in a commercial battery (e. g. , GS-Myca 5 by Japan Storage Battery Company) in 1968. This battery was used for new cars in 1969.
In 1971 another new type of battery, GS-7 (Japan Storage Battery Company)
which needs no water addition was introduced on the market. It has far
more electrolyte in the battery from the beginning and uses an electrode
material which does not decrease the charge voltage after long use.

Another battery of this type is YSB-Neo Perfector (Yuasa Battery
Company), which needs no water addition for 1 year.

In September 1973, another new replacement battery, NSB-Panasonic
Carec, was introduced by Matsushita Electric Industrial Company. The
battery used calcium alloy grids for the negative electrode and had also a
large amount of electrolyte in the battery. Ordinary battery plates use Pb-
Sb alloy grid and upon repeated overcharges the grid is oxidized to PbO_2
and Sb is dissolved in the electrolyte. The antimony deposits on the negative plate and reduces the hydrogen overpotential. Therefore under the
present constant-voltage charging system, hydrogen evolution increases
with time and water additions become more frequent toward the end of the
battery life.

In the United States, a calcium grid battery was already used for GM's
new cars in 1971. Recently (1974) Gould, Inc. , manufactured a maintenance-
free battery which needs no water addition and is sold through J. C. Penney
stores. By using calcium-lead or low-antimony-lead alloy as grid, the self-
discharge on standing is reduced considerably. The extended shelf life competes with the dry-charged battery (which has the inconvenience of the filling
procedure to be done by the customer or store).

4.2.7. Other New Batteries

4.2.7.1. Heated Batteries

Matsushita Electric Industrial Company introduced the Cell Heat battery NS60(W) having a 45 Ah capacity. It is recommended in extremely cold areas such as Hokkaido. The battery contains heaters (total of 60 W) distributed in the cells and the battery temperature is controlled by a thermostat. The battery can always be charged, even on extremely cold days.

4.2.7.2. Y-110-5-type Battery

Yuasa Battery Company introduced a new battery which uses Yumicron, a microporous thin vinyl chloride film, as separator. Since the separator has significantly low resistance, the battery has a much better engine-starting capability at low temperatures. In Fig. 4 a comparison is made between ordinary batteries (N50Z, NS70, N100Z using regular paper separators) and a Y110-5 battery using a Yumicron separator at -15°C at 300-A discharge.

4.2.7.3. G110-5-type Battery

Japan Storage Battery Company introduced this high-capability battery in 1975. The battery uses a new polyolefin separator, Polymion, which not only has low electrical resistivity similar to Yumicron but also is easy to recycle, since no chloride is involved in the separator material. The engine-starting capability at low temperatures is equivalent to the Yumicron batteries.

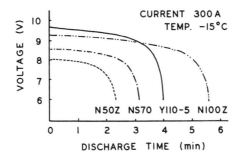

FIG. 4. Comparison of batteries with ordinary separators (broken lines) and a battery with Yumicron separator (full line).

5. RAW MATERIALS FOR LEAD-ACID BATTERIES

The important raw materials for lead-acid batteries are lead, antimony, ABS resin, and polypropylene resin. Lead to be used for batteries is usually electrolytic lead which has a purity of 99.99% or better. The grid of the electrode plate is usually cast from a lead-antimony alloy containing several percent antimony. Pure lead is too soft to use for the grid. The function of the added antimony is to improve mechanical strength.

Table 11 shows distribution of weight of the raw material for a typical modern automobile battery. Sixty-eight percent of the raw material cost is the cost of lead, 11% for antimony, 11% for the separator, 8% for the casing, and 2% for the electrolyte. The price of lead and antimony fluctuate considerably, not only from year to year, but also even within a year. Lately (1975) the U.S. price of lead dropped due to the surplus created by the elimination of the lead additive in gasoline.

Fifty percent of the total lead consumption in Japan is used for batteries. Ninety percent of the lead comes from Japanese sources.

Table 12 shows the lead production of the world in 1972 and 1973. The production and consumption of lead in the major battery-producing countries

TABLE 11

Weight of Components of a Typical

U.S. Automotive Battery Weighing 47 lb (21.3 kg)

	(a) Weight (kg)	(b) Wt %
1. Grid (lead + antimony)	5.0	23.5
2. Active material		
Anode (Pb)	3.2	15.0
Cathode (PbO_2)	3.7	17.3
3. Strap and terminal	1.02	4.9
4. Electrolyte	7.00	33.0
5. Plastic (polypropylene) case	1.30	6.0
6. Separator	0.08	0.4
	21.5	100%

TABLE 12

World Production and Consumption of Lead (Unit: 1,000 Metric Tons)

	Ore produced 1972		Ore produced 1973		Lead produced 1972		Lead produced 1973		Lead consumed 1972		Lead consumed 1973	
United States	585	16.4%	564	15.5%	1,076	24.8%	1,089	24.4%	1,252	28.8%	1,289	28.1%
Canada	385	10.8	404	11.0	189	4.4	190	4.3	59	1.3	62	1.3
Mexico	178	5.0	181	4.9	159	3.7	170	3.8	71	1.6	90	2.0
Peru	160	4.5	187	5.1	75	1.7	75	1.7	5	0.1	11	0.2
Other	117	3.2	114	3.1	87	2.0	90	2.0	119	2.7	144	3.1
Total of America	1,425	39.9	1,450	39.6	1,586	36.6	1,614	36.2	1,506	34.6	1,596	34.7
Belgium	27	0.8	24	0.7	93	2.1	95	2.1	48	1.1	51	1.1
France	40	1.1	38	1.0	187	4.3	188	4.2	202	4.7	214	4.7
West Germany	34	1.0	28	0.8	273	6.3	299	6.7	274	6.3	280	6.1
Italy	73	2.0	75	2.0	69	1.6	50	1.1	186	4.3	190	4.1
Sweden	69	1.9	68	1.9	45	1.0	45	1.0	34	0.8	32	0.7
Great Britain					271	6.2	275	6.2	278	6.4	285	6.2
Spain	106	3.0	110	3.0	104	2.4	114	2.6	107	2.5	125	2.7
Yugoslavia	93	2.6	108	2.9	88	2.0	101	2.3	54	1.2	64	1.4
Other					82	1.9	85	1.9	187	4.3	196	4.3
Total of Europe	442	12.4	451	12.3	1,212	27.8	1,252	28.1	1,370	31.6	1,437	31.3
Japan	63	1.8	54	1.5	223	5.1	227	5.1	231	5.3	256	5.5
Other	64	1.8	62	1.7	18	0.4	19	0.4	119	2.7	141	3.1
Total of Asia	127	3.6	116	3.2	241	5.5	246	5.5	350	8.0	397	8.6
Morocco	94	2.6	110	3.0	—	—	—	—	3	0.1	4	0.1
South Africa	59	1.7	65	1.8	68	1.5	71	1.6	26	0.6	26	0.5
Other	72	2.0	74	2.0	55	1.3	50	1.1	9	0.2	12	0.3
Total of Africa	225	6.3	249	6.8	123	2.8	121	2.7	38	0.9	42	0.9
Australia	386	10.8	402	11.0	209	4.8	225	5.0	63	1.5	69	1.5
Other	2	0.1	2	0.1	—	—	—	—	7	0.1	7	0.2
Total of Oceania	388	10.9	404	11.1	209	4.8	225	5.0	70	1.6	76	1.7
Total of Free World	2,607	73.0	2,670	73.0	3,371	77.6	3,458	77.5	3,334	76.8	3,548	77.2
USSR	544	15.2	577	15.8	544	12.5	577	13.0	490	11.3	510	11.1
Other	420	11.8	412	11.2	429	9.9	425	9.5	517	11.9	539	11.7
Total of Communist World	964	27.0	989	27.0	973	22.4	1,002	22.5	1,007	23.2	1,049	22.8
Total of World	3,571	100.0	3,659	100.0	4,344	100.0	4,460	100.0	4,341	100.0	4,597	100.0

(United States, Germany, Great Britain, France, Japan) is approximately
balanced. This means that in these countries they produce the amount of
lead they need. The countries in which production is far greater than consumption are Canada, Australia, Mexico, Peru, etc., which are lead-
exporting countries. Italy imports considerable amounts of lead. The total
lead production of the world is about 4,460,000 tons/year. This is somewhat smaller than the total zinc production (5,380,000 tons/year). Sometimes people raise the question, "Can we have sufficient lead for batteries
in the future?" According to experts on resources, we can produce sufficient lead for some time even when 10 to 20% of our present automobiles
become electric cars. Since we can efficiently recycle used batteries, the
material supply situation is very good as far as the lead-acid battery is concerned.

In Japan, recycling used lead-acid batteries as a source of lead was
poor, but is increasing. Since the recovered lead contains about 2% antimony, it is mostly used for grid casting after adding more antimony. The
rate of lead recovery was about 20% in 1966, 28% in 1970, and has now
probably reached 35 to 40% in 1975. In the United States it is far higher.

6. TYPICAL PROCESS FOR MANUFACTURING BATTERIES

In this section a typical process for manufacturing lead-acid batteries in
Japan is described. Lead-acid batteries can be classified into two types,
depending on the structure of the positive plate: (1) pasted or flat-plate
type and (2) tubular or clad type. All negative electrodes are paste-type
plates. The life of a lead-acid battery is generally determined by the life
of the positive electrodes. The tubular-type battery has three to five times
longer life than the paste type.

All of the Japanese automobile batteries, which account for more than
80% of the total lead-acid batteries, use paste-type plates for both the positive and negative electrode. In Japan, industrial batteries, such as stationary batteries, forklift, and traction batteries, use a tubular-type plate for
the positive electrode.

FIG. 7. The Shimazu ball mill for lead powder production used today (Japan Storage Battery Company).

Red lead (Pb_3O_4): Some manufacturers use red lead to prepare the paste. This material can be prepared by heating the lead powder in an air stream at about 450°C. If the temperature is raised once to form litharge, it is hard to obtain red lead.

6.2. Grid Casting

The functions of the grid are to hold the active material (a paste) and also to collect current upon charge and discharge. The casting is done by automatic machines. See Fig. 8 for the view of a modern Japanese grid casting plant.

6.3. Preparation of the Paste

The paste is prepared by blending the lead powder and dilute sulfuric acid by a mixing machine. In place of the lead powder, litharge or red lead is

FIG. 8.   Grid casting plant (Japan Storage Battery Company).

used totally or partially.   The amount of sulfuric acid depends on the property of the lead powder and also depends on the type of battery to be made.
In order to understand the general property of the paste preparation, Fig. 9
(the paste diagram) is helpful.   The paste diagram shows the properties of
various pastes which are obtained when 100 g of lead oxide are mixed with
0 to 50 cm^3 of various sulfuric acid solutions (sp gr 1.00 to 1.30).   The
paste diagram varies considerably depending on the composition and particle size and surface area of the component material, but the diagram represents the physical states of the produced paste.   When the amount of sulfuric acid is not sufficient, the blended material is brittle or too loose and
does not form a paste.   When the amount of sulfuric acid is right, it has
the property of a paste which can be spread on a grid properly.   When the
amount of sulfuric acid is too high, it becomes soft and muddy and no longer
has the property of a paste.   A good paste which is useful for making battery plates is obtained only under the conditions enclosed by the three areas:

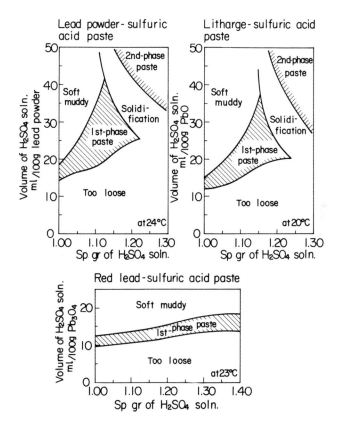

FIG. 9. Some examples of paste diagrams.

soft muddy area, too loose area, and solidification area. The enclosed
area is called the first-phase paste. When a relatively large amount of
sulfuric acid having a relatively large specific gravity is used, the lead
powder reacts with the sulfuric acid violently and the material solidifies
rapidly. This condition does not produce a usable paste. When the specific gravity and/or the amount of sulfuric acid is further increased, a
paste called second-phase paste is obtained. This paste is never used for
practical battery plates. The first-phase paste prepared with a relatively
low-density sulfuric acid solution tends to produce a long-life battery, and
those with a high-density sulfuric acid solution tend to produce a high-
performance battery. Extensive studies on that subject were made in Japan
[1-3].

6.4. Curing or Drying

The drying process is also called "curing" and is very important in order
to obtain a good electrode. During the curing, oxidation of metallic lead
powder takes place. When excessive long curing is applied, the paste ab-
sorbs carbon dioxide from air and part of the material converts to basic
lead carbonate 2 $PbCO_3 \cdot Pb(OH)$ [4].

6.5. Formation

Formation is a process in which the cured plates are electrolyzed (charged)
to form PbO_2 for the positive electrode and sponge lead for the negative
electrode. There are two types of forming processes. In "tank formation,"
the positive and the negative plates are placed alternately in a large tank
containing sulfuric acid solution of sp gr 1.05 to 1.10. In "container formation," the cured plates are assembled in the final battery case and the formation is carried out using a sulfuric acid solution of sp gr 1.10 or greater
(see Fig. 7). After formation the sulfuric acid, of sp gr 1.26 to 1.38, is
filled in or corrective adjustments are made. A view of a modern Japanese
formation plant is shown in Fig. 10.

6.6. Manufacturing Process for Separators

The 1974 Japanese separator production was about 6,000 to 7,000 sheets of
the NS battery site (116 × 134 × 0.7 mm). In Japan there are three major
manufacturers at this time: Nippon Mukiseni Kogyo Company, Matsubayashi
Kogyo Company, and Abegawa Kogyo Company. Most of the separators are
phenol resin-impregnated cellulose separators. The shapes may be flat,
waved, or have ribs. As mentioned earlier, most of the Japanese separators are laminated with a glass mat layer. The glass mat layer is produced
by Nippon Glass Fiber Company and Nippon Mukiseni Kogyo Company. The
glass mat layer protects the positive active material from shedding and also
protects the paper separator from oxidation by the positive electrode.

FIG. 10.   Formation plant (Japan Storage Battery Company).

Recent trend in separators: A sintered polyvinyl chloride separator is popular in Europe, but in Japan no polyvinyl chloride separator is used for lead-acid batteries. The reason for this is not only the performance and the cost, but also the fact that there is difficulty in the recovery process of lead from the used batteries when vinyl chloride is used as separator. Non-woven polypropylene sheet separators are being considered, since the thickness is 0.1 to 0.2 mm and the pores are less than $1\,\mu$m in diameter. New separator material needs a long life-test period before reaching wide practical use.

7. MAJOR MANUFACTURERS OF LEAD-ACID STORAGE
    BATTERIES IN JAPAN

It may be helpful to list the major manufacturers of lead storage batteries in Japan. As is the custom with U.S. and European manufacturers, Japanese firms publish special engineering handbooks describing the properties

TABLE 13

Major Lead-Acid Battery Manufacturers in Japan and Their Brand Names

Company	Japan Storage Battery Company	Yuasa Battery Company	Furukawa Battery Company	Shinkobe Denki Company	Matsushita Electric Industrial Company (Storage Battery Division)
Address	Inobaba-cho, Kisshoin, Minami-Ku, Kyoto Japan 601	6-6 Josai-cho, Takatsuki-shi Osaka, Japan 569	2-246, Hoshikawa Hodogaya-Ku Yokohama-shi Japan 240	2-1-1, Nishi-Shinjyuku, Shinjyuku, Mitsui Bldg. Tokyo, Japan 160	1006 Kadoma Kadoma-shi Osaka, Japan 571
Total sales ($) 1974	$148 million	$150 million	$55 million	$75.3 million	$3,283 million for the whole company
Main products	Lead-acid batteries, alkaline batteries, rectifiers, illumination equipment, other	Lead-acid batteries, alkaline batteries, rectifiers, other	Lead-acid batteries, alkaline batteries, rectifiers	Lead-acid batteries, alkaline batteries, electrical equipment, plastic resin products	Lead-acid batteries, various industrial and consumer electrical and electronic products
Brand names					
1. Automobile batteries	Tafner Myca 5	Perfector	Hidashi	Dry Pack	Big Power Carec
2. Stationary batteries	Myty Power Myty Flash	Highrater (HS) Golden Clad (CS)	Furukawa Clad	Super Clad Highlong Power (HS)	Hyper Clad
3. Small sealed batteries	Portalac	Noyper		Sealac	Panalloid
4. Cycle service batteries	Myty power-1	Ultra Gold Clad	Furukawa Clad	Super Clad	Pyper Clad

of their products. Table 13 represents such a list. The product brand names are included to serve as a guide for identifying them in the market; the size of the companies can be estimated from their sales volume.

ACKNOWLEDGMENT

This work was compiled for the U.S. Office of the Electrochemical Society of Japan, based on material submitted by Yuasa Battery Co. (Mr. S. Tsuji), Storage Battery Division of Matsushita Electric Industrial Co. (Mr. M. Ozeki), Furukawa Battery Co. (Dr. K. Shimizu), Shin Kobe Electric Co. (Dr. S. Ikari), and Japan Storage Battery Co. (Mr. I. Okazaki).

REFERENCES

1. T. Takagaki, J. Electrochem. Soc. Japan, 23, 449, 567, 637 (1955).
2. T. Takagaki, J. Electrochem. Soc. Japan, 25, 492 (1957).
3. T. Takagaki, J. Electrochem. Soc. Japan, 26, 278, 320, 354 (1958).
4. T. Takagaki, J. Electrochem. Soc. Japan, 23, 173, 232, 308, 399 (1955).

Numbers in parentheses are reference numbers and indicate that an author's work is referred to although his name is not cited in the text. Underlined numbers give the page on which the complete reference is listed.